AF355962

Micromachined Acoustic Transducers for Audio-Frequency Range

Micromachined Acoustic Transducers for Audio-Frequency Range

Guest Editor

Libor Rufer

Basel • Beijing • Wuhan • Barcelona • Belgrade • Novi Sad • Cluj • Manchester

Guest Editor
Libor Rufer
MEMS
Advanced Design & Technology
Rives
France

Editorial Office
MDPI AG
Grosspeteranlage 5
4052 Basel, Switzerland

This is a reprint of the Special Issue, published open access by the journal *Micromachines* (ISSN 2072-666X), freely accessible at: www.mdpi.com/journal/micromachines/special_issues/ Micromachined_Acoustic_Transducers.

For citation purposes, cite each article independently as indicated on the article page online and using the guide below:

Lastname, A.A.; Lastname, B.B. Article Title. *Journal Name* **Year**, *Volume Number*, Page Range.

ISBN 978-3-7258-3266-8 (Hbk)
ISBN 978-3-7258-3265-1 (PDF)
https://doi.org/10.3390/books978-3-7258-3265-1

Contents

About the Editor

Libor Rufer

Libor RUFER received his Engineer and PhD degrees from Czech Technical University in Prague. Between 1976 and 1993, he served as an Assistant Professor at the university's Faculty of Electrical Engineering. From 1993 to 1994, he conducted research at the Acoustics Laboratory of École Centrale de Lyon, France. Subsequently, from 1994 to 2019, he worked as a Senior Scientist at the TIMA Laboratory, University Grenoble Alpes, France. Since 2021, he has been managing ADT MEMS (Advanced Design and Technology for MEMS), a consulting company specializing in microelectromechanical systems.

His expertise spans the following:

Modeling MEMS through analytical approaches, behavioral modeling, and finite element analysis.

Electromechanical systems utilizing piezoelectric, electrostatic, piezoresistive, and electromagnetic transduction.

Applications of microfabrication technologies, focusing on energy harvesting, acoustic/ultrasonic transducers, physical sensors, and electro-active polymer structures.

Measurement techniques associated with these applications.

He has contributed to numerous European projects, including MEDEA+, LEONARDO, EUREKA, and FP7, as well as national and industrial collaborations. He led the ANR project SIMMIC, which developed a wide-band silicon measurement microphone for high-frequency acoustics (10 kHz–1 MHz) and was recognized as a CNRS "Key Event". He also spearheaded international mobility projects fostering collaborations with institutions in Hong Kong, China, Italy, Slovakia, and the Czech Republic.

He has authored over 150 publications in peer-reviewed international journals and conferences. He has been a committee member of competitive international conferences, a reviewer for leading journals, and an independent expert for the European Commission on FP7 and H2020 programs. He is a former member of the French Acoustical Society, the Audio Engineering Society (USA), the Acoustical Society of America, and a Senior Member of IEEE.

Preface

In the 21st century, microelectromechanical systems (MEMSs) and acoustics have become inseparable disciplines. Microsystem technology, which emerged nearly four decades ago, paved the way for the development of miniature acoustic transducers for the audio-frequency range. The drive to create smaller, more efficient, and cost-effective acoustic sensors and actuators for consumer devices (such as cell phones and hearing aids) and industrial applications (such as acoustic source localization or imaging using compact sensor arrays) has been the main force behind this evolution.

Following years of innovation and overcoming technical challenges, MEMS microphones have become one of the most successful commercial products in the history of microsystem technology. Commercialized since 2004, capacitive MEMS microphones are now indispensable components of cell phones. In recent years, MEMS speakers have garnered significant interest from both research and industry. The demand for miniaturized speakers has increased in applications such as headsets, smartphones, hearing aids, voice assistants, and the Internet of Things (IoT). However, current MEMS speakers face challenges, including limited sound pressure output, sound quality, and linearity in frequency response. Recent advancements are primarily focused on piezoelectric MEMS speakers, which are considered a promising solution due to their ability to achieve large actuation forces with relatively low driving voltages.

This Special Issue is dedicated to showcasing works on micromachined acoustic transducers for the audio-frequency range and providing a state-of-the-art overview of this dynamic field as it pertains to audio applications.

The articles in this Special Issue cover a range of topics within the domain of MEMS acoustic devices, including microphones, speakers, and hearing aids. They address key research questions related to the design, modeling, evaluation, and applications of these devices. The diverse origins of the contributing authors—from industry, academia, and research institutions—bring a multidisciplinary perspective that ensures a comprehensive and balanced exploration of the field. This diversity also guarantees the practical relevance and scientific rigor of the presented works, offering valuable insights for researchers and engineers alike.

Libor Rufer
Guest Editor

Editorial

Editorial for the Special Issue on Micromachined Acoustic Transducers for Audio-Frequency Range

Libor Rufer [1,*], Shubham Shubham [2], Haoran Wang [3], Tom Miller [4], Petr Honzík [5] and Vittorio Ferrari [6]

[1] Advanced Design & Technology for MEMS, 38140 Rives, France
[2] Syntiant Corporation, Itasca, IL 60143, USA; shubham.shubham@syntiant.com
[3] Applied Materials Inc., Santa Clara, CA 95054, USA; wanghaoranuf@gmail.com
[4] Knowles, Electronics LLC, Itasca, IL 60143, USA; tom.miller@knowles.com
[5] Department of Radioelectronics, Faculty of Electrical Engineering, Czech Technical University in Prague, 166 27 Prague, Czech Republic; honzikp@fel.cvut.cz
[6] Department of Information Engineering, University of Brescia, 25121 Brescia, Italy; vittorio.ferrari@unibs.it
* Correspondence: liborrufer01@gmail.com

1. Introduction

Microelectromechanical systems (MEMS) refer to miniaturized mechanical and electromechanical elements that are fabricated through microelectronic processes. The development of MEMS technology dates back to the 1960s, along with the development of silicon-based integrated circuits and advancements in micromachining techniques. Early MEMS research primarily focused on sensors and actuators, where microfabrication methods such as photolithography, etching, and deposition enabled the creation of highly precise structures at microscopic dimensions.

In the 1980s and 1990s, MEMS technology gained traction as advances in semiconductor fabrication processes made it possible to manufacture complex structures at a very small scale. These developments enabled MEMS devices to operate with high accuracy and low power consumption, sparking innovations across a range of industries, including automotive, biomedical, telecommunications, and consumer electronics.

One of the most significant and impactful applications of MEMS technology is in the field of acoustics. MEMS-based acoustic devices, such as microphones, speakers, and sensors, have revolutionized the way sound is captured, processed, and reproduced. MEMS microphones, for example, leverage the small size, low power requirements, and high sensitivity of MEMS technology to create miniature microphones that are widely used in smartphones, hearing aids, and other portable devices. These microphones provide high performance and durability in compact packages, making them ideal for consumer electronics that demand miniaturization without degrading the main parameters. The integration of MEMS technology in acoustic applications has also led to innovations in noise-cancelling headphones, smart speakers, and acoustic sensing systems for industrial and medical purposes.

The future of MEMS in acoustics looks promising, with ongoing research pushing the boundaries of device performance, miniaturization, and integration. As this technology continues to evolve, we can expect further innovations in audio processing, smart devices, and immersive acoustic experiences, transforming industries ranging from entertainment to healthcare. The integration of MEMS with the internet of things (IoT) will accelerate this transformation, enabling smarter, interconnected devices that can seamlessly communicate and adapt to their environments. MEMS-based acoustic sensors in IoT applications, for

check for updates

Received: 23 December 2024
Revised: 2 January 2025
Accepted: 3 January 2025
Published: 8 January 2025

Citation: Rufer, L.; Shubham, S.; Wang, H.; Miller, T.; Honzík, P.; Ferrari, V. Editorial for the Special Issue on Micromachined Acoustic Transducers for Audio-Frequency Range. *Micromachines* **2025**, *16*, 67. https://doi.org/10.3390/mi16010067

example, will enhance real-time audio analysis, environmental monitoring, and personal health tracking, leading to more efficient, responsive, and intelligent systems.

This Special Issue featured a total of ten contributions, comprising nine original research articles and one review paper. Each contribution integrates one of the main topics in the field, which are detailed in the following paragraphs.

2. MEMS Microphones

2.1. Overview of Recent Developments

MEMS microphones are widely used in various applications, including IoT devices such as smartphones, speakers, tablets, laptops, smartwatches, and true wireless earphones, as well as other consumer appliances in automobiles, industrial goods, and machinery. Due to their miniaturized size, low cost, high-quality acoustic performance, exceptional integration compatibility with application-specific integrated circuits (ASICs), and improved reliability against environmental factors, there has been a continuous demand to further improve the signal-to-noise ratio (SNR) performance and other key parameters related to MEMS microphones, including sensitivity, noise, total harmonic distortion (THD), and dynamic range.

The authors of [1] presented a new differential MEMS capacitive microphone transducer design based on a sealed-dual membrane (SDM); they significantly reduced the fluidic noise component caused by air inside the MEMS transducer by eliminating the air between the two membranes. An SNR of 72 dBA was reported, with a sensitivity of −38 dBFS and an acoustic overload point (AOP) of 130 dB; this was achieved with latest generation digital ASIC and a reduced power consumption. This SNR performance marked a significant improvement on the current state-of-the-art MEMS microphones on the market, which have an SNR in the range of 65 to 70 dBA [2–4]. In 2021, a single corrugated membrane design with a single-backplate architecture and capacitive transduction mechanism was proposed, delivering a 73 dBA SNR with an analog ASIC, closely matching the performance of the SDM MEMS with a localized vacuum [5]. Again, the focus was on minimizing the MEMS acoustic damping losses to achieve a high SNR performance. The MEMS sensitivity achieved was −34 dBV/Pa. Steady progress has also been made in the piezoelectric transduction mechanism, which offers the advantages of low power consumption and makes the sensor less susceptible to particle contamination. In 2023, a novel design based on a fully clamped corrugated aluminum nitride (AlN) membrane was presented, offering very effective intrinsic stress relaxation and enabling differential signal generation by exploiting a pseudo-bimorph effect [6]. A sensitivity of −47.1 dBV/Pa and an SNR of 62.6 dBA were reported. However, the authors left room for further design improvements to achieve higher mechanical compliance, while maintaining a system resonance above 20 kHz, and enhance the overall performance. To address the impending fundamental physical, design, and material limitations for MEMS microphones, several works have demonstrated the use of graphene as a diaphragm. In 2024, a novel approach to integrating multi-layer graphene into condenser MEMS microphones, without the need for transfer or polymer support, was presented [7]. This approach addressed the previous limitations associated with fabricating wafer-scale graphene microphones without polymer supports, potentially opening new possibilities for future microphone technology. The sensitivity achieved was 24.3–321 mV/Pa, which is more than twice that of the reported state-of-the-art MEMS microphones. Further exploration is needed to enhance other critical MEMS microphone performance metrics. On the microphone modeling front, a behavioral model to capture the nonlinear phenomena in MEMS microphones was developed using a circuit simulator tool and hardware description languages (Verilog-A) [8]. The proposed model improved the accuracy of predicting sensitivity, SNR, and THD performance. A constrained MEMS microphone architecture was used to validate the findings with measurements. Shubham et al. [9] also demonstrated a similar behavioral model for capturing nonlinearities using circuit modeling tools to represent blocks with couplings between the acoustic, mechanical, and electrical domains. Although the approach and intent were similar, the modeling method presented was generalized to cover devices for which there were no closed-form solutions regarding the diaphragm shape or variable capacitance of the microphone. This became increasingly important when considering the boundary conditions of the diaphragm and its deformed shape under applied bias or acoustic pressure. These models excel in precisely predicting pull-in, sensitivity, SNR, and THD performances over wide acoustic pressure ranges.

2.2. Future Research Areas to Address Knowledge Gaps in the Field

There has been a continuous push to expand the boundaries of physics in order to further improve the SNR performance, enhance the sensitivity, and reduce the overall noise of MEMS microphones. This will require focusing on minimizing other components, such as noise related to the thermal boundary layer. The noise floor in the MEMS microphone design is limited by the thermal noise generated by the heat exchange with the isothermal boundary condition at the walls in the back cavity. This typically results in increased noise as the microphone package size decreases. A diffusion-dominated design regime has been explored that can yield high SNR with very small back volumes. In such a design, the challenge becomes producing enough output signal, which often requires the use of very-high-bias voltages in the sensor [10,11]. The need for good MEMS microphone performance has opened the door for the further exploration of other transduction mechanisms, such as optical technology. Although the proof-of-concept dates back to 2003, it was not until 2016 that a Norwegian company, SensiBel, demonstrated a breakthrough with an 80 dBA microphone SNR and 146 dB AOP using this technology [12]. However, the typical challenges related to power consumption, size, and the high cost of optical components remain unresolved. To overcome the design challenges and improve the performance of MEMS microphones, this Special Issue also included novel design features. Shubham et al. [13] proposed a semi-constrained diaphragm design with flexible springs, supported by center and peripheral protrusions originating from the back plate. This helped to significantly increase the effective area of the diaphragm by 85% under bias, thereby improving the linearity, sensitivity, and SNR performance of the microphone.

All the modeling explorations assume a rigid (clamped at all boundaries) perforated backplate. However, in this Special Issue, Šimonová and Honzík address this gap by including the movement effect of the perforated moving electrode to accurately capture the damping effects related to thermal and viscous boundary layers; they compare the analytical results with finite element modeling [14]. In another study [15], the authors considered the backplate as a vibrating structure and extended the lumped model to introduce coupling between the mechanical and electrical domains, showcasing the impact on the microphone sensitivity and device pull-in. In this Special Issue, Rufer et al. address concerns related to the poor AOP performance of MEMS microphones under high sound pressure levels and in harsh environmental conditions [16]. A novel approach to improving the AOP performance is through piezoelectric coupling. With a modeling and optimization approach for microphone sensitivity and AOP tuning in the audio band, the piezoelectric MEMS microphone can be used in applications that require high sound pressure level detection, such as aircraft design, rocket-launching vehicles, gunshot detection, and urban security systems. The issue also demonstrates a new approach to enabling sensitivity control by placing one electrode at the diaphragm center and one annular electrode closer to the clamped edge of the diaphragm. The first electrode serves as a sensor to provide the output signal, whereas the second acts as an actuator that mechanically pre-stresses the diaphragm and alters the microphone sensitivity.

3. MEMS Speakers

In the past few years, MEMS speakers, another type of micromachined acoustic transducer, have drawn significant interest from researchers and businesses. With the flourishing development of consumer electronics and artificial intelligence (AI), there are increasing demands for miniaturized speakers in headsets, smartphones, hearing aids, voice assistants, and internet of things devices. Benefitting from micromachining technologies, MEMS speakers offer numerous advantages; they have a small carbon footprint, a light weight, low power consumption, low-cost batch fabrication compatibility, and an improved compatibility with application-specific integrated circuits.

However, as a consequence of their small device size, most of the existing MEMS speakers face the challenge of limited sound pressure output [17]. Sound quality and linearity in frequency response, which are typically evaluated according to the total harmonic distortion (THD), are also important considerations in the development of MEMS speakers [18]. Significant progress has been made in addressing challenges related to materials, device structure designs, electrode configuration, and driving methods [19,20]. Lumped element modeling and acoustic simulation are also important methods that have been explored to effectively optimize MEMS speaker design and performance [21].

According to transduction mechanisms, MEMS speakers are classified as electrodynamic, electrostatic, piezoelectric, and thermoacoustic. Breakthroughs have been made in all types of MEMS speakers, including in terms of the various materials and fabrication approaches, special device designs, and improved sound pressure levels (SPLs) over wide bandwidths [22,23]. Due to their ability to achieve large actuation forces with relatively small driving voltages, piezoelectric MEMS speakers show the most promising results and have become a research hotspot. In terms of recent developments, there are more and more studies dedicated to piezoelectric MEMS speakers.

In this Special Issue, we include three such studies. In particular, Wang et al. presented a comprehensive review of MEMS speakers, including their recent development, remaining challenges, and future outlook [24]. Liechti et al. proposed a method to evaluate the THD generated by a cantilever-based MEMS speaker inside an acoustic coupler [25]. Teuschel et al. investigated the potential of using epitaxial grown PZT with an imprint for stable bipolar operation, demonstrating a strong piezoelectric response for MEMS actuator applications such as speakers [26].

In the future, more research is expected on how to improve the SPL of MEMS speakers. Obtaining large driving forces with low power consumption will continue to be a key developmental challenge. A balance between small diaphragm sizes and relatively large SPL outputs can be achieved by exploring new materials, novel structure designs, and special electrode configurations. With a more comprehensive understanding of the modeling, special structures and acoustic enclosures will also be explored in order to optimize the overall frequency response of MEMS speakers, including further reduced THD and high SPL over wide bandwidths. In addition to the optimized device performance, easier fabrication techniques with low costs, high yields, and good compatibility with other materials and processes will continue to be studied before MEMS speakers can become mainstream and large-scale commercial products.

4. MEMS in Hearing Aids and Implantable Devices

MEMS microphones first saw mass market acceptance with their introduction into cell phones in 2003 and have been dominant in hearing health products since 2019. They offer improved electroacoustic performance, reliability and repeatability at a lower cost than electret microphones. Devices currently offer noise levels below 25 dBA, handling sound levels over 135 dB in a package less than 3.5 mm long [27]. Knowles Electronics pioneered the creation of MEMS microphones for the hearing health and commercial markets and continues to be a key supplier. Sonion is now also a significant supplier to the hearing health market. InvenSense is a key supplier of MEMS silicon to many makers of MEMS-based microphones. The latter face continued pressure to provide ever greater dynamic range at both the softest and loudest extremes while further reducing size. Future research goals include further reducing nonlinearity, greater resistance to contamination from exposure to dust and liquids, and selectively reducing sensitivity to ultrasound signals emitted by proximity sensors. Research in the ASICs for these products is seeking to reduce noise and distortion, reduce power consumption, and enable a growing list of programmable options such as gain, filtering, and power consumption.

Commercial applications of MEMS microphones are virtually all air-conduction. There has been research into using implanted sensors as microphones, with particular interest in their use with cochlear implants. However, their SNR and directivity are inherently much poorer than in air-conduction devices, preventing their widespread adoption for many hearing health applications. There is also potential for using implanted MEMS audio devices to monitor voice production and other audio signals within the body. One major hurdle to using MEMS implanted devices is protecting the delicate sensor from contact with bodily tissues and fluids while avoiding attenuating the signals the sensor should be receiving. In this Special Issue, Prochazka et al. describe a method for such protection that offers a reduced attenuation of the desired signal [28].

MEMS audio-frequency vibration-sensing devices attached to the outside of the body, particularly near the jaw and ears, are seeing increasing adoption. These are often referred to as bone-conduction sensors and the sensors are placed on the skin over areas of bone or cartilage that offer strong signals. Audio band vibration sensors can also be used to convert any vibrating surface in a room or on a car into a microphone [29,30]. Human body applications include speaker identification and speech capture augmentation in high wind conditions, for which high SNR and low power consumption are critical.

While body-contact MEMS sensors are in development, they do not yet make very effective actuators. The most significant issue is the very large impedance mismatch between the MEMS actuator and body tissues. The actuator must also be protected from the forces applied by the tissues. The transducers currently used in bone-conduction speakers use a moving mass that is many orders of magnitude larger than could fit into a MEMS device. It will be interesting to see if there are niche applications where a MEMS actuator could be useful. One possibility would be applying force directly to the tiny bones of the middle ear or to the eardrum.

MEMS speakers are beginning to be used in earphones, but have not yet seen acceptance in hearing health devices. MEMS devices struggle to produce the desired 120 dB sound levels in the ear canal. They also need high-voltage electronics to power them and hence significantly more battery power than traditional balanced-armature (BA) devices. This prevents them from meeting hearing aid designers' battery life requirements. For comparison, the Knowles balanced armature driver package size is $5.3 \times 3.1 \times 2.7$ mm, a size that is widely used in hearing instruments. This device can produce a sound level of 130 dB at its 3 kHz resonance and produces 112 dB at 1 mW of drive power (rising to 122 dB/mW at 3 kHz) [31]. The size of the BA device also includes the air enclosure behind the transducer. In hearing health applications, the sound radiating from the rear of the transducer must be contained in a sound-proof enclosure, so that it cannot leak to the microphone and cause feedback. MEMS speakers often require a high-compliance acoustic load behind the radiator, which can be larger than the BA housing.

5. Measurement Methods for MEMS Devices

Although most MEMS acoustic devices currently in use operate on the same principles and using the same methods as their traditional millimeter-scale counterparts, applying the same evaluation methods as those used for traditional devices is not straightforward. This paragraph highlights the extraction of MEMS parameters presented in this Special Issue, and provides examples of two additional methods specifically tailored for MEMS microphone characterization.

5.1. Extraction of MEMS Transducer Mechanical Properties

Extracting the mechanical properties of complex three-dimensional MEMS transducers through acoustic transmission measurements offers a novel approach that bypasses the limitations of traditional characterization methods. Becker et al. focus on folded diaphragms with buried in-plane vibrating structures, which expand the active area while maintaining a compact form factor [32]. Conventional techniques such as bulge tests and atomic force microscopy often require destructive preparation or rely on optical accessibility, which is infeasible for structures hidden within high-aspect-ratio trenches [33–36]. The presented methodology employs a passive acoustic transmission setup where diaphragms are actuated by an acoustic signal and analyzed using lumped-element modeling (LEM). This non-invasive method effectively captures dynamic mechanical behaviors such as compliance, even for complex geometries, without altering boundary conditions.

The experimental framework utilizes two acoustically isolated chambers separated by the diaphragm under test. Sound pressures on both sides of the diaphragm are measured to extract mechanical properties by fitting experimental data to LEM simulations [37,38]. The results demonstrate high accuracy, with the measured compliances differing by less than 4% from numerical simulations for diaphragms exceeding 1 mm in length.

By eliminating extensive sample preparation and ensuring boundary conditions mimic practical applications, this method advances the characterization of MEMS transducers with complex three-dimensional geometries. It provides valuable insights into the mechanical properties critical for optimizing the design and performance of MEMS devices in acoustic applications.

5.2. High-Frequency Calibration of MEMS Microphones

Due to their compact dimensions, MEMS microphones can achieve frequency responses extending into several hundreds of kilohertz. Characterizing the frequency response of such microphones and pressure sensors presents a significant challenge. To address this, a calibration method has been developed utilizing a spherical weak shock acoustic pulse generated by a spark source. The pressure waves, in the form of asymmetric N-waves with a duration of approximately 40 µs and a front shock rise time of around 0.1 µs, were characterized using an optical interferometer. By accounting for the nonlinear propagation of weak shockwaves, it became possible to estimate the incident pressure wave and derive the frequency response of MEMS microphones across a range of 10 kHz to 1 MHz [39,40]. This approach is independent of the transduction principle or sensor mounting configuration.

5.3. Measurement of MEMS Microphone Nonlinearities

Measuring the nonlinearities of MEMS microphones has become an important research area, particularly due to the widespread use of these microphones in consumer electronics and measurement applications. MEMS microphones are popular for their small size, low cost, high sensitivity, low inherent noise, and low power consumption. However, nonlinear distortions, such as harmonic and intermodulation distortions, can impact precision, making their investigation crucial. These distortions from multiple factors, including electrostatic transduction, the mechanical properties of the diaphragm, air gap damping, and amplifier behavior [41,42]. In particular, the electrostatic transduction mechanism is often identified as the primary source of distortion in condenser microphones, producing both harmonic and intermodulation distortion [43–45]. According to [46], the second harmonic contributes approximately 90% of the total harmonic distortion.

In a typical experimental setup used to study MEMS microphone nonlinearities, a reference low-distortion microphone and a loudspeaker can be employed to generate a sinusoidal acoustic pressure signal with minimized higher harmonics. Achieving a perfectly linear acoustic signal source is a challenging task, as most sources inherently introduce some degree of distortion. Therefore, harmonic correction [47] is advantageous in effectively reducing these distortions. The distortion of the nonlinear source (loudspeaker) is reduced to the noise level using this adaptive correction method, assuming that the nonlinearities of the reference microphone are negligible. This setup allows for a direct measurement of the harmonic distortion of MEMS microphones under testing at different frequencies and sound pressure levels. For example, the authors of [48] describe such measurements conducted at frequencies of 20 Hz, 200 Hz, and 2 kHz, and sound pressure levels ranging from 90 to 128 dB. A simple nonlinear model based on the electrostatic transduction principle was used to compare the measured results for single-backplate condenser MEMS microphones. The results indicate that the second harmonic behaves predictably and is well described by the model. However, the third harmonic distortion product shows deviations, suggesting the influence of nonlinearities not included in the simple analytical description.

To conclude, measuring the nonlinearities of MEMS microphones is feasible, especially with the use of advanced harmonic correction techniques that minimize source distortion. The electrostatic transduction mechanism is identified as the primary source of distortion, mainly affecting the second harmonic. However, challenges remain, particularly in modeling and accounting for higher-order nonlinear effects and other sources of distortion, such as diaphragm deflection and dynamic changes in air gap thickness [49]. Future work should focus on two main areas: refining models to include these additional nonlinearities for better prediction accuracy, and achieving effective reduction in nonlinear distortion, particularly by minimizing the level of the second harmonic in single-backplate condenser MEMS microphones.

Piezoelectric MEMS microphones can exhibit nonlinearities due to material properties and fabrication imperfections. The former occur when piezoelectric materials like aluminum nitride (AlN) respond nonlinearly under high electric fields or mechanical stresses, leading to harmonic distortion in the output signal [50]. Geometric nonlinearities arise from fabrication-induced imperfections, such as initial curvature or residual stresses in the diaphragm; these result in hardening or softening behaviors that affect device performance [51]. Understanding and mitigating these nonlinearities is crucial for the accurate and reliable operation of piezoelectric MEMS microphones, particularly in applications requiring high fidelity and precision.

6. Applications of MEMS Acoustic Devices

MEMS devices' small size, low power consumption, ability to integrate with sophisticated electronics, and cost-effective mass production enable their widespread application across industrial, military, and consumer sectors. This paragraph presents two papers featured in this Special Issue, along with additional papers, illustrating some of their potential applications.

6.1. Acoustic Source Localization and Ranging

Acoustic source localization, ranging, and imaging are essential techniques for determining the position and characteristics of sound sources in various environments. Acoustic source localization involves pinpointing the location of a sound source by analyzing the time-of-arrival (TOA) or phase shifts of sound waves reaching multiple microphones or sensors. Since the distance is the product of the signal speed and the time of flight of the signal traveling from the source, the ranging accuracy depends on the signal speed and the precision of TOA measurements. Compared to other signal types, such as light or radio, acoustic signals are often better suited for achieving high accuracy due to their relatively slow speed. However, ensuring precise TOA measurements remains a significant challenge in system implementation [52].

Acoustic source localization is widely used in underwater sonar systems and applications such as speech tracking and wildlife monitoring [53]. Another important area of research is gunshot acoustic localization, which is employed in military and civilian security systems. In this context, microphone array data are used to identify the event time, the direction of arrival (DOA), and to calculate the shooter's position [54].

Acoustic ranging is a technique for estimating the distance between two objects. It plays a critical role in applications such as motion tracking, gesture and activity recognition, and indoor localization. Although numerous ranging algorithms have been developed, their performance often degrades significantly under conditions of strong noise, interference, and hardware limitations. Various ranging applications using MEMS microphones have been explored, spanning simple methods with three sensors for sound source localization [55], gesture recognition [56,57], and indoor positioning systems for public use in smart city-related applications [58].

6.2. MEMS Microphone Arrays for Acoustic Imaging

Acoustic imaging takes acoustic source localization and ranging a step further, creating a spatial map or visual representation of sound waves as they interact with objects. This technique is widely used in medical imaging and non-destructive testing. Acoustic imaging is crucial in fields like marine biology, seismic exploration, autonomous vehicle navigation, and aeroacoustics, providing vital information about the environment through sound. Due to their small size, MEMS microphones are inherently well suited for integration into compact arrays, enabling high spatial resolution and controlled directional sensitivity in detecting acoustic fields [59]. In particular, aeroacoustic testing is an application where MEMS microphone arrays are both attractive and promising [60,61].

In this context, in this Special Issue, Ahlefeldt et al. report the results of an experimental study demonstrating that MEMS microphones assembled into an array allowed for detecting a plane fuselage's surface pressure fluctuations during flight [62]. While the state-of-the-art sensors for measuring such fluctuating pressure distributions are small piezoresistive pressure transducers, MEMS devices offer smaller dimensions and reduced costs. The article describes testing MEMS microphones for in-flow measurements. Commercially available MEMS microphones have a limited AOP of up to a maximum of 135 dB, which is too low for in-flow operation. To overcome this limitation, a covering

Kapton foil of 25 μm in thickness was applied on the microphones to attenuate the pressure by approximately 38 dB. The cover layer also introduces a change in the frequency response, which was taken into account. The results of both wind-tunnel and flight tests indicate that the Kapton covering damped the pressure fluctuations as expected, while still preserving the critical phase information required for wavenumber analysis.

6.3. Electrical Tuning of MEMS Acoustic Transducer Resonant Frequency

MEMS acoustic transducers can function as sound receivers (microphones) or emitters (speakers) within the audio-frequency range. Additionally, they can operate in the ultrasonic frequency range as micromachined ultrasonic transducers (MUTs), which are further categorized into piezoelectric (PMUTs) and capacitive (CMUTs) types.

The acoustic response and performances of the transducers depend, among other parameters, on the mechanical conditions in the microstructure; these conditions can affect its elastic properties, such as the degree of static stress in vibrating diaphragms and plates. Static stress can be influenced or determined by DC bias voltages purposely applied to suitable electrodes, particularly with piezoelectric MEMS transducers. This opens the possibility for the electrical adjustment/tuning of transducer properties relevant to acoustic response, such as resonant frequencies and corresponding quality factors. This approach has been investigated in the ultrasonic frequency range in PMUTs [63].

In such a scenario, in this Special Issue, Nastro et al. report the results of an experiment exploring and validating a technique to obtain an electrically tunable matching between the series and parallel resonant frequencies of a piezoelectric MEMS acoustic transducer [64]. This can increase the effectiveness of acoustic emission/detection in voltage-mode driving and sensing.

The adopted piezoelectric MEMS transducer is based on an aluminum nitride (AlN) active layer on top of a square diaphragm. By applying an adjustable DC bias voltage between pairs of properly connected electrodes, a planar static compressive or tensile stress in the diaphragm is electrically induced, depending on the voltage sign, thereby shifting its resonant frequency. The piezoelectric MEMS transducer operates at its first flexural resonance of about 5 kHz. The results of the experimental tests run in both receiver and transmitter modes showed that the resonance can be electrically tuned in the bias voltage range between −8 V and +8 V with estimated tuning sensitivities of about 9 Hz/V and 8 Hz/V in transmitter and receiver modes, respectively. The reported device can be employed in pulsed-echo mode as a proximity/presence or gesture detector. The proposed technique can be transferred to down-scaled structures to obtain tunable PMUTs.

A similar concept for controlling microphone sensitivity is presented in another paper in this Special Issue, that by Rufer et al. [16]. The proposed approach utilizes a MEMS piezoelectric microphone design featuring two electrodes: a circular electrode positioned at the diaphragm's center and an annular electrode located near its clamped edge. One electrode operates as a sensor, generating the microphone's output signal, while the other serves as an actuator. By applying a DC bias voltage to the actuator, the diaphragm can be mechanically pre-stressed, thereby altering the microphone's sensitivity. This sensitivity control approach holds significant potential for applications involving high acoustic loads, where reduced microphone sensitivity can be electronically achieved when the acoustic level exceeds a predefined threshold.

Author Contributions: Writing, structuring, review and editing the paper: L.R.; Section 1: L.R.; Section 2: S.S.; Section 3: H.W.; Section 4: T.M.; Section 5: P.H. and L.R.; Section 6: L.R. and V.F. All authors have read and agreed to the published version of the manuscript.

Acknowledgments: We would like to take the opportunity to thank all the authors for submitting their papers to this Special Issue and all the reviewers for dedicating their time in helping to improve the quality of the submitted papers.

Conflicts of Interest: The authors declare no conflicts of interest. They have obtained consent to publish this paper of the companies to which they are affiliated.

References

1. Sant, L.; Füldner, M.; Bach, E.; Conzatti, F.; Caspani, A.; Gaggl, R.; Baschirotto, A.; Wiesbauer, A. A 130 dB SPL 72dB SNR MEMS Microphone Using a Sealed-Dual Membrane Transducer and a Power-Scaling Read-Out ASIC. *IEEE Sens. J.* **2022**, *22*, 7825–7833. [CrossRef]

2. Bach, E.; Gaggl, R.; Sant, L.; Buffa, C.; Stojanovic, S.; Straeussnigg, D.; Wiesbauer, A. 9.5 A 1.8 V true-differential 140 dB SPL full-scale standard CMOS MEMS digital microphone exhibiting 67 dB SNR. In Proceedings of the IEEE International Solid-State Circuits Conference (ISSCC), San Francisco, CA, USA, 5–9 February 2017; pp. 166–167. [CrossRef]

3. SmartSound™ Digital Microphones. Available online: https://invensense.tdk.com/smartsound/digital/ (accessed on 27 November 2024).

4. SmartSound™ Analog Microphones. Available online: https://invensense.tdk.com/smartsound/analog/ (accessed on 27 November 2024).

5. Naderyan, V.; Lee, S.; Sharma, A.; Wakefield, N.; Kuntzman, M.; Ma, Y.; Da Silva, M.; Pedersen, M. MEMS microphone with 73 dBA SNR in a 4 mm × 3 mm × 1.2 mm Package. In Proceedings of the 21st International Conference on Solid-State Sensors, Actuators and Microsystems (Transducers), Orlando, FL, USA, 20–25 June 2021; pp. 242–245. [CrossRef]

6. Bosetti, G.; Bretthauer, C.; Bogner, A.; Krenzer, M.; Gierl, K.; Timme, H.J.; Heiss, H.; Schrag, G. A Novel High-SNR Full Bandwidth Piezoelectric MEMS Microphone Based on a Fully Clamped Aluminum Nitride Corrugated Membrane. In Proceedings of the 22nd International Conference on Solid-State Sensors, Actuators and Microsystems (Transducers), Kyoto, Japan, 25–29 June 2023; pp. 366–369.

7. Pezone, R.; Anzinger, S.; Baglioni, G.; Wasisto, H.S.; Sarro, P.M.; Steeneken, P.G.; Vollebregt, S. Highly-sensitive wafer-scale transfer-free graphene MEMS condenser microphones. *Microsyst. Nanoeng.* **2024**, *10*, 27. [CrossRef] [PubMed]

8. Anzinger, S.; Wasisto, H.S.; Basavanna, A.; Fueldner, M.; Dehé, A. Non-Linear Behavioral Modeling of Capacitive MEMS Microphones. In Proceedings of the 36th IEEE International Conference on Micro Electro Mechanical Systems (MEMS), Munich, Germany, 15–19 January 2023; pp. 973–976. [CrossRef]

9. Shubham, S.; Nawaz, M.; Song, X.; Seo, Y.; Zaman, M.F.; Kuntzman, M.L.; Pedersen, M. A behavioral nonlinear modeling implementation for MEMS capacitive microphones. *Sens. Actuators A Phys.* **2024**, *371*, 115294. [CrossRef]

10. Pedersen, M.; Naderyan, V.; Loeppert, P. Novel design method for sub-miniature MEMS microphones with enhanced acoustic SNR. *J. Acoust. Soc. Am.* **2023**, *153*, A144. [CrossRef]

11. Naderyan, V.; Pedersen, M.; Loeppert, P.V. Sub-Miniature Microphone. Available online: https://patents.google.com/patent/US11509980B2/en (accessed on 2 January 2025).

12. White Paper: How Optical Technology Enables a Generation Shift in MEMS Microphone Performance. Available online: https://sensibel.com/introducing-sensibels-optical-mems-white-paper/ (accessed on 2 January 2025).

13. Shubham, S.; Seo, Y.; Naderyan, V.; Song, X.; Frank, A.; Johnson, J.; da Silva, M.; Pedersen, M. A Novel MEMS Capacitive Microphone with Semi-Constrained Diaphragm Supported with Center and Peripheral Backplate Protrusions. *Micromachines* **2022**, *13*, 22. [CrossRef]

14. Šimonová, K.; Honzík, P. Modeling of MEMS Transducers with Perforated Moving Electrodes. *Micromachines* **2023**, *14*, 921. [CrossRef]

15. Peng, T.H.; Huang, J.H. The Effect of Compliant Backplate on Capacitive MEMS Microphones. *IEEE Sens. J.* **2024**, *24*, 17803–17811. [CrossRef]

16. Rufer, L.; Esteves, J.; Ekeom, D.; Basrour, S. Piezoelectric Micromachined Microphone with High Acoustic Overload Point and with Electrically Controlled Sensitivity. *Micromachines* **2024**, *15*, 879. [CrossRef] [PubMed]

17. Garud, M.; Pratap, R. MEMS Audio Speakers. *J. Micromech. Microeng.* **2023**, *34*, 013001. [CrossRef]

18. Cheng, H.; Fang, W. THD improvement of piezoelectric MEMS speakers by dual cantilever units with well-designed resonant frequencies. *Sens. Actuators A Phys.* **2024**, *377*, 115717. [CrossRef]

19. Cheng, H.H.; Lo, S.C.; Wang, Y.J.; Chen, Y.C.; Lai, W.C.; Hsieh, M.L.; Wu, M.; Fang, W. Piezoelectric Microspeaker Using Novel Driving Approach and Electrode Design for Frequency Range Improvement. In Proceedings of the 33rd International Conference on Micro Electro Mechanical Systems (MEMS), Vancouver, BC, Canada, 18–22 January 2020; pp. 513–516.

20. Wang, Y.J.; Lo, S.C.; Hsieh, M.L.; Wang, S.D.; Chen, Y.C.; Wu, M.; Fang, W. Multi-Way In-Phase/Out-of-Phase Driving Cantilever Array for Performance Enhancement of PZT MEMS Microspeaker. In Proceedings of the 34th International Conference on Micro Electro Mechanical Systems (MEMS), Gainesville, FL, USA, 25–29 January 2021; pp. 83–84.

21. Gazzola, C.; Zega, V.; Cerini, F.; Adorno, S.; Corigliano, A. On the Design and Modeling of a Full-Range Piezoelectric MEMS Loudspeaker for In-Ear Applications. *J. Microelectromech. Syst.* **2023**, *32*, 626–637. [CrossRef]

22. Bhuiyan, M.E.H.; Palit, P.; Pourkamali, S. Electrostatic MEMS Speakers With Embedded Vertical Actuation. *J. Microelectromech. Syst.* **2024**, *33*, 446–455. [CrossRef]

23. Kaiser, B.; Schenk, H.A.G.; Ehrig, L.; Wall, F.; Monsalve, J.M.; Langa, S.; Stolz, M.; Melnikov, A.; Conrad, H.; Schuffenhauer, D.; et al. The push-pull principle: An electrostatic actuator concept for low distortion acoustic transducers. *Microsyst. Nanoeng.* **2022**, *8*, 125. [CrossRef] [PubMed]

24. Wang, H.; Ma, Y.; Zheng, Q.; Cao, K.; Lu, Y.; Xie, H. Review of Recent Development of MEMS Speakers. *Micromachines* **2021**, *12*, 1257. [CrossRef]

25. Liechti, R.; Durand, S.; Hilt, T.; Casset, F.; Poulain, C.; Le Rhun, G.; Pavageau, F.; Kuentz, H.; Colin, M. Total Harmonic Distortion of a Piezoelectric MEMS Loudspeaker in an IEC 60318-4 Coupler Estimation Using Static Measurements and a Nonlinear State Space Model. *Micromachines* **2021**, *12*, 1437. [CrossRef] [PubMed]

26. Teuschel, M.; Heyes, P.; Horvath, S.; Novotny, C.; Rusconi Clerici, A. Temperature Stable Piezoelectric Imprint of Epitaxial Grown PZT for Zero-Bias Driving MEMS Actuator Operation. *Micromachines* **2022**, *13*, 1705. [CrossRef]

27. Knowles, MEMS Microphone MM20-33366-000, Datasheet. Knowles Electronics LLC: Itasca, IL, USA, 2003; pp. 1–12. Available online: https://www.knowles.com/docs/default-source/default-document-library/mm20-33366-000.pdf?Status=Master&sfvrsn=5bb171b1_0 (accessed on 2 January 2025).

28. Prochazka, L.; Huber, A.; Schneider, M.; Ghafoor, N.; Birch, J.; Pfiffner, F. Novel Fabrication Technology for Clamped Micron-Thick Titanium Diaphragms Used for the Packaging of an Implantable MEMS Acoustic Transducer. *Micromachines* **2022**, *13*, 74. [CrossRef]

29. Knowles, Multimode Digital Vibration Sensor V2S200D, Datasheet. Knowles Electronics LLC: Itasca, IL, USA, 2003; pp. 1–12. Available online: https://www.knowles.com/docs/default-source/default-document-library/kno_v2s200d-datasheet.pdf (accessed on 2 January 2025).

30. Clemens, P. Sonion Voice Pick Up (VPU) Sensor. In Proceedings of the Humanizing the Digital Experience: TDK Developers Conference, Santa Clara, CA, USA, 17–18 September 2018; pp. 1–38.

31. Knowles, Balanced Armature Driver RAQ-33862-000, Datasheet. Knowles Electronics LLC: Itasca, IL, USA, 2003; pp. 1–2. Available online: https://www.knowles.com/docs/default-source/default-document-library/receiver-datasheet-raq-33862-000_woprelim.pdf (accessed on 2 January 2025).

32. Becker, D.; Littwin, M.; Bittner, A.; Dehé, A. Acoustic Transmission Measurements for Extracting the Mechanical Properties of Complex 3D MEMS Transducers. *Micromachines* **2024**, *15*, 1070. [CrossRef]

33. Neggers, J.; Hoefnagels, J.P.M.; Geers, M.G.D. On the validity regime of the bulge equations. *J. Mater. Res.* **2012**, *5*, 1245–1250. [CrossRef]

34. Martins, P.; Delobelle, P.; Malhaire, C.; Brida, S.; Barbier, D. Bulge test and AFM point deflection method, two technics for the mechanical characterization of very low stiffness freestanding films. *Eur. Phys. J. Appl. Phys.* **2009**, *45*, 10501. [CrossRef]

35. Petitgrand, S.; Bosseboeuf, A. Simultaneous mapping of out-of-plane and in-plane vibrations of MEMS with (sub)nanometer resolution. *J. Micromech. Microeng.* **2004**, *14*, S97–S101. [CrossRef]

36. Ennen, M.; Lherbette, M.L. In-Plane and Out-of-Plane Analyses of Encapsulated Mems Device by IR Laser Vibrometry. In Proceedings of the 35th IEEE International Conference on Micro Electro Mechanical Systems Conference (MEMS), Tokyo, Japan, 9–13 January 2022; pp. 814–817.

37. Del Rey, R.; Alba, J.; Rodríguez, C.J.; Bertó, L. Characterization of New Sustainable Acoustic Solutions in a Reduced Sized Transmission Chamber. *Buildings* **2019**, *9*, 60. [CrossRef]

38. Urbán, D.; Roozen, N.B.; Jandák, V.; Brothánek, M.; Jiříček, O. On the Determination of Acoustic Properties of Membrane Type Structural Skin Elements by Means of Surface Displacements. *Appl. Sci.* **2021**, *11*, 10357. [CrossRef]

39. Ollivier, S.; Desjouy, C.; Yuldashev, P.; Koumela, A.; Salze, E.; Karzova, M.; Rufer, L.; Blanc-Benon, P. Calibration of high frequency MEMS microphones and pressure sensors in the range 10 kHz–1 MHz. *J. Acoust. Soc. Am.* **2015**, *138*, 1823. [CrossRef]

40. Ollivier, S.; Desjouy, C.; Salze, E.; Yuldashev, P.; Koumela, A.; Rufer, L.; Blanc-Benon, P. Caractérisation de la réponse capteurs de pression et de microphones MEMS. In Proceedings of the 12th Congrès Français d'Acoustique (CFA) Joint avec le Colloque Vibrations, Shocks, and Noise (VISHNO), Le Mans, France, 11–15 April 2016; pp. 833–837.

41. Dessein, A. Modelling Distortion in Condenser Microphones. Ph.D. Thesis, Technical University of Denmark, DTU, Lyngby, Denmark, 2009.

42. Chowdhury, S.; Ahmadi, M.; Miller, W.C. Nonlinear effects in MEMS capacitive microphone design. In Proceedings of the International Conference on MEMS, NANO, and Smart Systems, Banff, AB, Canada, 23 July 2003; pp. 297–302. [CrossRef]

43. Frederiksen, E. Reduction of non-linear distortion in condenser microphones by using negative load capacitance. In Proceedings of the International Congress on Noise Control Engineering, Liverpool, UK, 30 July–2 August 1996.

44. Pastillé, H. Electrically manifested distortions of condenser microphones in audio frequency circuits. *J. Audio Eng. Soc.* **2000**, *48*, 559–563.

45. Abuelma'atti, M.T. Improved analysis of the electrically manifested distortions of condenser microphones. *Appl. Acoust.* **2003**, *64*, 471–480. [CrossRef]

46. Djuric, S.V. Distortion in microphones. In Proceedings of the IEEE International Conference on Acoustics, Speech, and Signal Processing, Philadelphia, PA, USA, 12–14 April 1976; pp. 537–539.

47. Novak, A.; Simon, L.; Lotton, P. A simple predistortion technique for suppression of nonlinear effects in periodic signals generated by nonlinear transducers. *J. Sound Vib.* **2018**, *420*, 104–113. [CrossRef]

48. Novak, A.; Honzík, P. Measurement of nonlinear distortion of MEMS microphones. *Appl. Acoust.* **2021**, *175*, 107802. [CrossRef]

49. Pedersen, M.; Olthuis, W.; Bergveld, P. Harmonic distortion in silicon condenser microphones. *J. Acoust. Soc. Am.* **1997**, *102*, 1582–1587. [CrossRef]

50. Boales, J.A.; Erramilli, S.; Mohanty, P. Measurement of nonlinear piezoelectric coefficients using a micromechanical resonator. *Appl. Phys. Lett.* **2018**, *113*, 20. [CrossRef]

51. Arora, N.; Singh, P.; Kumar, R.; Pratap, R.; Naik, A. Mixed Nonlinear Response and Transition of Nonlinearity in a Piezoelectric Membrane. *ACS Appl. Electron. Mater.* **2024**, *6*, 155–162. [CrossRef]

52. Peng, C.; Shen, G.; Zhang, Y.; Li, Y.; Tan, K. BeepBeep: A high accuracy acoustic ranging system using COTS mobile devices. In Proceedings of the 5th ACM Conference on Embedded Networked Sensor Systems (SenSys), Sydney, Australia, 6–9 November 2007; pp. 1–14. [CrossRef]

53. Hedley, R.W.; Joubert, B.; Bains, H.K.; Bayne, E.M. Acoustic detection of gunshots to improve measurement and mapping of hunting activity. *Wildl. Soc. Bull.* **2022**, *46*, e1370. [CrossRef]

54. Astapov, S.; Berdnikova, J.; Ehala, J.; Kaugerand, J.; Preden, J.-S. Gunshot acoustic event identification and shooter localization in a WSN of asynchronous multichannel acoustic ground sensors. *Multidim. Syst. Sign. Process.* **2018**, *29*, 563–595. [CrossRef]

55. Zhou, Z.; Rufer, L.; Salze, E.; Yuldashev, P.; Ollivier, S.; Wong, M. Bulk micro-machined wide-band aero-acoustic microphone and its application to acoustic ranging. *J. Micromech. Microeng.* **2013**, *23*, 1094–1100. [CrossRef]

56. Wang, Y.; Shen, J.; Zheng, Y. Push the Limit of Acoustic Gesture Recognition. *IEEE Trans. Mob. Comput.* **2020**, *21*, 1798–1811. [CrossRef]

57. Pan, J.; Bai, C.; Zheng, Q.; Xie, H. Review of Piezoelectric Micromachined Ultrasonic Transducers for Rangefinders. *Micromachines* **2023**, *14*, 374. [CrossRef] [PubMed]

58. Zhang, Z.; Yu, Y.; Chen, L.; Chen, R. Hybrid Indoor Positioning System Based on Acoustic Ranging and Wi-Fi Fingerprinting under NLOS Environments. *Remote Sens.* **2023**, *15*, 3520. [CrossRef]

59. Chowdhury, S.; Ahmadi, M.; Miller, W.C. Design of a MEMS Acoustical Beamforming Sensor Microarray. *IEEE Sens. J.* **2002**, *2*, 617–627. [CrossRef]

60. Shams, Q.A.; Graves, S.S.; Bartram, S.M.; Sealey, B.S.; Comeaux, T. Development of MEMS microphone array technology for aeroacoustic testing. *J. Acoust. Soc. Am.* **2004**, *116*, 2511. [CrossRef]

61. del Val, L.; Izquierdo, A.; Villacorta, J.J.; Suárez, L. Using a Planar Array of MEMS Microphones to Obtain Acoustic Images of a Fan Matrix. *J. Sens.* **2017**, *2017*, 3209142. [CrossRef]

62. Ahlefeldt, T.; Haxter, S.; Spehr, C.; Ernst, D.; Kleindienst, T. Road to Acquisition: Preparing a MEMS Microphone Array for Measurement of Fuselage Surface Pressure Fluctuations. *Micromachines* **2021**, *12*, 961. [CrossRef] [PubMed]

63. Wu, Z.; Liu, W.; Tong, Z.; Cai, Y.; Sun, C.; Lou, L. Tuning Characteristics of AlN-Based Piezoelectric Micromachined Ultrasonic Transducers Using DC Bias Voltage. *IEEE Trans. Electron Devices* **2022**, *69*, 729–735. [CrossRef]

64. Nastro, A.; Ferrari, M.; Rufer, L.; Basrour, S.; Ferrari, V. Piezoelectric MEMS Acoustic Transducer with Electrically-Tunable Resonant Frequency. *Micromachines* **2022**, *13*, 96. [CrossRef] [PubMed]

 *micromachines*

Article

A Novel MEMS Capacitive Microphone with Semiconstrained Diaphragm Supported with Center and Peripheral Backplate Protrusions

Shubham Shubham *, Yoonho Seo, Vahid Naderyan, Xin Song, Anthony J. Frank II, Jeremy Thomas Morley Greenham Johnson, Mark da Silva and Michael Pedersen

Knowles Corporation, LLC., Itasca, IL 60143, USA; yoonho.seo@knowles.com (Y.S.);
vahid.naderyan@knowles.com (V.N.); xin.song@knowles.com (X.S.); tony.frank@knowles.com (A.J.F.II);
jeremy.johnson@knowles.com (J.T.M.G.J.); mark.dasilva@knowles.com (M.d.S.);
michael.pedersen@knowles.com (M.P.)
* Correspondence: shubham.shubham@knowles.com

Abstract: Audio applications such as mobile phones, hearing aids, true wireless stereo earphones, and Internet of Things devices demand small size, high performance, and reduced cost. Microelectromechanical system (MEMS) capacitive microphones fulfill these requirements with improved reliability and specifications related to sensitivity, signal-to-noise ratio (SNR), distortion, and dynamic range when compared to their electret condenser microphone counterparts. We present the design and modeling of a semiconstrained polysilicon diaphragm with flexible springs that are simply supported under bias voltage with a center and eight peripheral protrusions extending from the backplate. The flexible springs attached to the diaphragm reduce the residual film stress effect more effectively compared to constrained diaphragms. The center and peripheral protrusions from the backplate further increase the effective area, linearity, and sensitivity of the diaphragm when the diaphragm engages with these protrusions under an applied bias voltage. Finite element modeling approaches have been implemented to estimate deflection, compliance, and resonance. We report an 85% increase in the effective area of the diaphragm in this configuration with respect to a constrained diaphragm and a 48% increase with respect to a simply supported diaphragm without the center protrusion. Under the applied bias, the effective area further increases by an additional 15% as compared to the unbiased diaphragm effective area. A lumped element model has been also developed to predict the mechanical and electrical behavior of the microphone. With an applied bias, the microphone has a sensitivity of −38 dB (ref. 1 V/Pa at 1 kHz) and an SNR of 67 dBA measured in a 3.25 mm × 1.9 mm × 0.9 mm package including an analog ASIC.

Keywords: MEMS; capacitive microphone; finite element modeling; reduced order modeling; effective area; peripheral protrusion; center protrusion; serpentine spring

 check for updates

Citation: Shubham, S.; Seo, Y.; Naderyan, V.; Song, X.; Frank, A.J., II; Johnson, J.T.M.G.; da Silva, M.; Pedersen, M. A Novel MEMS Capacitive Microphone with Semiconstrained Diaphragm Supported with Center and Peripheral Backplate Protrusions. *Micromachines* **2022**, *13*, 22. https://doi.org/10.3390/mi13010022

Academic Editor: Libor Rufer

Received: 4 December 2021
Accepted: 22 December 2021
Published: 25 December 2021

Publisher's Note: MDPI stays neutral with regard to jurisdictional claims in published maps and institutional affiliations.

1. Introduction

After the commercialization of the first microelectromechanical system (MEMS) microphone by Knowles in 2002 [1], the microphone market has witnessed a giant leap toward high-performance audio applications meant for consumer electronics, automotive, hearing aids, military, and aerospace markets. The capacitive MEMS microphone industry has attracted significant interest because these designs facilitate small size, reduced cost, low power consumption, high sensitivity, low noise, flat frequency response, and good stability with respect to temperature and humidity [2,3]. The silicon micromachining technology enables high-volume production of MEMS microphone devices with an outstanding level of miniaturization that can be achieved within an area less than 1 mm^2 [4]. Alternate transduction mechanisms, i.e., piezoelectric [5], piezoresistive [6,7], and optical [8,9], have also been demonstrated to convert the sound pressure waves into the electrical signal.

In order to design a MEMS microphone and predict its behavior, simplified network modeling is typically used due to the complex nature of the coupled acoustical, mechanical, and electrical components. Extensive network modeling approaches have been discussed over the past years and have been published [10,11]. To improve the sensitivity, a diaphragm having slits at the edges was proposed by Yoo et al. [12]. Circular diaphragms with and without slits at the edges of the diaphragm have been investigated, in which the diaphragm with slits shows a higher displacement. A deflected membrane causes a change in the effective area of the diaphragm, and the capacitance is decreased, causing lower sensitivity. In order to overcome this problem, the authors in [13] suggested two coupled membranes in which the first membrane closer to the port experiences the sound pressure and the other membrane attached at the center of the first membrane acts as a moving electrode.

Thin-film diaphragms usually are affected by residual stresses of the film, which can cause issues related to high diaphragm deflection and, even in some cases, buckling of the diaphragm after the release process. The mechanical compliance of the membranes is further limited by the residual stresses in the released layer, potentially altering sensitivity. Different techniques have been used in the past to reduce the residual stresses in the thin films, such as corrugations in the diaphragms, slits on the diaphragm, spring-supported structures, and controlling the deposition parameters in the fabrication process. With the increase in the number of corrugations in the diaphragm, the mechanical sensitivity can be increased [14]. However, this trend is not always observed to be true, as the mechanical sensitivity tends to reduce after a certain number of corrugations as it begins to stiffen up the diaphragm further to acoustic level pressure loads. A strong function of corrugation width with number of corrugations is discussed in [15]. In [16], finite element analysis was used to optimize the number of corrugations and corrugation depth. The diaphragm with different types of springs was analyzed analytically, and FEA simulations were performed to check and compare their sensitivities. In [17], a frog-arm-supported diaphragm is presented, which shows higher sensitivities as compared to that with a simple clamped diaphragm. Sui et al. [18] analyze how a spring-supported diaphragm can achieve improved sensitivity with a reduction in diaphragm radius as compared to the simple clamped design [18]. Instead of using a flexible diaphragm, a microphone consisting of a rigid diaphragm supported by flexible springs is proposed [19], which helped reduce the diaphragm deformation caused by thin-film residual stress. This design was further improved in terms of sensitivity, signal-to-noise ratio (SNR), and bandwidth by using a flexible V-shaped spring, silicon nitride electrical isolation, and the ring-type oxide/Polysilicon mesa, respectively [20].

Other research studies (on dual backplate designs) have also shown advantages in sensitivity and SNR compared to a single-backplate microphone [21–24]. SNR, a key indicator of performance, is the most important parameter of MEMS microphones and is defined as the difference between a microphone's sensitivity and its noise floor and is expressed in decibels (dB, A-weighted). Current MEMS microphones with analog application-specific integrated circuits (ASIC) can be found in the range of about 60 dB to 73 dB SNR [25] by balancing high sensitivity and low self-noise of the microphone. Improving the SNR can also be achieved by designing a fully differential architecture double-diaphragm [26] or double-backplate microphone with performance benefits [27].

In this paper, we will focus on improving the SNR by increasing the effective area of the diaphragm with peripheral and center protrusions on the backplate acting like a simply supported boundary condition for the diaphragm. The polysilicon diaphragm is suspended using serpentine springs to minimize the residual film stress of the diaphragm. In addition, analytical, finite element, and lumped element models are established to predict the electroacoustic behavior of the proposed microphone whose performance is in good agreement with the measurements.

2. Microphone Structure and Design

A generic Knowles capacitive MEMS microphone is represented in Figure 1a. The cross-section of the structure consists of a polysilicon diaphragm suspended on springs that are constrained at the perimeter region, as shown in Figure 1b. When a bias voltage is applied, the diaphragm is electrostatically attracted to the backplate but simply supported with a center and eight peripheral protrusions (or posts) that extend from the backplate. A variable capacitor is formed by the relatively rigid backplate and the movable diaphragm. In response to acoustic pressure changes, the capacitance of the diaphragm changes, and this change is sensed and converted into a proportional electrical voltage through the ASIC.

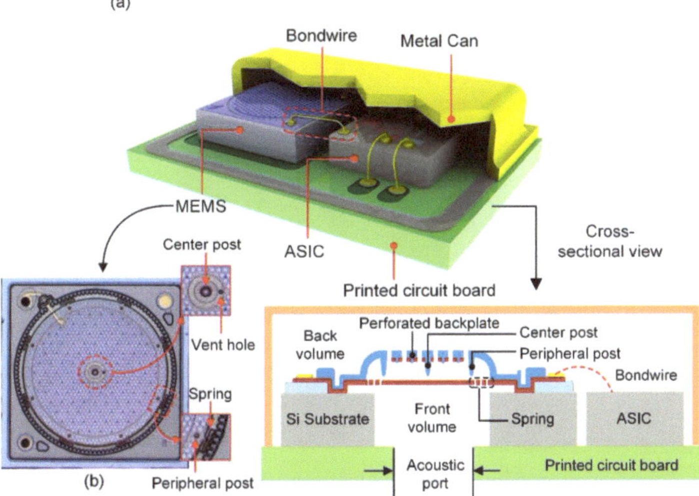

Figure 1. (**a**) A schematic cross-sectional view of a generic Knowles MEMS microphone package. (**b**) A top–down optical image of the MEMS die collected using a microscope. The movable diaphragm is suspended with eight compliant springs along the perimeter. The center and eight peripheral posts on the backplate prevent the movable diaphragm from electrostatic collapse onto the backplate.

An optical microscope image of the MEMS die is shown in Figure 1b. It clearly shows the eight serpentine spring structures of the diaphragm which are aligned with the peripheral posts. The serpentine springs are anchored with the perimeter anchor structure of the backplate. A small vent hole is located near the center post to allow ventilation of the pressure between the front and back volume of the microphone. This is used to control the low-frequency roll-off (LFRO) by forming an acoustical high pass filter for sound. The bias voltage is applied to the diaphragm bond pad, and the change in capacitance as output signal is measured across the backplate bond pad.

Details of the MEMS structure are visible in Figure 2a–d. Figure 2a shows the top–down view of the MEMS die. Figure 2b is a close-up view of the serpentine spring that comprises spring arms with different lengths. The long arm connected to anchor fingers, as shown in Figure 2b, provides enough mechanical compliance to achieve the desired high sensitivity, whereas the short arm connected to the diaphragm provides torsional rigidity to the diaphragm and more uniform lift-off under bias voltage. Figure 2b also shows the antistiction dimple structure present underneath the diaphragm.

Figure 2. SEM images of critical features of the proposed design. (**a**) Top view of the MEMS microphone die, (**b**) serpentine spring design, (**c**) perforated backplate with antistiction bumps, and (**d**) backplate anchor design to substrate.

Figure 2c presents perforation holes formed in the backplate. Figure 2d shows the backplate anchor design to the substrate.

3. Microfabrication

The schematic in Figure 3 shows a simplified MEMS fabrication process for the MEMS device. The first sacrificial layer deposition is followed by polysilicon diaphragm deposition and etch (Figure 3a). The next sacrificial layer deposition will form a gap between the diaphragm and backplate electrode and also define the shape of the peripheral and center posts (Figure 3b). After that, the polysilicon is deposited and patterned to form an electrode for a backplate formed by the subsequent Si_3N_4 deposition (Figure 3c). The backplate is patterned to form acoustic holes, and gold is deposited to form metal contact pads. The back cavity hole is formed by a silicon deep-reactive ion etch process. Then, the sacrificial oxide is removed (Figure 3d).

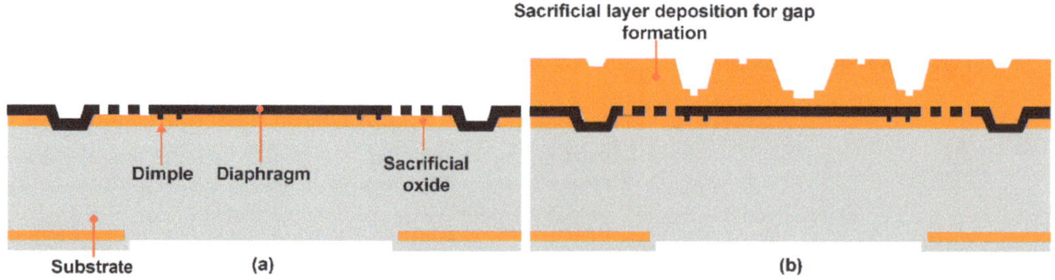

Figure 3. *Cont.*

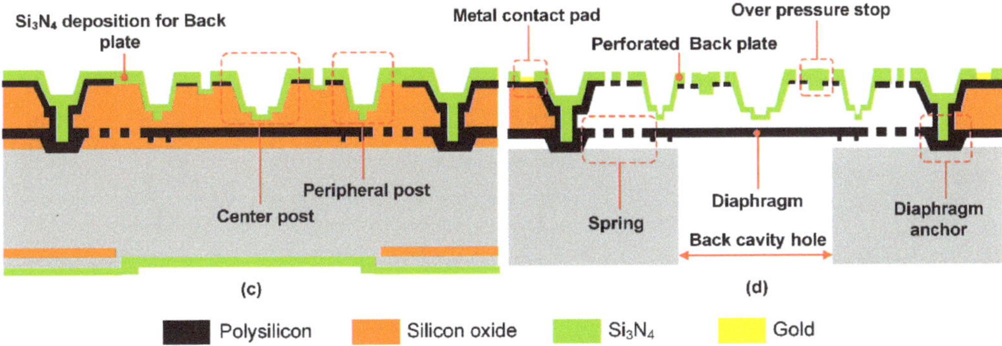

Figure 3. Proposed design fabrication sequence: (**a**) 1st sacrificial layer deposition and polysilicon deposition for diaphragm with dimples. (**b**) Sensing gap formation with sacrificial layer deposition and patterning. (**c**) Si$_3$N$_4$ deposition for formation of backplate with polysilicon electrode with a center and peripheral protrusions. (**d**) Backplate patterning, formation of the back cavity, and gold metal pads with sacrificial layer release.

4. Reduced Order Modeling

A proposed reduced order model for the MEMS microphone is shown in Figure 4. The diaphragm mechanical model is coupled with acoustic and electrical domains on its left- and right-hand side, respectively. This coupling is represented using transformers, which account for flow and effort variables in each domain, as shown in Table 1. The flow of **air** or sound pressure represented in the acoustic domain creates mechanical movement of the diaphragm, which further drives the electric circuit by changing the capacitance between the motor and electrode.

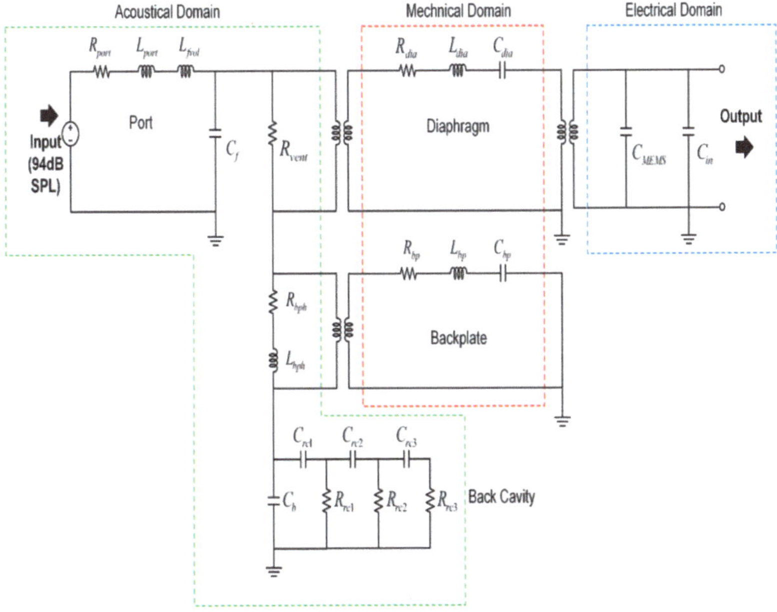

Figure 4. A reduced order model for a single-motor MEMS microphone.

Table 1. Description of domains in an acoustomechanical lumped model.

Domain	Flow Variable	Effort Variable	Impedance	Units of Impedance
Acoustical	volume velocity, U	pressure, P	$Z_a = \frac{P}{U}$	$\frac{Pa \cdot s}{m^3}$ or $\frac{N \cdot s}{m^5}$
Mechanical	velocity, v	force, F	$Z_m = \frac{F}{v}$	$\frac{N \cdot s}{m}$
Electrical	current, i	voltage, V	$Z_e = \frac{V}{i}$	Ω

In terms of the elements such as mass, spring, and damper, the lumped equivalent mechanical, electrical, and acoustical models are represented in Table 2.

Table 2. Lumped elements in a microphone model.

Element	Mechanical Equivalent Model	Electrical Equivalent Model	Acoustical Equivalent Model
Mass	M_m. (kg)	Inductance L (H)	$M_a = M_m / A_{d\_eff}^2$ (kg/m^4)
Spring	C_m (m/N)	Capacitance, C (F)	$C_a = C_m A_{d\_eff}^2$ (m^5/N)
Damper	B_m (N.s/m)	Resistance, R (Ω)	$R_a = B_m / A_{d\_eff}^2$ (N·s/m^5)

The description of each element for the proposed reduced order model is presented in Table 3.

Table 3. Description of the lumped elements used in Figure 4.

Category	Parameter	Description	Unit
Acoustical (Port and back cavity)	R_{port}	Acoustic port resistance	N·s/m^5
	L_{port}	Acoustic port inductance	N·s^2/m^5
	L_{fvol}	Front volume inductance	N·s^2/m^5
	C_f	Front volume compliance	m^5/N
Acoustical (Vent)	R_{vent}	Vent flow resistance	N·s/m^5
Acoustical (Backplate)	R_{bph}	Backplate hole resistance	N·s/m^5
	L_{bph}	Backplate hole inductance	N·s^2/m^5
Acoustical (Back cavity volume)	C_b	Back volume compliance	m^5/N
	C_{rc1}, C_{rc2}, C_{rc3}	TBL * compliance term #1, #2, and #3	m^5/N
	R_{rc1}, R_{rc2}, R_{rc3}	TBL damping term #1, #2, and #3	N·s/m^5
Mechanical (Diaphragm)	R_{dia}	Diaphragm mechanical damper	N·s/m
	L_{dia}	Diaphragm mechanical mass	kg
	C_{dia}	Diaphragm mechanical compliance	$\frac{m}{N}$
Mechanical (Backplate)	R_{bp}	Backplate mechanical damper	N·s/m
	L_{bp}	Backplate mechanical mass	kg
	C_{bp}	Backplate mechanical compliance	$\frac{m}{N}$
Electrical (MEMS and ASIC)	C_{MEMS}	Motor Capacitance (including MEMS parasitic capacitances)	pF
	C_{in}	ASIC input Capacitance	pF

* TBL is abbreviated for thermal boundary layer.

The acoustical lumped parameters are related to mechanical lumped parameters using the Equation (1):

$$Z_a = Z_m / A_{d\_eff}^2 \tag{1}$$

where Z_m is the mechanical impedance, Z_a is the acoustical impedance, and A_{d_eff} is the effective diaphragm area.

4.1. Diaphragm Effective Area (Acoustic-to-Mechanical Transformer Ratio)

The transformer that turns the ratio for conversion between the acoustical and mechanical domains is the diaphragm area. An effective area is typically used instead of a geometric one to better represent a distributed system as the network of lumped elements, as shown in Figure 5. The effective area is defined so that the air volume displaced by the lumped element system is equal to the air volume displaced by the distributed system.

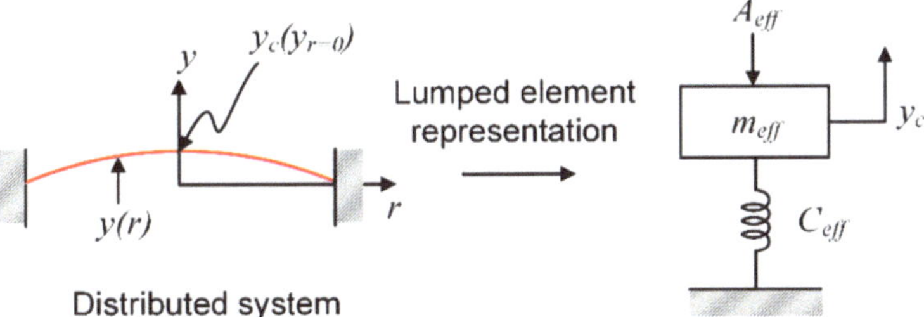

Figure 5. Lumped element representation of a distributed system.

Mathematically it can be stated as,

$$A_{d\_eff} \cdot y_c = \int_0^a y(r)2\pi r dr \tag{2}$$

where the left side of the Equation (2) is the volume displaced by the lumped piston model, and the right side of the Equation (2) is the volume displaced by the distributed system, where y_c is the maximum deflection of the diaphragm, a is the diaphragm radius, and $y(r)$ is the diaphragm deflection as a function of radius.

The Equation (2) can be rearranged to define a lumped area coefficient so that

$$A_{d\_eff} = \beta A_d \tag{3}$$

where A_d is the geometric area of the diaphragm.

The effective area coefficient is defined as

$$\beta = \frac{A_{d\_eff}}{A_d} = \frac{\int_0^a y(r)2\pi r dr}{A_d y_c} \tag{4}$$

The effective area coefficient, β, can be obtained by integrating analytical equations for diaphragm displacement or by numerical integration of actual displacement data for the diaphragm.

To determine β, one must know $y(r)$, the deflected shape of the diaphragm. In this section, we use analytical equations for clamped and simply supported plates to obtain β for each case and then compare the results to ones obtained using FEA.

The expression for obtaining the small deflection shape of a simply supported boundary condition is given by [28]:

$$y(r) = \frac{q(a^2 - r^2)}{64D}\left(\left(\frac{5+v}{1+v}\right)a^2 - r^2\right) \tag{5}$$

where q is the uniform pressure difference across the diaphragm, D is the flexural rigidity for the diaphragm as determined by material properties and diaphragm dimensions, and v is the Poisson's ratio.

However, the constrained diaphragm or clamped boundary condition with no residual stress is given by [28]:

$$y(r) = \frac{q}{64D}\left(a^2 - r^2\right)^2 \tag{6}$$

Substituting Equations (5) and (6) above into the Equation (4) for β and solving the integral yields:

$\beta = 0.46$ for a simply supported plate and $\beta = \frac{1}{3} \approx 0.33$ for a constrained diaphragm. Both calculations are for a generic polysilicon diaphragm with assumed elastic properties.

An FEA approach has been used to verify these analytical results and the effective area coefficient, β, for peripheral posts and center post boundary conditions for free plate diaphragm without springs and a semiconstrained diaphragm with springs. Results of the FEA simulations are shown in Figure 6 and summarized in Table 4.

The FEA is set up with a 45-degree symmetric boundary condition, and a small signal load of 1Pa pressure is applied to make the diaphragm deflect. The effective area coefficient, β, is calculated based on Equation (4), defined above.

The diaphragm area, A_d, is calculated to be 5.671×10^{-7} m^2. For a perfect piston-shaped diaphragm deflection, β is ~1. The higher β is, the closer deflection of the diaphragm is to the piston-shaped motion. The diaphragm with the peripheral and center post configuration has an 85% increase in the effective area as compared to the constrained diaphragm. The center post configuration combined with the peripheral posts has a 48% higher effective area compared to the peripheral post configuration without any center post. Under an applied bias voltage, the effective area additionally increases by 15% as compared to unbiased case for the proposed configuration since the diaphragm is more stiffened under bias voltage. This increase is also observed in the case of free plate with peripheral and center post.

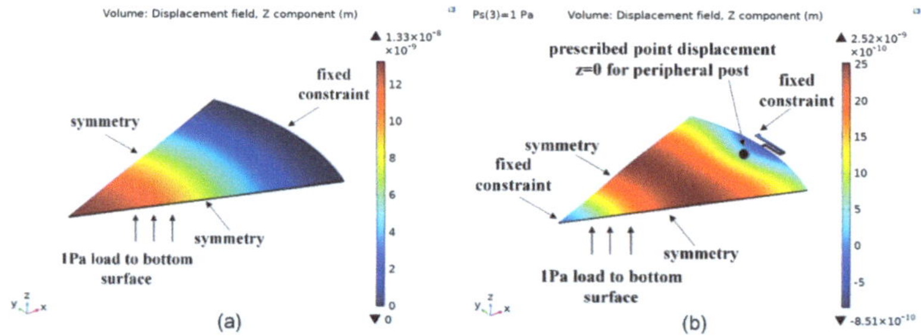

Figure 6. Defined boundary conditions and $z-$displacement field for: (**a**) constrained diaphragm and (**b**) spring$-$supported diaphragm with peripheral and center post under bias.

Table 4. FEA simulated values of effective area coefficient for different diaphragm boundary conditions.

Diaphragm Boundary Condition	Volume Displacement, $\int_0^a y(r)2\pi r dr$ (m^3)	Maximum Displacement, y_c (m)	Simulated Effective Area Coefficient, β
Constrained	2.51×10^{-15}	1.33×10^{-8}	0.33
Simply supported	1.45×10^{-14}	5.57×10^{-8}	0.46
Free plate with Peripheral Post	1.01×10^{-14}	4.29×10^{-8}	0.41
Free plate with Peripheral and Center Post	1.33×10^{-15}	3.86×10^{-9}	0.61
Semiconstrained with Peripheral Post	9.22×10^{-15}	3.95×10^{-8}	0.41
Semiconstrained with Peripheral Post and Center Post	1.3×10^{-15}	3.74×10^{-9}	0.61
Semiconstrained with Peripheral Post and Center Post under bias	1×10^{-15}	2.52×10^{-9}	0.7

4.2. Electrostatic Coupling Coefficient (Mechanical-to-Electrical Transformer Ratio)

The electrostatic coupling coefficient, φ, can be calculated directly from the equations governing ideal transformers and the definitions of the effort and flow variables. Assuming the diaphragm is blocked so that it cannot move, the electrostatic force can be expressed as

$$F = \varphi V_{ac} = \frac{\varepsilon A V^2}{2g_0^2} \tag{7}$$

where expression on the right-hand side is the force between the plates of a parallel plate capacitor. Note that V^2 consists of both a DC component (the bias voltage) and an AC component (the signal). Because we are interested in a linear, small-signal model, the V^2_{bias} (DC) and V^2_{ac} (nonlinear) terms can be dropped, and the expression becomes

$$F = \varphi V_{ac} = \frac{\varepsilon A \left(V^2_{bias} + V^2_{ac} + 2 V_{bias} V_{ac} \right)}{2 g^2_o} \approx \frac{\varepsilon A V_{bias} V_{ac}}{g^2_o} \tag{8}$$

Solving for the coupling coefficient yields

$$\varphi = \frac{\varepsilon A V_{bias}}{g^2_o} = \frac{C_o}{C_{em}} \tag{9}$$

where C_o is the MEMS capacitance at the bias voltage, and $C_{em} = \frac{g_o}{V_{bias}}$ is the reactance of the transducer [29], where g_o is the gap after bias.

4.3. Effective Mass of the Diaphragm

The effective mass of the diaphragm accounts for the nonuniform deflection profile of the diaphragm (not all of the mass is moving by the same amount). Referring to Figure 7, the effective mass used in the lumped model is defined such that kinetic energy is equal for the lumped and distributed representations. The effective mass can be expressed as

$$m_{d\_eff} = \frac{2\pi \rho t}{y_c{}^2} \int_0^a y^2(r) r \, dr \tag{10}$$

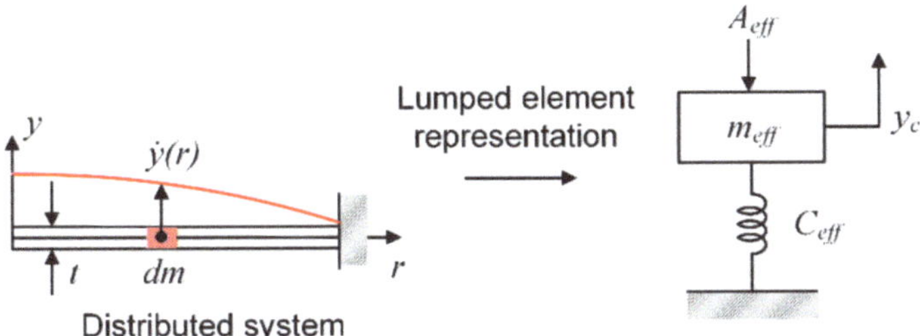

Figure 7. Lumped element representation of a system with distributed mass.

An effective mass coefficient, α, can be defined such that $m_{d\_eff} = \alpha m_d$, which results in

$$\alpha = \frac{m_{d\_eff}}{m_d} = \frac{2\pi \rho t}{m_d y_c{}^2} \int_0^a y^2(r) r \, dr \tag{11}$$

The values of α for free plate and clamped diaphragms are calculated for each case as shown below. Analytical expressions for $y(r)$ are given in Equations (5) and (6) for simply supported and clamped boundary conditions, respectively. Substituting each into the Equation (13) for the effective mass coefficient and solving yields, $\alpha = \frac{m_{d\_eff}}{m_d} = \frac{2\pi \rho t}{m_d y_c{}^2} \int_0^a y^2(r) r \, dr = 0.296$ for a simply supported plate, and $\alpha = \frac{m_{d\_eff}}{m_d} = \frac{2\pi \rho t}{m_d y_c{}^2} \int_0^a y^2(r) r \, dr = 0.2$ for a clamped plate.

Another technique that can be used to extract the effective mass is by measuring the resonance frequency in vacuum, which accounts for the unloaded diaphragm resonance. The effective mass equation is given by

$$m_{d,eff} = \frac{1}{\omega^2_{0,vac} C_{d,mech}} = \frac{A_{d,eff}}{\omega^2_{0,vac} C_{d,cp}} \tag{12}$$

where $\omega_{0,vac}$ is the measured resonance frequency in vacuum and $C_{d,mech}$ is the mechanical compliance of the diaphragm in units of m/N. The mechanical compliance of the diaphragm can be also expressed in terms of effective area, $A_{d,eff}$, of the diaphragm and acoustic compliance, $C_{d,\ cp}$, in units of m/Pa. This technique is used for direct measurements for the semiconstrained MEMS with peripheral posts and center post boundary conditions.

4.4. Diaphragm Compliance

The diaphragm compliance is measured under an applied bias voltage. There is often a negative mechanical compliance included in the lumped models for electrostatic transducers that accounts for a phenomenon commonly referred to as electrostatic softening, whereby the mechanical compliance is effectively increased by the tendency of the electrostatic force to counteract the diaphragm restoring force when the diaphragm is displaced toward the counter electrode [29]. However, the negative compliance is not included in the model presented here, since the MEMS is biased when the compliance is measured. The softening term is already accounted for during the measurement, and no separate term is needed in the model. The compliance values obtained are only valid for the bias at which they are measured. If the bias is changed, the compliance needs to be remeasured.

The compliance of the diaphragm was measured here directly using a Laser Doppler Vibrometer (LDV). The device under test (DUT) is mounted on a pressure cavity, as shown in Figure 8. A small speaker provides an actuation pressure, while the LDV monitors the maximum diaphragm displacement. A reference microphone is used to calibrate the actuation signal to 1Pa. The resulting measurement gives the diaphragm compliance in units of m/Pa, corresponding to specific acoustic impedance. To convert to units of mechanical impedance, multiply by the diaphragm effective area, as shown in Table 5.

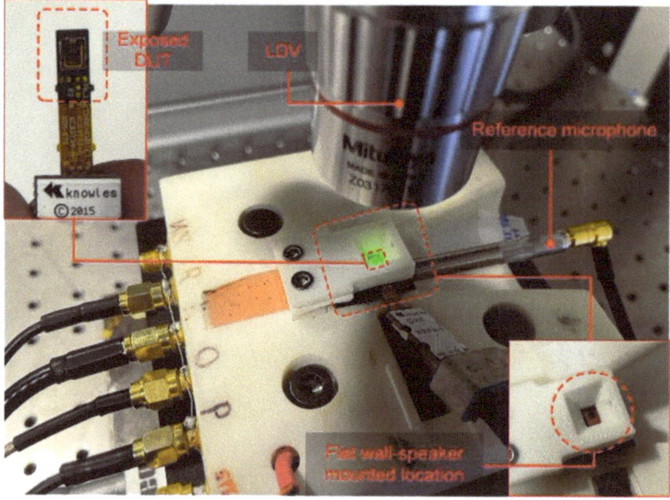

Figure 8. Experimental setup for the diaphragm compliance measurement using a scanning LDV.

Table 5. Summary of techniques for measuring diaphragm compliance.

Expression	Unit	Description
$C_{d,sa} = \frac{\delta_{max}}{p_{ref}}$	m/Pa	Specific Acoustic Impedance
$C_d = \frac{\delta_{max}}{p_{ref} A_{deff}}$	m/N	Mechanical Impedance

An interesting observation was reported with the diaphragm acoustic compliance under bias voltage with the simply supported conditions using peripheral posts and center post design. The acoustic compliance decreased with increasing bias voltage in this configuration as opposed to increasing acoustic compliance with increasing bias voltage from electrostatic spring softening for constrained diaphragms. This particular phenomenon in the simply supported design is usually termed as the electrostatic spring stiffening effect. The apparent stiffening of the diaphragm is driven by the nonlinear static displacement of the diaphragm due to the electrostatic force. The FEA model is used to capture this behavior for the described model. The diaphragm acoustic compliance is plotted against the bias voltage in Figure 9.

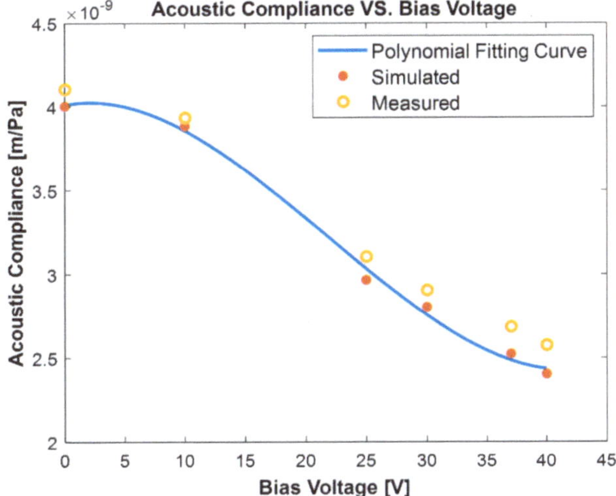

Figure 9. Measurement and simulation of the diaphragm acoustic compliance vs. bias voltage for the semiconstrained with peripheral post and center post boundary condition. The polynomial fitting curve is for the simulation.

5. Diaphragm FEA Model

FEA models have been developed to predict the mechanical response of the MEMS microphone diaphragm considering serpentine spring architecture with peripheral posts and a center post. To capture the fringe field effect due to the perforation holes on the backplate, a separate model was established that can be fed into the electromechanical model to simulate the coupled effect.

All FEA models are obtained through COMSOL Multiphysics® software and a thermoviscous acoustic model was developed to calculate the acoustic damping of the MEMS.

5.1. Unit Cell Capacitance Model

We begin by calculating the correction factor of the electrostatic force and capacitance for the perforated backplate by taking advantage of the uniform hexagonal distribution across the electrode surface, as shown in Figure 10. For the unit cell model setup, we define the air gap and the perfectly matched layer (PML) on the top, as shown in Figure 11a. The permittivity of vacuum for air, Si_3N_4, and polysilicon materials is defined in the material properties section of the module. The electric potential of 1 V is applied to the electrode, and the ground is set to 0 V, as shown in Figure 11b. A fringing field effect can be captured using the 1/6th unit cell FEA. The fringe field distribution near the acoustic hole is as shown in Figure 11c.

The acoustic hole perforation ratio is approximately 54% of the defined electrode area. The electrostatic force and capacitance functions are shown in Figure 12a,b and are represented by fitting polynomials as a function of gap. These functions are used in the mechanical FEA model in the form of correction factors to calculate the motor capacitance and collapse voltage. For a gap of 5 μm, the force correction factor is determined to be 0.7, meaning 30% reduction in force with respect to the solid plate. Similarly, the capacitance correction factor is estimated to be 0.84, meaning 16% reduction in capacitance with respect to the solid plate.

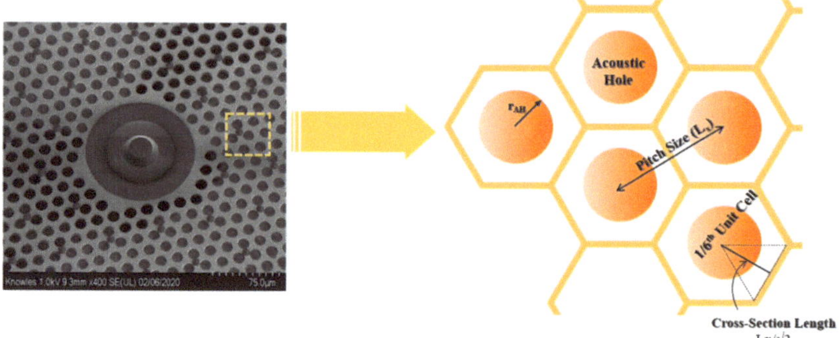

Figure 10. Hexagonal symmetry of the perforated backplate.

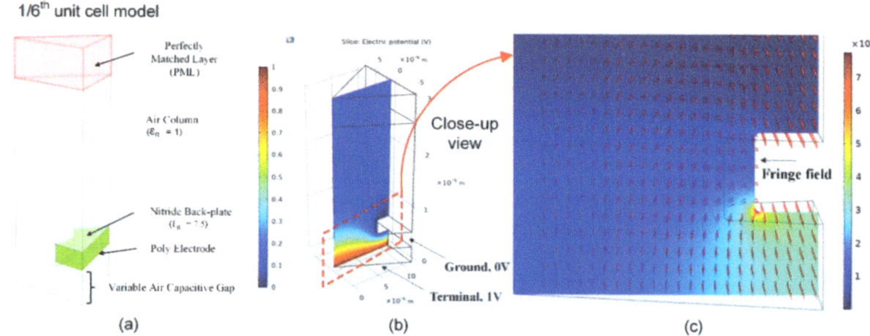

Figure 11. (**a**) Defined boundary condition for the 1/6th unit cell model with Si_3N_4 backplate and underlying polysilicon electrode. (**b**) Electric potential distribution with ground and terminal voltage. (**c**) Electric field distribution with fringing field effect.

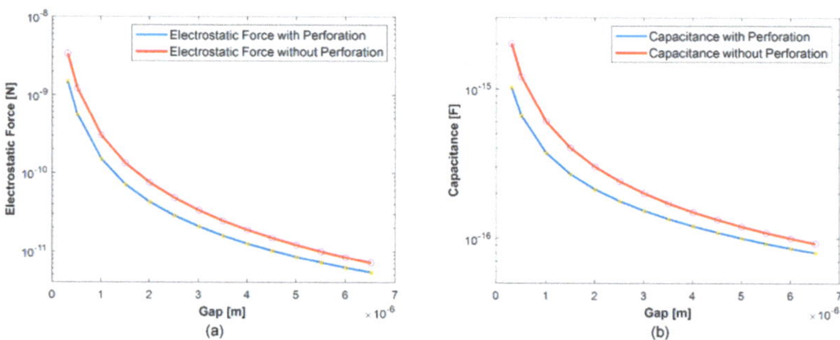

Figure 12. (**a**) Electrostatic force as a function of the gap between the electrodes for solid and perforated plates. (**b**) Capacitance as a function of the gap for solid and perforated plates.

5.2. Electromechanical Model

Symmetry of the diaphragm with a 45-degree segment is utilized to capture deflection, capacitance, and resonance modes under bias voltage. This model is obtained using COMSOL Multiphysics® software with MEMS module. The electrostatic force F is fed into the model as per the coupling Equation (13), defined as

$$F = \frac{\varepsilon A V^2}{2(g_0 - w)^2} \tag{13}$$

where A is the surface area of the diaphragm, g_0 is the initial gap between the diaphragm and backplate electrode, and w is the diaphragm deflection under bias voltage, V. Furthermore, the above electrostatic force equation is multiplied by an analytical correction function to account for the effect of perforations. The material properties and dimensions of the polysilicon diaphragm are tabulated in Table 6, below.

Table 6. Dimensions of the MEMS microphone diaphragm.

Property	Value
Diaphragm diameter	850 (μm)
Diaphragm thickness	1.4 (μm)
Spring long arm length	80 (μm)
Spring short arm length	45 (μm)
Spring width	8 (μm)
Gap between springs	4 (μm)
Spring count	8

The mesh and deflection shape of the diaphragm under bias voltage are shown in Figure 13a–d. Since there is a presence of a rigid center post on the backplate, the deflection shape of the diaphragm looks like that of a 'donut' under the bias voltage. The diaphragm capacitance is also plotted as a function of bias voltage until the diaphragm collapse happens with the backplate at 50 V. The presence of the small kink in Figure 13e at 25 V bias indicates engagement of the diaphragm with the peripheral posts. Figure 13f shows the deflected cross-section shape of the diaphragm under different bias voltages. At diaphragm radius $r = 0$ μm, the diaphragm is deflected to 0.6 μm at the center and is further restricted to move at center due to the presence of a rigid center post.

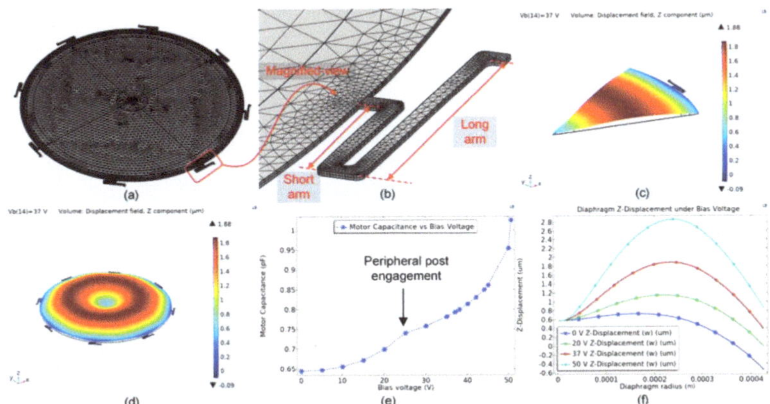

Figure 13. (**a**) Developed fine mesh for the full FEA model. (**b**) Developed mesh for the spring. (**c**) A 45-degree symmetry deflection model of the diaphragm under bias. (**d**) Full diaphragm deflection model showing a donut deflected shape under bias. (**e**) Capacitance as a function of bias voltage with collapse 50 V, parasitic capacitance not included. (**f**) Diaphragm deflection along the radius cross-section at different bias voltages.

5.3. Resonance Modes

The resonance modes of the diaphragm under bias are extracted using the prestressed eigenfrequency study in COMSOL Multiphysics®. The first four vibration modes are as shown in Figure 14a–d. The first vibration mode is an ideal z-displacement vibration mode of the diaphragm. The first vibration mode (54.5 kHz) and the second mode (61.2 kHz) are high enough to not create any interference with the normal operating frequency range of microphone in the 20–20 kHz. The third and fourth vibration modes are even higher than

100 kHz. Note that the resonance obtained here is under vacuum condition, and no air damping has been used in the model. The structural diaphragm resonance under ambient pressure loading is discussed in Section 7.1, with the first resonance mode measured in air to be 39 kHz.

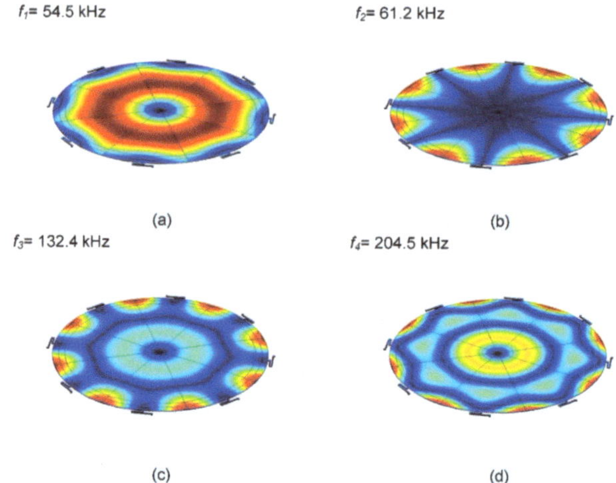

Figure 14. The first four vibration modes of the FEA model of the microphone diaphragm: (a) $f_1 = 54.5$ kHz, (b) $f_2 = 61.2$ kHz, (c) $f_3 = 132.4$ kHz, and (d) $f_4 = 204.5$ kHz. The red and blue surface profiles indicate regions of maximum and minimum deflections, respectively.

The deflection of the diaphragm under bias voltage is plotted against the applied sound pressure, as shown in Figure 15. For the small signal applied acoustic pressure, the diaphragm acoustic compliance can be calculated using the linear slope of the curve to be 2.5 nm/Pa. The acoustic compliance is the deflection averaged over the surface of the diaphragm under the applied sound pressure. The mechanical compliance of the diaphragm is calculated to be 7.154×10^{-3} m/N.

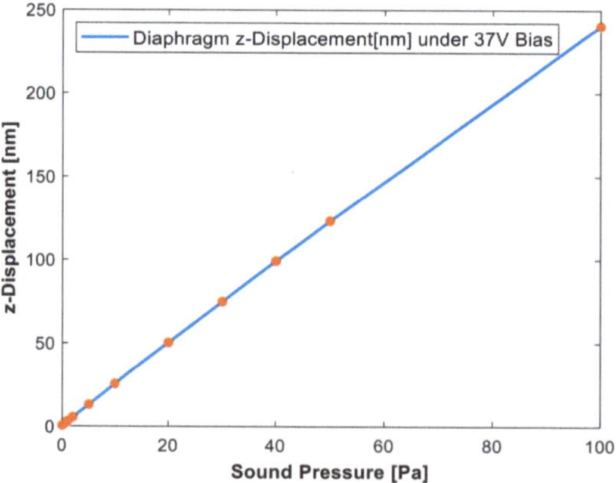

Figure 15. Simulation result of diaphragm deflection under the bias voltage as a function of sound pressure.

6. Noise Sources in MEMS Microphone

The four major acoustic sources of noise in MEMS microphones are acoustic port damping and vent resistance, which includes slit flow (around the perimeter of the diaphragm since it is not fully constrained) in parallel with flow through the pierce (hole punched on the diaphragm near the center post). The four major noise impedances are highlighted in red in the Figure 16. Z^a_{rad}, Z^a_{fv}, Z^a_e, and Z^a_{bv} are the port, front volume, diaphragm, and back volume impedances, respectively. The analytical or FEA approach to obtain these parameters will be discussed in this section.

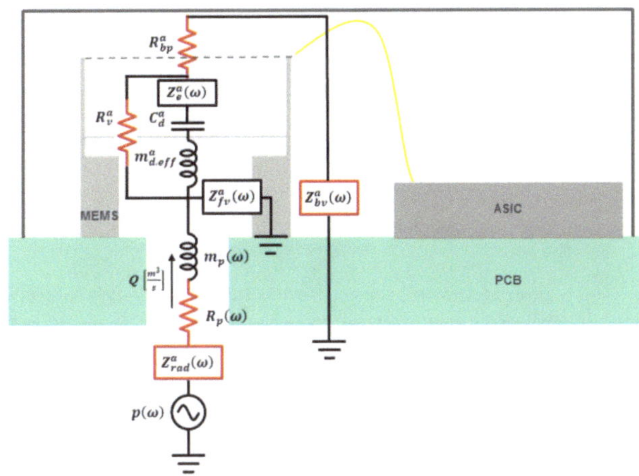

Figure 16. Simplified electroacoustic lumped model representing noise sources.

6.1. Analytical Calculation of Port and Cavity Acoustic Parameters

For the noise model, the dimensions and properties shown in Table 7 were used.

Table 7. Microphone lumped parameter values.

Category	Parameter	Description	Value	Unit
Package	V_b	Back cavity volume	1.676×10^{-9}	m^3
	l_p	Acoustic port length	250×10^{-6}	m
	a_p	Acoustic port radius	175×10^{-6}	m
	V_f	Front cavity volume	0.88×10^{-9}	m^3
MEMS	r_d	Diaphragm radius	425×10^{-6}	m
	t_d	Diaphragm thickness	1.4×10^{-6}	m
	r_{AH}	Acoustic hole radius	8.25×10^{-6}	m
	g_o	Average gap after bias over electrode region	4×10^{-6}	m
	V_{bias}	Bias Voltage	37	V
	$C_{mems,tot}$	MEMS total capacitance	0.9	pF
	C_p	MEMS parasitic capacitance [30]	0.12	pF
	$\omega_{o,air}$	Diaphragm resonance in characterization package, measured in air	39	kHz
	C^a_d	Diaphragm compliance	2.5	nm/Pa
	f_c	Low-frequency corner	35	Hz
	R^a_{bp}	Backplate damping	1.74×10^8	Pa·s/m^3
	$m^a_{d\_eff}$	Effective acoustic diaphragm mass	1.13×10^{-10}	kg
	α	Effective mass coefficient for simply supported plate	0.296	dimensionless
	A_{d_eff}	Effective diaphragm moving area	3.52×10^{-7}	m^2
	β	Effective area coefficient	0.7	dimensionless
	φ	Electrostatic coupling coefficient (Transduction factor)	1.201×10^5	V/N

The acoustical–domain lumped parameters for the acoustic port and cavities can be calculated using well-known analytical expressions, as shown in Table 8.

Table 8. Calculation of port and cavity parameters.

Symbol	Description	Expression	Unit
m_p	Port mass	$m_p = \dfrac{\rho_o\left(l_p+1.7a_p\right)}{\pi a_p^2}$	$\dfrac{\text{N·s}^2}{\text{m}^5}$
R_p	Port resistance	$R_p = \dfrac{8\eta l_p}{\pi a_p^4}$	$\dfrac{\text{N·s}}{\text{m}^5}$
C_f	Front cavity compliance	$C_f = \dfrac{V_f}{\rho_o c^2}$	m^5/N
C_b	Back cavity compliance	$C_b = \dfrac{V_b}{\rho_o c^2}$	m^5/N

Where l_p is the length of the acoustic port (m), a_p is the radius of the acoustic port (m), V_f is the volume of the front cavity (m^3), and V_b is the volume of the back cavity (m^3). The physical constants represented here are density of air, ρ_0; viscosity of air, η; and speed of sound in air, c.

For calculating the port impedance, Beranek's analytical equations are used [31]. The radiation port impedance is given by Equation (14):

$$Z_{rad} = \frac{\rho_o c}{\pi a_p^2}\left[\left(1 - \frac{2J_1\left(2ka_p\right)}{2ka_p}\right) - j\frac{2H_1\left(2ka_p\right)}{2ka_p}\right] \tag{14}$$

where $H_1(z) \approx \frac{2}{\pi} - J_0(z) + \left(\frac{16}{\pi} - 5\right)\frac{\sin z}{z} + \left(12 - \frac{36}{\pi}\right)\frac{1-\cos z}{z^2}$, and $J_1(x)$ is the first-order Bessel function. For an approximation for the Sturve function of the first kind, $H_1(x)$ is used as given by [32]. The plots for port damping and port mass as a function of frequency, obtained using the lumped parameter model, are shown in Figure 17a,b.

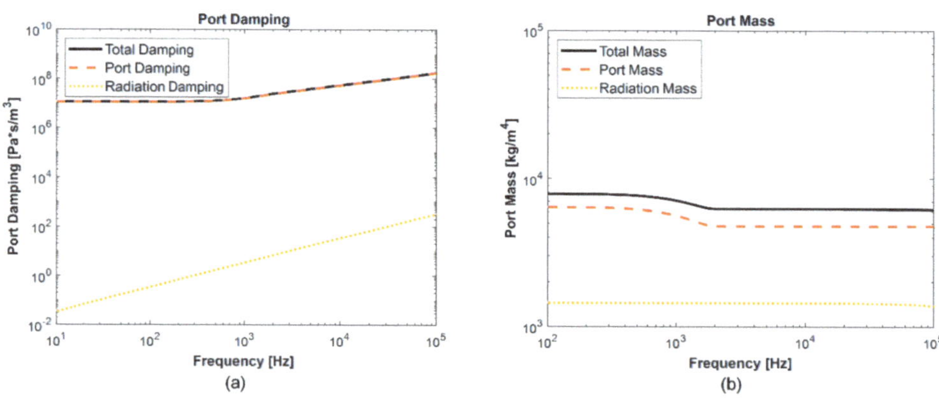

Figure 17. (**a**) Total damping obtained using port damping and radiation damping as function of frequency, and (**b**) total mass obtained using port mass and radiation mass as a function of frequency.

6.2. Calculation of Enclosure Impedance

The thermal boundary layer thickness, δ_t, as a function of frequency is given by

$$\delta_t = \sqrt{\frac{2\kappa}{\omega\rho_0 C_p}} \tag{15}$$

Equation (16) assumes the wall to be an isothermal boundary and ignores the influence of adjacent walls. The thermally corrected impedance for a thin parallelepiped enclosure with a height 2a is given by Equation (16) [33]:

$$Z = \frac{1}{j\omega C_a\left[1 + (\gamma - 1)\left(\frac{\tanh(\beta a)}{\beta a}\right)\right]} \tag{16}$$

where C_a is the adiabatic compliance of the air volume, $\beta = \sqrt{\frac{j\omega\rho_0 C_p}{k}}$, ρ_0 is the density, k is the thermal conductivity, γ is the ratio of specific heats for gas inside the enclosure, and C_p is the specific heat at constant pressure of the gas inside the enclosure. For a more detailed discussion on this topic, see [33].

6.3. Vent Resistance

The LFRO of the microphone is set, effectively, by an RC high-pass filter formed between the total vent resistance, $R_{v,tot}$, and the back volume compliance, $C_{b,tot}$. Because the LFRO can be easily measured and the back volume compliance is generally well-known, the vent resistance can be calculated using Equation (17):

$$R_{v,tot} \approx \frac{1}{2\pi f_{LFRO} C_{b,tot}} \tag{17}$$

6.4. Backplate Damping

Analytical solutions for the acoustic damping in perforated MEMS with a piston-like motion of the diaphragm and also a clamped circular diaphragm [34] have been developed. However, for the presented microphone, due to the complicated shape of the diaphragm under bias, developing a closed form analytical formula for the damping would be challenging. The damping of flexible structures can be accurately estimated with thermoviscous FEM simulations [35].

The Thermoviscous Acoustic (TA) module of the COMSOL Multiphysics® software is used for three-dimensional FEM frequency domain simulations of the backplate acoustic damping. Due to the symmetric pattern of the posts, a 22.5 deg section of the plate is modeled, and the symmetry condition is applied to the edges. Appropriate no-slip and isothermal wall boundary conditions are applied to all pertinent included model surfaces. The deflected shape of the diaphragm under bias voltage from the diaphragm model in Section 5.2 is used for the static shape of the diaphragm in the damping model. The velocity of the diaphragm, as a function of the radius and angular coordinate of the diaphragm, is extracted from the diaphragm model. In the damping model, the diaphragm surface is driven with this velocity function.

A column of air and a PML are included at the top of the backplate to guarantee no reflection from the top.

The cross-section of the velocity field from the FEM model is shown in Figure 18. The acoustic damping (resistance) is calculated from Equation (18):

$$R_{bp} = \frac{P_{diss}}{\frac{1}{2}|U|^2} \tag{18}$$

where P_{diss} is the total dissipated power calculated from volume integral of the total power dissipation density, and U is the diaphragm volume velocity. The simulated value for the damping was $R_{bp} = 1.74 \times 10^8 \ (\text{Pa·s/m}^3)$.

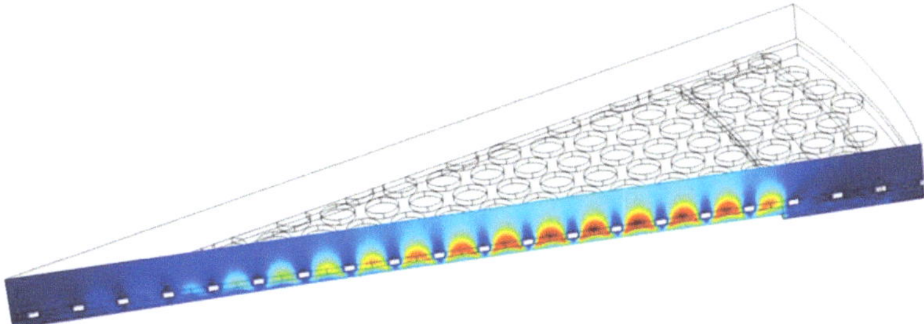

Figure 18. Cross-section of the velocity field in the FEM domain at 1 kHz. Red is associated with high velocity, and blue is associated with low velocity.

7. Microphone Performance

7.1. Microphone Sensitivity

The sensitivity under bias voltage is defined by Equation (19):

$$S = 20 \log_{10}\left(\frac{\Delta C \cdot V_B}{(C_m + C_p + C_i) \cdot \Delta P}\right) \tag{19}$$

where C_m is the diaphragm capacitance, C_p is the parasitic capacitance, C_i is the ASIC input capacitance of 0.215 pF, ΔC is the change in capacitance under the applied acoustic pressure, and ΔP is the pressure change accounting for the back volume compliance change due to diaphragm movement.

The frequency response of the microphone was measured in an anechoic chamber and was compared to the simulated frequency response from the lumped element model shown in Figure 19. A good correlation between the measured and the simulated curves is established. The measured sensitivity at 1 kHz is −38 dBV/Pa. The microphone resonance in air is 39 kHz.

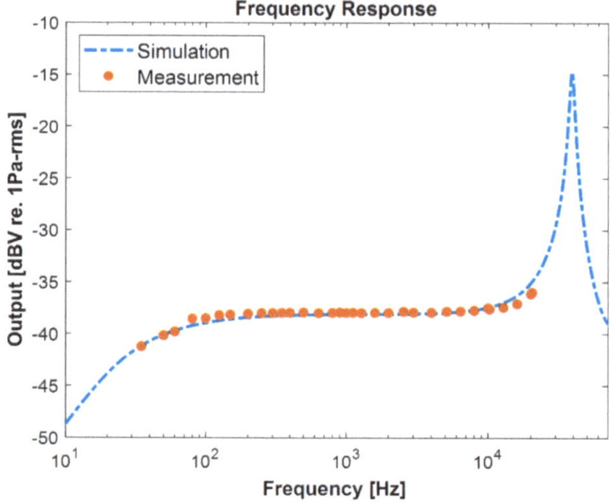

Figure 19. Measured frequency response along with the simulation.

7.2. Noise

The noise spectrum of the microphone can be obtained by measuring the power spectral density in a sound isolation chamber. The measured noise spectral density is shown in Figure 20 with the simulation results from the lumped element model from Figure 16. Good agreement is observed between the simulated noise spectral density and the measured noise spectral density. The total noise spectrum is the incoherent superposition of all the noise sources. By solving the lumped circuit as per Figure 16 for responses from each noise source individually, the total noise can be calculated.

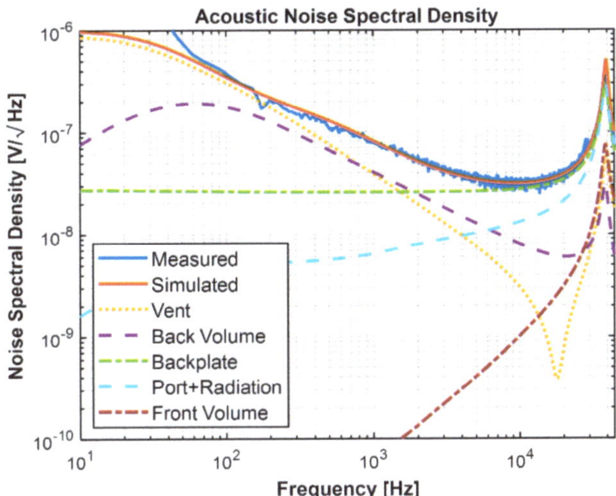

Figure 20. Measured and simulated noise spectral density. The component-wise noise spectra are plotted together to show their contributions to the simulated noise spectrum in total.

By subtracting the ASIC noise from the total microphone noise, the acoustic SNR was calculated to be 69 dBA. The acoustic SNR is a measure of the MEMS and package noise contributions, which is calculated by excluding the ASIC noise from the total noise. The typical A-weighted input-referred ASIC noise is approximately 3.4 µVrms. The microphone SNR obtained is 67 dBA in a 3.25 mm × 1.9 mm × 0.9 mm package.

As shown in Figure 21a, the backplate damping with ~50% is the highest contributor for the acoustic noise source, followed by the back volume, vent, and port noises, respectively. At the microphone level, including ASIC, as shown in Figure 21b, the analog ASIC noise with 38% is the highest contributor to the noise source followed by backplate damping, back volume, vent, and port noises.

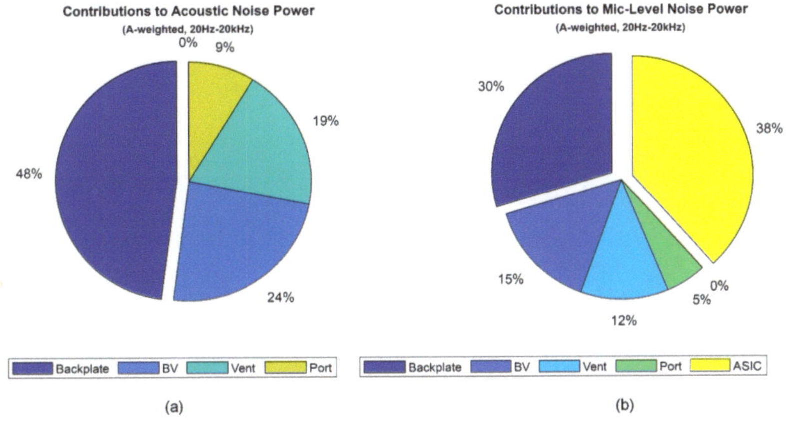

Figure 21. Pie charts for the total acoustic noise contributors from (**a**) MEMS only and (**b**) a package including MEMS and ASIC.

7.3. Resonance Peak

The noise spectral density is plotted in Figure 22 against the frequency by measuring in vacuum and air. The resonance in the vacuum is purely the mechanical diaphragm resonance owing to no air damping and is reported to be 55.5 kHz, which is 2% higher than the predicted diaphragm resonance (54.5 kHz) using the FEA model in Figure 14. Due to the air loading effect, the microphone resonance in air is measured to be 39 kHz, which is in excellent agreement with the lumped element model prediction, as shown in Figure 20.

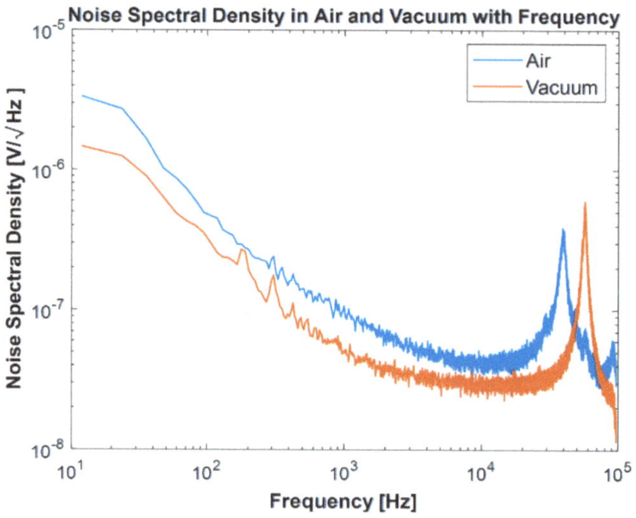

Figure 22. Noise spectral density in air and vacuum with respect to frequency.

7.4. Total Harmonic Distortion

Figure 23 shows the measured total harmonic distortion (THD) of the microphone, showing a distortion of 1.4% at 130 dB SPL and 7.2% at 134 dB SPL. The acoustic overload point (AOP) of 10% THD is higher than 134 dB SPL, at which point the electrical signal clips in the ASIC.

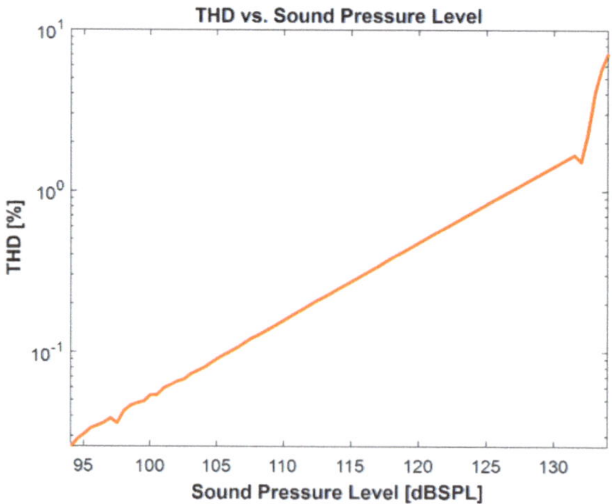

Figure 23. Measured THD curve.

In Table 9, the electroacoustic performance of the proposed MEMS microphone is summarized. In addition, Table 10 shows a comparison with other bottom port MEMS microphones available in the market with a similar package size. Our proposed MEMS microphone compares favorably with the available microphones in the market in terms of sensitivity, SNR, and THD performance in a relatively smaller package.

Table 9. Summary of the electroacoustic performance of the proposed MEMS microphone.

Property	Value
Sensitivity	−38 dBV/Pa
Signal-to-noise ratio (SNR)	67 dBA
Total harmonic distortion (THD)	7.2% at 134 dB SPL
Bandwidth	35–10 kHz
Motor capacitance	0.9 pF
Bias voltage	37 V
Pull-in voltage	50 V

Table 10. Performance comparison of the proposed design and commercially available MEMS microphones in market.

Microphone	ASIC Interface	Sensitivity	SNR (dBA)	AOP (dB SPL)	Package Size (mm × mm × mm)
Knowles, Proposed design	Analog	−38 dBV/Pa	67	>134	3.25 × 1.90 × 0.9
Infineon partner, MMA208-001 [36]	Analog	−38 dBV/Pa	67	135	3.35 × 2.50 × 0.98
Infineon partner, MMA208-W02 [36]	Analog	−38 dBV/Pa	66	136	3.35 × 2.50 × 0.98
Infineon partner, MA-ERA381-H43-1 [36]	Analog	−38 dBV/Pa	65.5	137	3.35 × 2.50 × 0.98
Infineon partner, S14OB381 [36]	Analog	−38 dBV/Pa	65	135	3.35 × 2.50 × 0.98
TDK InvenSense, ICS-40618 [37]	Analog	−38 dBV/Pa	67	132	3.5 × 2.65 × 0.98
ZillTek, ZTS6554 [38]	Analog	−37 dBV/Pa	67	120	3.35 × 2.50 × 0.98
ZillTek, ZTS6054 [38]	Analog	−38 dBV/Pa	65	125	3.35 × 2.50 × 0.98
TDK InvenSense, ICS-4078 [39]	Analog	−38 dBV/Pa	66	135	3.35 × 2.50 × 0.98

8. Conclusions

A semiconstrained diaphragm with a unique spring design supported with peripheral posts and a center post has been developed with performance of 67 dBA SNR in a 3.25 mm × 1.9 mm × 0.9 mm package. This is among the highest SNR that has been reported in this package as compared to other commercially available microphones. The microphone has a diaphragm with eight serpentine springs consisting of a shorter arm connected to the diaphragm side and a longer arm constrained on the anchor finger. The peripheral posts near the springs have been designed along with a center support structure, providing a simply supported boundary condition for the diaphragm under bias voltage. An 85% increase in the effective area of the diaphragm in this configuration was found with respect to a constrained diaphragm and a 48% increase with respect to a simply supported diaphragm without the center post architecture. Under the bias condition, the effective area further increases by 15%, as compared to the unbiased case. Detailed analytical, FEA, and lumped element simulations were utilized to predict and optimize the performance levels of the microphone. The results of the FEA and lumped element simulations show a good agreement with the measurement values obtained for the microphone.

Author Contributions: Conceptualization and methodology, S.S., Y.S., V.N. and X.S.; writing—original draft preparation and figures preparation, S.S., Y.S., V.N., X.S. and A.J.F.II; writing—review and editing, J.T.M.G.J., M.d.S. and M.P. All authors have read and agreed to the published version of the manuscript.

Funding: This research received no external funding.

Acknowledgments: The authors would like to sincerely acknowledge the contributions from other cross-functional teams at Knowles who helped in making this project a success.

Conflicts of Interest: The authors declare no conflict of interest.

References

1. Loeppert, P.V.; Lee, S.B. SiSonic™-The first commercialized MEMS microphone. In Proceedings of the Solid-State Sensors, Actuators, and Microsystems Workshop, Hilton Head, SC, USA, 5–9 June 2006; pp. 27–30.
2. Kim, B.H.; Lee, H.S. Acoustical-thermal noise in a capacitive MEMS microphone. *J. Sens. IEEE* **2015**, *15*, 6853–6860. [CrossRef]
3. Ganji, B.A.; Sedaghat, S.B.; Roncaglia, A.; Belsito, L.; Ansari, R. Design, modeling, and fabrication of crab-shape capacitive microphone using silicon-on-isolator wafer. *J. Micro Nanolithogr. MEMS MOEMS* **2018**, *17*, 015002. [CrossRef]
4. Scheeper, P.R.; van der Donk, A.G.H.; Olthius, W.; Bergveld, P. A review of silicon microphones. *Sens. Actuators A Phys.* **1994**, *44*, 1–11. [CrossRef]
5. Tadigadapa, S.; Mateti, K. Piezoelectric MEMS sensors: State-of-the-art and perspectives. *Meas. Sci. Technol.* **2009**, *20*, 092001. [CrossRef]
6. Papila, M.; Haftka, R.T.; Nishida, T.; Sheplak, M. Piezoresistive microphone design Pareto optimization: Tradeoff between sensitivity and noise floor. *J. MEMS* **2006**, *15*, 1632–1643. [CrossRef]
7. Kon, S.; Oldham, K.; Horowitz, R. Piezoresistive and piezoelectric MEMS strain sensors for vibration detection. In Proceedings of the Sensors and Smart Structures Technologies for Civil, Mechanical, and Aerospace Systems 2007, Vancouver, BC, Canada, 10 April 2007.
8. Hall, N.A.; Bicen, B.; Jeelani, M.K.; Lee, W.; Qureshi, S.; Degertekin, F.L. Micromachined microphones with diffraction-based optical displacement detection. *J. Acoust. Soc. Am.* **2005**, *118*, 3000–3009. [CrossRef]
9. Kuntzman, M.L.; Garcia, C.T.; Onaran, A.G.; Avenson, B.; Kirk, K.D.; Hall, N.A. Performance and modeling of a fully packaged micromachined optical microphone. *J. MEMS* **2011**, *20*, 828–833. [CrossRef]
10. Bernstein, J. A micromachined condenser hydrophone. In Proceedings of the Technical Digest IEEE Solid-State Sensor and Actuator Workshop, Hilton Head, SC, USA, 22–25 June 1992; pp. 161–165.
11. Scheeper, P.R.; Olthuis, W.; and Bergveld, P. Improvement of the performance of microphones with a silicon nitride diaphragm and backplate. *Sens. Actuators A Phys.* **1994**, *40*, 179–186. [CrossRef]
12. Yoo, I.; Sim, J.; Yang, S.; Kim, H. Development of capacitive MEMS microphone based on slit-edge for high signal-to-noise ratio. In Proceedings of the IEEE Micro Electro Mechanical Systems (MEMS), Belfast, UK, 21–25 January 2018; pp. 1072–1075.
13. Gharaei, H.; Koohsorkhi, J. Design and characterization of high sensitive MEMS capacitive microphone with fungous coupled diaphragm structure. *Microsyst. Technol.* **2016**, *22*, 401–411. [CrossRef]
14. Fu, M.; Dehe, A.; Lerch, R. Analytical analysis and finite element simulation of advanced membranes for silicon microphones. *J. Sens. IEEE* **2005**, *5*, 857–863. [CrossRef]
15. Shubham, S. Silicon Nitride Corrugated Membrane with High-Width-Aspect-Ratio for MEMS Microphones. In Proceedings of the COMSOL Conference, Boston, MA, USA, 7–8 October 2020.
16. Ying, M.; Zou, Q.; Yi, S. Finite-element analysis of silicon condenser microphones with corrugated diaphragms. *Finite Elem. Anal. Des.* **1998**, *30*, 163–173. [CrossRef]
17. Sedaghat, S.B.; Ganji, B.A. A novel MEMS capacitive microphone using spring-type diaphragm. *Microsyst. Technol.* **2019**, *25*, 217–224. [CrossRef]
18. Sui, W.; Zhang, W.; Song, K.; Cheng, C.H.; Lee, Y.K. Breaking the size barrier of capacitive MEMS microphones from critical length scale. In Proceedings of the 19th International Conference on Solid-State Sensors, Actuators and Microsystems (Transducers), Kaohsiung, Taiwan, 18–22 June 2017; pp. 946–949.
19. Chan, C.K.; Lai, W.C.; Wu, M.; Wang, M.Y.; Fang, W. Design and implementation of a capacitive-type microphone with rigid diaphragm and flexible spring using the two poly silicon micromachining processes. *J. Sens. IEEE* **2011**, *11*, 2365–2371. [CrossRef]
20. Lo, S.C.; Yeh, S.K.; Wang, J.J.; Wu, M.; Chen, R.; Fang, W. Bandwidth and SNR enhancement of MEMS microphones using two poly-Si micromachining processes. In Proceedings of the 2018 IEEE Micro Electro Mechanical Systems (MEMS), Belfast, UK, 21–25 January 2018; pp. 1064–1067.
21. Rombach, P.; Müllenborn, M.; Klein, U.; Rasmussen, K. The first low voltage, low noise differential silicon microphone, technology development and measurement results. *Sens. Actuators A Phys.* **2002**, *95*, 196–201. [CrossRef]
22. Martin, D.T.; Liu, J.; Kadirvel, K.; Fox, R.M.; Sheplak, M.; Nishida, T. A micromachined dual-backplate capacitive microphone for aeroacoustic measurements. *J. MEMS* **2007**, *16*, 1289–1302. [CrossRef]
23. Martin, D.T. Design, Fabrication, and Characterization of a MEMS Dual-Backplate Capacitive Microphone. Ph.D. Thesis, University of Florida, Gainesville, FL, USA, 2007.
24. Martin, D.T.; Kadirvel, K.; Liu, J.; Fox, R.M.; Sheplak, M.; Nishida, T. Surface and bulk micromachined dual backplate condenser microphone. In Proceedings of the 18th IEEE International Conference on Micro Electro Mechanical Systems, Miami Beach, FL, USA, 30 January–3 February 2005; pp. 319–322.
25. Naderyan, V.; Lee, S.; Sharma, A.; Wakefield, W.; Kuntzman, M.; Ma, Y.; da Silva, M.; Pedersen, M. MEMS microphone with 73dBA SNR in a 4 mm × 3 mm × 1.2 mm Package. In Proceedings of the 21st International Conference on Solid State Sensors, Actuators and Microsystems (Transducers), Orlando, FL, USA, 20–24 June 2021.
26. Bay, J.; Hansen, O.; Bouwstra, S. Design of a silicon microphone with differential read-out of a sealed double parallel-plate capacitor. *Sens. Actuators A Phys.* **1996**, *53*, 232–236. [CrossRef]
27. Wang, Z.; Zou, Q.; Song, Q.; Tao, J. The era of silicon MEMS microphone and look beyond. In Proceedings of the 2015 Transducers—2015 18th International Conference on Solid-State Sensors, Actuators and Microsystems (Transducers), Anchorage, AK, USA, 21–25 June 2015; pp. 375–378.

28. Timoshenko, S. *Theory of Plates and Shells*; McGraw-Hill: New York, NY, USA, 1959.
29. Hunt, F.V. *Electroacoustics: The Analysis of Transduction, and Its Historical Background*; American Institute of Physics: New York, NY, USA, 1982.
30. Shubham, S.; Nawaz, M. Estimating Parasitic Capacitances in MEMS Microphones using Finite Element Modeling. In Proceedings of the COMSOL Conference, Boston, MA, USA, 2–4 October 2019.
31. Beranek, L. *Acoustics*; American Institute of Physics: New York, NY, USA, 1954.
32. Aarts, R.M.; Janssen, A.J.E.M. Approximation of the Struve function H_1 occurring in impedance calculations. *J. Acoust. Soc. Am.* **2003**, *113*, 2635–2637. [CrossRef] [PubMed]
33. Kuntzman, K.; LoPresti, J.; Du, Y.; Conklin, W.; Naderyan, V.; Lee, S.; Schafer, D.; Pedersen, M.; Loeppert, P. Thermal boundary layer limitations on the performance of micromachined microphones. *J. Acoust. Soc. Am.* **2018**, *144*, 2838–2846. [CrossRef] [PubMed]
34. Naderyan, V.; Raspet, R.; Hickey, C. Thermo-viscous acoustic modeling of perforated micro-electro-mechanical systems (MEMS). *J. Acoust. Soc. Am.* **2020**, *150*, 2749–2756. [CrossRef] [PubMed]
35. Naderyan, V.; Raspet, R.; Hickey, C. Analytical, computational, and experimental study of thermoviscous acoustic damping in perforated micro-electro-mechanical systems with flexible diaphragm. *J. Acoust. Soc. Am.* **2021**, *148*, 2376–2385. [CrossRef] [PubMed]
36. Infineon Inside MEMS Microphone Partners. Available online: https://www.infineon.com/cms/en/product/sensor/mems-microphones/mems-microphones-for-consumer/infineon-inside/ (accessed on 8 November 2021).
37. High SNR Microphone with Differential Output and Low Power Mode. Available online: https://invensense.tdk.com/products/analog/ics-40618/ (accessed on 8 November 2021).
38. Analog MEMS Microphone. Available online: http://www.zilltek.com/en-us/Product (accessed on 8 November 2021).
39. Wide Dynamic Range, High SNR, Small Package Analog Microphone. Available online: https://invensense.tdk.com/products/analog/t4078/ (accessed on 8 November 2021).

Article

Piezoelectric Micromachined Microphone with High Acoustic Overload Point and with Electrically Controlled Sensitivity [†]

Libor Rufer [1,*], Josué Esteves [2], Didace Ekeom [3] and Skandar Basrour [2]

1   ADT MEMS, 360 Rue Taillefer, F-38140 Rives, France
2   University Grenoble Alpes, CNRS, Grenoble INP, TIMA, 46 Av. Félix Viallet, F-38000 Grenoble, France; josue.esteves@univ-lyon1.fr (J.E.); skandar.basrour@univ-grenoble-alpes.fr (S.B.)
3   Microsonics, 39 Rue Des Granges Galand, F-37550 Saint Avertin, France; didace.ekeom@microsonics.fr
*   Correspondence: liborrufer01@gmail.com; Tel.: +33-6-61-72-90-93
†   Forum Acusticum 2023 (10th Convention of the European Acoustics Association).

Abstract: Currently, the most advanced micromachined microphones on the market are based on a capacitive coupling principle. Capacitive micro-electromechanical-system-based (MEMS) microphones resemble their millimetric counterparts, both in function and in performance. The most advanced MEMS microphones reached a competitive level compared to commonly used measuring microphones in most of the key performance parameters except the acoustic overload point (AOP). In an effort to find a solution for the measurement of high-level acoustic fields, microphones with the piezoelectric coupling principle have been proposed. These novel microphones exploit the piezoelectric effect of a thin layer of aluminum nitride, which is incorporated in their diaphragm structure. In these microphones fabricated with micromachining technology, no fixed electrode is necessary, in contrast to capacitive microphones. This specificity significantly simplifies both the design and the fabrication and opens the door for the improvement of the acoustic overload point, as well as harsh environmental applications. Several variations of piezoelectric structures together with an idea leading to electrically controlled sensitivity of MEMS piezoelectric microphones are discussed in this paper.

Keywords: aeroacoustics; piezoelectric transducer; microphone; AOP; design; sensitivity control

Citation: Rufer, L.; Esteves, J.; Ekeom, D.; Basrour, S. Piezoelectric Micromachined Microphone with High Acoustic Overload Point and with Electrically Controlled Sensitivity. *Micromachines* **2024**, *15*, 879. https://doi.org/10.3390/mi15070879

Academic Editor: Huikai Xie

Received: 31 May 2024
Revised: 29 June 2024
Accepted: 1 July 2024
Published: 3 July 2024

1. Introduction

The field of micromachined microphones has seen significant advancements in recent years with the growing demand for miniaturization and high-performance sensing capabilities in various applications. For a long time, micro-electromechanical-system-based (MEMS) acoustic sensors have been the focus of academic and industrial research teams. The first developments of micromachined microphones were enabled by the progress in material science, fabrication technologies, miniaturization, and sensor techniques. Examples of these preliminary steps are the invention of the electret microphone [1] and progress in silicon-based static pressure sensors [2]. Further developments of micromachined microphones were achieved by many research teams and focused on the most common general approach using a diaphragm as an active microphone element converting the acoustic signal to the mechanical signal, and then converting the mechanical signal to the electric signal through known transduction principles. This effort resulted in the first microphones using piezoelectric [3], capacitive [4], and piezoresistive [5] couplings. Later, the FET (field effect transistor) microphone using a new principle, enabled only by silicon micromachining, was invented [6]. Finally, an optical microphone was invented, in which a diaphragm and a rigid structure form an optical waveguide with the geometry, and thus the transmission properties dependent on the diaphragm deflections modify the intensity and the phase of the transmitting optical signal [7].

From these early demonstrated micromachined devices, a capacitive microphone has been adopted as the dominant microphone type for further development for several reasons. The first reason was its much lower noise floor compared to piezoelectric or piezoresistive microphones. The other reason was the applicability of currently used industrial microtechnologies for its fabrication, with no requirement for additional structural layers or process steps. With strong industrial support, the micromachined capacitive microphone reached a commercial form after more than twenty years of incubation and became one of the most successful commercial MEMS products in the history of microsystem technology [8]. The continuing research of this kind of microphone resulted in key performance parameters such as sensitivity, signal to noise ratio, and distortion, meeting high requirements for microphones for mobile applications [9].

More recently, the availability of aluminum nitride (AlN) layers in industrial fabrication processes brought an increased interest in piezoelectric thin-plate-based micromachined devices. The following two kinds of acoustic devices using this new AlN-based technology were developed almost simultaneously: piezoelectric micromachined ultrasonic devices (PMUTs) and piezoelectric microphones.

PMUTs have been presented as counterparts of already mature capacitive micromachined ultrasonic devices (CMUTs). These kinds of thin-film-based micromachined ultrasonic transducers (MUTs) have been investigated as an alternative to conventional bulk, or thick-film, or piezocomposite ultrasonic transducers due to the advantages offered by microsystem technologies, namely small size, low power consumption, easy interconnexion, batch fabrication, and low cost. The working principle of CMUTs, like that of capacitive microphones, relies on the conversion of mechanical energy into electrical energy through an electrostatic field between two electrodes of a device. The energy conversion can take place in both ways, so the device can both transmit and receive acoustic signals. CMUTs have been demonstrated to work efficiently for both air and immersion applications and their main use is in medical imaging, but also in underwater imaging and nondestructive evaluation [10]. PMUTs exploit a 31-mode of a piezoelectric layer to generate or detect an acoustic signal. In cases of detection, an acoustic signal deforms the device diaphragm containing the piezoelectric layer and thus creates in-plain mechanical strain (1-direction), which results in an out-of-plane electric field (3-direction) obtained through the 31-transverse piezoelectric effect. The main applications of PMUTs are rangefinders [11], fingerprint sensors [12], and implantable micro-devices [13].

A piezoelectric microphone works, like a PMUT, in a 31-transverse mode. The main difference between the two devices consists of the useful frequency range definition. PMUTs are resonant devices and their useful frequency range is spread in a relatively narrow band around the resonant frequency. The useful frequency range of a microphone is located under its resonant frequency. The frequency range is limited at its high end by the resonant frequency and its low end is characterized by the leaks due to electrical and acoustic resistances. The piezoelectric microphone structure, thanks to the absence of a fixed electrode, offers a unique advantage due to its fabrication simplicity. The piezoelectric microphone can bring improvements in ruggedness and in moisture tolerance that are important in harsh environmental applications. Compared to the capacitive microphone, the piezoelectric structure enables higher diaphragm excursions, limited only by its nonlinear behavior, and thus higher acoustic overload point (AOP). This feature has attracted attention for industrial and aerospace applications with extremely high acoustic levels. One of the first piezoelectric microphones designed for aeroacoustic applications reached an AOP of 172 dB, which is substantially higher than in the currently available typical micromachined capacitive microphones [14].

The aim of this paper is to introduce modeling approaches and main design considerations related to MEMS piezoelectric microphones, and to compare and discuss simulation results for piezoelectric microphones with a circular diaphragm with a piezoelectric layer located in two specific diaphragm regions (one is close to the center and the other is close to the clamped edge). Based on the parametric optimization, the effects of several design parameters, such as the piezoelectric layer thickness and width, the sensing electrode localization, and the pressure equalization hole dimensions, on the microphone sensitivity and signal-to-noise ratio (SNR) are demonstrated in a typical case study. The optimization towards a high AOP is presented and the expected performance of this parameter is estimated.

The ability of a piezoelectric layer to serve not only as a sensor but also as an actuator opens the space to advanced designs of acoustic devices aimed at the control of various parameters such as resonant frequency, stiffness, or sensitivity. Various approaches using piezoelectric layers have been applied to acoustic and ultrasonic micromachined devices to achieve resonant frequency tuning [15,16] or sensitivity improvement [17]. A configuration with a piezoelectric layer with two distinguished sections, one used as a sensor and the other as an actuator, is also proposed in this paper. To the best of the authors' knowledge, a similar solution demonstrating the diaphragm stiffening due to bias voltage, enabling microphone sensitivity and AOP tuning, has not been applied in the audio frequency range.

After this short overview of micromachined microphones, Section 2 presents the piezoelectric materials that are considered in the study, the microphone mechanical structure corresponding to the silicon on insulator (SoI) fabrication process, and the associated acoustic elements. Modeling approaches are described in Section 3, and optimization processes are explained in Section 4. Finally, Section 5 is devoted to the proposed microphone structure aimed at electrically controlled sensitivity.

2. Piezoelectric Microphone Structure

The microphone structure involves mechanical elements, neighboring acoustic elements and components introducing electromechanical coupling, enabling its main function—sensitivity to acoustic stimulation. The microphone's performance is determined by the response of such a complete structure to an acoustic stimulus applied on its diaphragm.

The number and dimensions of each layer composing the microphone structure, as well as their material constants and associated stresses, depend heavily on the used fabrication technologies. The choice of the piezoelectric material was made in agreement with the technology applicable for its deposition and it also determined the appropriate materials for the bottom and top electrode layers.

2.1. Piezoelectric Layer

The function of a piezoelectric device requires a capacitor structure with a piezoelectric layer sandwiched between its top and bottom electrodes. If the main deformation of the piezoelectric layer obtained in diaphragm-based microphones is considered, the general tensor constitutive equation, coupling the electrical and mechanical domains, can be reduced to the following equations [18]:

$$S_1 = s_{11}^E T_1 + d_{31} E_3, \tag{1}$$

$$D_3 = d_{31} T_1 + \epsilon_{33}^T E_3, \tag{2}$$

where S_1 and T_1 are the mechanical strain and stress in axis *1*, E_3 and D_3 are the electric field and the electric density displacement in axis *3*, $s_{11}{}^E$ is the compliance constant at a constant electric field, d_{31} is the piezoelectric constant, and $\varepsilon_{33}{}^T$ is the permittivity of the piezoelectric material at a constant mechanical stress. The components of Equations (1) and (2) correspond to the fact that the mechanical strain and stress are applied in the lateral dimension, which is perpendicular to the polarization, to the electric field and to the electric density displacement axis. Such a situation is described by the '*31*' components of the piezoelectric matrix and by the corresponding so-called '*31*' coupling mode of operation for piezoelectric materials.

In most of the applications using the bending of a piezoelectric thin film, the total elastic properties of a bending structure are often dominated by a substrate, which brings the main difference to the evaluation of the piezoelectric activity compared to bulk materials. The anisotropic interaction between the piezoelectric film and the substrate results in identical strains along in-plane directions (S_1 and S_2), and the stress perpendicular to the film surface is $T_3 = 0$. Such a situation enables the derivation of effective piezoelectric coefficients and an example is shown below [19]:

$$e_{31,f} = \frac{d_{31}}{s_{11}^E + s_{12}^E}.$$

(3)

Compared to its intrinsic bulk value, the absolute value of the effective e-coefficient is always larger than e_{31}. The effective piezoelectric coefficients can be conveniently used as an evaluation index for the piezoelectric characteristics of thin films. These coefficients can also be measured directly by unimorph cantilever-based methods [20]. Nevertheless, for numerical simulation-based modeling, all materials building the microphone structure must be described with complete matrices of intrinsic elastic, electric, and piezoelectric coefficients.

In this paper, an AlN layer is considered as a basic piezoelectric component of a microphone structure. In specific arrangements requiring an actuation function, the ferroelectric material lead zirconate titanate (Pb(Zr,Ti)O$_3$), abbreviated as PZT, is also taken into account. The main matrix components describing piezoelectric materials involved in our study related to the '31' coupling mode of operation are listed in Table 1.

Table 1. Main physical properties of piezoelectric materials used in the simulations.

Property	AlN	PZT
Compliance s_{11}^E [TPa^{-1}]	3.53	12.8
Compliance s_{12}^E [TPa^{-1}]	−1.01	−3.7
Compliance s_{13}^E [TPa^{-1}]	−0.77	−5.8
Material density [kg/m^3]	3260	7700
Piezoelectric constant d_{31} [pm/V]	−2.65	−118
Piezoelectric constant e_{31} [C/m^2]	−0.58	−4.1
Relative permittivity ϵ_{33}^T [-]	9.5	1160
Dielectric loss angle (tanδ) [%]	0.3	2
Figure of merit M$_N$ [10^5 Pa$^{1/2}$]	20.9	9

The material parameters displayed in Table 1 depend heavily on physical parameters applied during the fabrication process and good knowledge of these parameters is critical for obtaining accurate simulation results. For the modeling purpose of this paper, focusing various simulation cases, Table 1 was drawn from the reference literature [14,21–23].

Various figures of merit focusing on various criteria have been adopted for piezoelectric transducers with bending elements. If the transmission of the acoustic signal is focused, the effective transverse piezoelectric constant can be used as a figure of merit for the transmission case.

$$M_T = \left| e_{31,f} \right|.$$

(4)

Materials with a higher value of M_T can produce larger sound pressure at the same driving voltage, or the required driving voltage becomes lower to obtain the same pressure level [24].

In a sense mode, when the membrane is deflected due to an impinging acoustic wave, a piezoelectric g constant is important [24]. A corresponding figure of merit for a sense mode is shown below.

$$M_S = \left| \frac{e_{31,f}}{\varepsilon_{33}\varepsilon_0} \right|.$$

(5)

Another figure of merit representing the intrinsic signal-to-noise ratio of the material has been defined as follows [23]:

$$M_N = \left| \frac{e_{31,f}}{\sqrt{\varepsilon_{33}\varepsilon_0 tan\delta}} \right|. \tag{6}$$

If both materials considered in this study are compared with the aid of the figures of merit (4) and (5), AlN could be effectively applied to sensors, whereas PZT is better suited for actuation purposes. The fast and simple comparison using the figures of merit was confirmed by using multi-criteria decision-making (MCDM) material selection techniques [25]. From the currently available piezoelectric materials, AlN clearly stands out as the best candidate for use as a microphone sensitive layer. For voltage detection, aluminum nitride leads in the quantitative parameters with its low dielectric constant, high resistivity, low loss tangent, and high SNR values. Moreover, AlN has good compatibility with complementary metal oxide semiconductor (CMOS) processing and good process quality control in manufacturing, which is important for device scaling and commercial applications. However, if current detection is preferred for sensors, or if force performance is required in actuating applications, the PZT appears as a clear leader among piezoelectric materials [21].

2.2. Mechanical Body

The purpose of the modeling and simulation work exposed in this paper is to present the effect of various design parameters on the microphone performance, and to propose a microphone structure aimed at electrically controlled sensitivity. For the sake of clarity, and to allow presenting main behavioral tendencies without secondary effects, a simplified basic wafer, shown schematically in Figure 1, was considered for microphone modeling in the first approach. Later, important effects due to additional layers required by a chosen fabrication process must have been included in the model.

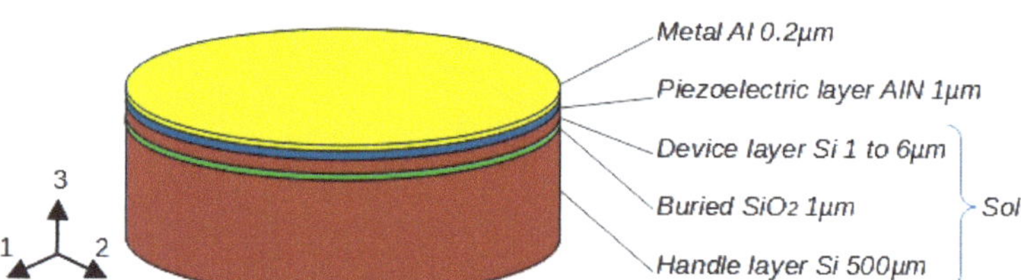

Figure 1. Basic SoI wafer with additional layers and their dimensions used for piezoelectric microphone modeling.

The structure of Figure 1 consisting of the SoI wafer with deposited piezoelectric and metallic layers was chosen as a good alternative to the wafer fabrication process. The top silicon layer (device layer) can be chosen for the exact thickness, crystal orientation, and conductivity required by the application, and the buried oxide layer provides brilliant etch stop characteristics. The thickness of each wafer layer is freely adjustable and it can be, depending on a producer, in the range of 340 to 725 μm for the handle layer, 0.5 to 3 μm for the buried oxide layer, and 1 to 300 μm for the device layer. During the fabrication process of the microphone structure, the front side patterning is applied on the metallic and piezoelectric layers to form the microphone sensitive parts. These sensitive parts are formed by a sandwich composition in which the piezoelectric layer is placed between two electrodes. In this study, the top electrode is formed of a metallic layer (aluminum), and a silicon device layer serves as the bottom electrode. The microphone diaphragm can be constructed using backside etching of the silicon handle layer and buried silicon oxide layer.

High-quality piezoelectric films cannot be grown directly on silicon. Depending on the piezoelectric material and on its deposition process, inter-layers are necessary to provide an optimum nucleation rate or growth direction to prevent interdiffusion and oxidation reactions or to improve adhesion.

In the case when AlN is used as a piezoelectric layer, commonly used underlying materials include Pt, Ti, Al, and Mo. Platinum underlying layers have demonstrated their ability to grow high-quality AlN, due to their inertness to nitrogen. However, it is not used in most applications due to cost and patterning difficulties. Molybdenum is the most common material used as the underlying seed layer, which ensures the high quality of the piezoelectric layer and of the electrical contact [26].

The PZT films for most applications are grown on an electrode, which should neither oxidize nor become insulating. The most often reported materials include Pt, and metal oxides. Usually, the chemical barrier function is provided by two or more layers, including the electrode. $PZT/Pt/Ti/SiO_2/Si$ is the most widely applied sequence, in which titanium is needed as an adhesion layer [21,27].

The residual stress in structural layers appearing as one of the most common outcomes of the integration of distinctly different materials must be well controlled during the fabrication process. Even if strong consequences of residual stresses such as creep, deformation, fracture, or fatigue are avoided, they can still affect the elastic properties of the structure and have a strong influence on the final behavior of the device. Hence, the assessment and regulation of residual stress are one of the prime challenges to predicting the final performance of MEMS devices [28].

The residual stress in devices based on SoI-MEMS technology arises primarily from the residual stress in the SoI wafer itself and from the residual stress formed during the additional process to achieve the final device. Silicon direct bonding technology, used in the preparation of SoI wafers, involves annealing and thermal oxidation steps, inducing the residual stress generation within the wafer layers. The gradual release of the silicon handle layer and buried SiO_2 layer disrupts the original stress balance mechanism within the SoI structure and leads to the development of tensile residual stress in the released silicon device layer. A mechanical theoretical model for the residual stress in SoI wafers was established and verified through experimental characterization and gives values of residual stress in the device layer in the range of 30 MPa [29].

The residual stress of additional processes is dependent on the materials deposited and on deposition conditions, and its final stress levels are known only in a relatively large interval of values. The thin film stress in polycrystalline AlN can range from compressive to tensile stress levels depending on the deposition technique and the parameters used. As an example, the residual stress in AlN thin films sputter-deposited in identical conditions on Si substrates was found to be compressive and its values were in the range of -300 (±50) MPa to -730 (±50) MPa. The difference in residual stresses can be attributed to the microstructure of the films and mismatch between in-plane atomic arrangements of the film and substrates with various orientations, including (111), (100) or (110) [30].

It is important to minimize the residual stress generated inside the device structure to minimize its effect on the performance, reliability, and yield. Simple compensation techniques to lower the overall stress are not sufficient, as AlN often exhibits a stress gradient along the thickness of the thin film. An example of a modified sputter process exploiting the influence of varying sputter pressure during deposition on the intrinsic stress component is presented in [31]. In the process, AlN thin films were synthesized with a DC magnetron sputter system at a temperature below 100 °C on p-type (100) silicon wafers. The back pressure of the pure nitrogen atmosphere in the sputter chamber was applied in two specific phases to reliably fine tune the resulting stress to -170 MPa, while keeping a high piezoelectric coefficient.

In the previous paragraphs, it was demonstrated that detailed knowledge of all components of the MEMS structure including the residual stress is highly important for accurate device simulation and design phases. It was also shown that with stress engineering, structures with piezoelectric films with low residual stress can be attained, but their study and elaboration are beyond the scope of this paper. For the simulations, residual stresses were not considered, and the simplified structure of Figure 1 was used, which is well in line with the purpose of the paper to present various microphone configurations and their design parameters. Table 2 summarizes the main material constants values of passive layers that were used in the work.

Table 2. Physical properties of passive mechanical layers used in the simulations.

Property	Si	SiO$_2$	Al
Compliance s_{11}^E [TPa^{-1}]	7.67	13.7	14.3
Compliance s_{12}^E [TPa^{-1}]	−2.13	−2.33	−5
Compliance s_{13}^E [TPa^{-1}]	−2.13	−2.33	−5
Material density [kg/m^3]	2330	2200	2700

It can be noted that although isotropic materials are typically described with two engineering constants as Young's modulus and Poisson's ratio, here the compliance matrix elements are presented to keep uniformity with Table 1.

In this study, the simulation results of the three axisymmetric microphone structures shown in Figure 2 are compared. The colors of the structure layers used in Figure 2 are identical to those described by the legend of Figure 1. Firstly, a microphone (type A, shown in Figure 2a) with a circular diaphragm, as described in [14], with the sensing electrode in the proximity of the clamped diaphragm edge is studied. In another microphone structure (type B, shown in Figure 2b), the diaphragm has its sensing electrode located around its center. Finally, the study is completed with a microphone structure (type C, shown in Figure 2c), exploiting both electrodes, including a peripheral and a central electrode [32]. This configuration can be used in two ways. Firstly, both electrodes are used as sensing electrodes and the microphone output is obtained as a difference in both electrode signals. Another method described in the paper consists of using one of the electrodes as a sensor and the other one as an actuator helping to electrically control the sensitivity of the overall structure.

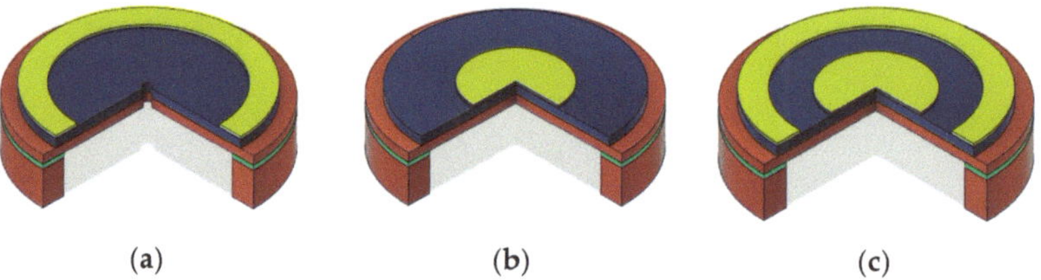

(a)   (b)   (c)

Figure 2. Mechanical structures of piezoelectric microphones used for the study schematically showing the electrodes (**a**) at the edge; (**b**) at the center; (**c**) both at the edge and at the center.

2.3. Acoustic Environment

Acoustic environment is a crucial part of the acoustic sensor. It is important not only as the propagation medium, ensuring the interaction between a sound source and the outer side of the diaphragm, representing the microphone input, but also for building elements placed inside the microphone body, communicating with the inner side of the diaphragm. Acoustic impedances of these elements must be taken into account together with mechanical impedances of structural parts of the microphone. Here, the main acoustic elements related to the microphone model are introduced, and simplified expressions corresponding to our design in terms of dimensions and the frequency range are presented. More detailed theory of this field can be found elsewhere [33].

The acoustic impedance Z_a is the ratio of sound pressure p (in Pa) to volume flow rate q (or volume velocity in m^3/s). Using the electroacoustic analogy, basic acoustic elements can be defined through simple structures presented by cavities and ducts.

If acoustic pressure p is applied in a small cavity with the volume V, with dimensions much smaller than the wavelength λ, the fluid (air) in the confined volume acts like a spring. In an analogy with mechanical compliance, a compact enclosed cavity is called an acoustic compliance with the following value [33]:

$$C_a = \frac{V}{\rho_0 c_0^2},\tag{7}$$

where ρ_0 is the fluid density and c_0 is the speed of the sound. A cavity placed underneath the diaphragm is an important part of a piezoelectric microphone. Its compliance can be fixed in a large range of values, and thus can tune the resulting resonance frequency of the microphone.

If a duct or pipe with a rectangular cross sectional area A and length $L \ll \lambda$ are considered, the fluid in the duct vibrates due to an acoustic pressure difference p applied across it. Such a component presents a complex acoustic impedance composed of the acoustic inertance (mass) with of the following value [33]:

$$M_a = \frac{\rho_0 L}{A},\tag{8}$$

and the acoustic resistance with the following value [33]:

$$R_a = \frac{12\mu L}{bh^3},\tag{9}$$

where μ is the viscosity of the fluid, and $b \gg h$ are the width and height of the duct. These elements are used to model the flow through the pressure equalization vent channel of the microphone.

The acoustic pressure p at the surface of the diaphragm, which can be approximated by a rigid piston, is the pressure due to the impedance Z_{rad} of the radiation field. The radiation impedance in the low-frequency approximation ($ka \ll 1$, where $k = \omega/c_0$, and a is the piston radius) can be simplified as a series combination of the radiation mass $M_{ad,rad}$ and the radiation resistance $R_{ad,rad}$ [33].

$$M_{ad,rad} = \frac{8\rho_0}{3\pi^2 a},\tag{10}$$

$$R_{ad,rad} = \frac{\rho_0 \omega^2}{2\pi c_0}.\tag{11}$$

In diaphragm-based microphones, thin film deformations induced by acoustic pressure play an important role in the device. It was verified that the corresponding stresses present in the microphone structure are significantly lower than the materials' tensile strengths.

3. Microphone Modeling

3.1. Finite-Elements Model

The modeling presented in the paper relies mainly on the following two approaches: finite-element modeling (FEM) and lumped-element modeling (LEM). The finite-element model was developed in the Ansys Workbench ver. 2022 R1. The proper definition of the boundary and symmetry conditions facilitates modelling of only a portion of the actual structure, which reduces the analysis run time and memory requirements with no losses in accuracy. Most of the simulations were carried out on one quarter of the structure, and in some cases, one eighth of the structure was used to speed up the calculation time. Structural layers (Si, SiO_2 and Al) are meshed with SOLID 186 elements; the piezoelectric layer (AlN) is meshed with SOLID 226 elements, allowing us to define piezoelectric properties. Meshing size varies with the dimensions of the structure during the optimization process. Nevertheless, working with the linear lumped-element model towards the complete microphone evaluation was preferred for speed and availability purposes. FEM was thus an important tool to define the lumped elements with high accuracy. It is briefly shown in Section 4.5 how FEM was used to evaluate the nonlinear behavior of the microphone diaphragm.

FEM also enables us to verify the stress situation in the structure and to predict the deformation and stress fields when intrinsic stress is present. It was shown in [34] that several restrictions exist in commercially available tools to combine a piezoelectric analysis with intrinsic stress in a harmonic response analysis. The arising difficulties could be overcome by employing a coupled thermo-electromechanical simulation in order to obtain consistent static and harmonic response results. A 3D FEM including intrinsic stress effects must be considered in advanced design, but this is not the focus in this paper.

3.2. Lumped-Element Model

Lumped-element modeling is a powerful and reliable method for predicting the multiphysics behavior of electroacoustic transducers. With this method, each element of the transducer is transformed to a circuit model thanks to the mechanical (mass–damping–stiffness) and electrical (inductor–resistor–capacitor) equivalence. Using this method requires characteristic lengths of the system smaller than the wavelength of the associated physical phenomena, which is satisfied in the audio-frequency range. Here, the LEM shown in Figure 3 already presented in [14] is used. This model allows evaluation of the microphone performance including its frequency range, sensitivity, noise, SNR, and minimum detectable pressure (MDP) values.

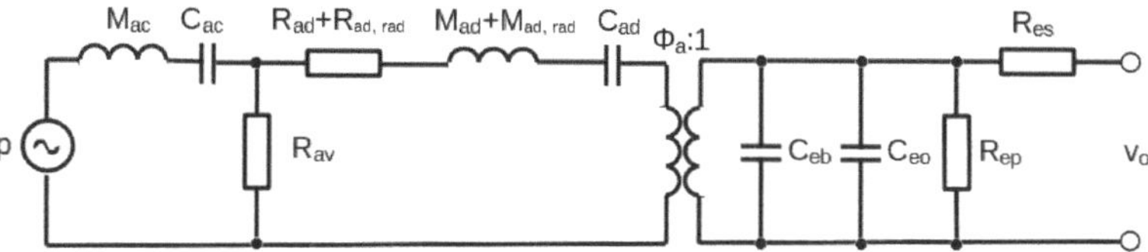

Figure 3. Lumped-element model of the piezoelectric microphone.

The model represents the mechanical elements transformed to its acoustic side (resistance, mass and compliance of the diaphragm, R_{ad}, M_{ad}, C_{ad}), acoustic elements (diaphragm radiation mass and resistance, $M_{ad,rad}$, $R_{ad,rad}$, mass and compliance of the back cavity, M_{ac}, R_{ac}, and the pressure equalization vent resistance, R_{av}), and the electrical elements (sensing and parasitic capacitance, C_{eb}, C_{e0}, resistance of the piezoelectric layer and of the connection lines, R_{ep}, R_{es}). The acoustic pressure at the microphone input p is transformed to the output electrical voltage v_0 with the transducer factor Φ_0.

The localized constants M_{ad}, C_{ad}, and C_{eb} are extracted from the finite element calculation, without making any assumptions about the geometry or the shape of the eigenmode. In the vicinity of the resonance, the effective mass M_{ad} of the microphone (in the z direction, normal to the membrane) is given by the following equation [35]:

$$M_{ad} = \frac{U^T[M]U_z}{U^T[M]U},$$

(12)

where $[M]$ designates the finite element global mass matrix; U is the eigenvector associated with the natural frequency f_r of the microphone. The displacement vector U_z is such that all its components in the z direction are equal to *1* and its other components are equal to *0* in the vicinity of the resonance frequency. The effective mass M_{ad} in the z direction is linked to the effective stiffness K_{ad} and to the resonance frequency f_r by the following relationship [35]:

$$\frac{K_{ad}}{M_{ad}} = \frac{U^T[K]U}{U^T[M]U} = (2\pi f_r)^2,$$

(13)

where $[K]$ is the global finite element stiffness matrix. It follows that K_{ad} is given by the following equation [35]:

$$K_{ad} = (2\pi f_r)^2 M_{ad} = \frac{1}{C_{ad}}. \tag{14}$$

The blocked capacitance of the transducer, C_{eb}, is obtained from the following equation [35]:

$$\frac{1}{2}\Phi^T[K_{\Phi\Phi}]\Phi = \frac{1}{2}C_{eb}V^2, \tag{15}$$

where $K_{\Phi\Phi}$ is the overall dielectric matrix, V is an arbitrary voltage applied to the hot electrode and Φ is the overall voltage vector, such that the voltage is equal to V for the nodes located on the electrode and 0 for the other nodes. The resulting relation for C_{eb} is as follows [35]:

$$C_{eb} = \Phi^T[K_{\Phi\Phi}]\Phi. \tag{16}$$

The remaining lumped elements, including M_{ac}, C_{ac}, R_{av}, $M_{ad,rad}$, and $R_{ad,rad}$, were obtained analytically from the expressions presented in Section 2.3. This work is focused on the presentation of the main structures applicable to piezoelectric microphones and on the comparison of their performance parameters. The parameters, such as the resonant frequency, the sensitivity, SNR, and AOP, used for the comparison are well known in the field of acoustic sensors, and thus they are not presented here in detail. More information on these parameters can be found elsewhere [14,36].

4. Parametric Optimization of the Microphone Structure

In piezoelectric microphones, thickness and lateral dimensions of all diaphragm layers, including piezoelectric and electrode materials, are key parameters that need to be optimized in the design loop. Typically, there is not a unique optimal solution satisfying the microphone specifications, and widely used optimization algorithms often lead to trivial solutions. For this reason, parametric optimization already used in [37] was applied. The optimization process is performed in two computing environments (Ansys 2022 R1 and Matlab R2023a) and involves two selection levels. This approach helps to limit the computational requirements by eliminating simulation cases not satisfying the condition of the first selection level, and thus to limit the number of numerical simulations entering the second selection level. The conditions corresponding to both selection levels depend on the focused parameters of the specifications.

By default, the cases discussed in this section are based on the structure presented in Figure 1 and on the variant (a) in Figure 2 with the use of the AlN piezoelectric layer. Any deviation from this plan will be announced when necessary. The results are presented with the aim of giving the main design tendency and for this reason, not all dimensions and details are strictly listed.

4.1. Optimization of the Piezoelectric Layer Thickness

In order to show the effect of the piezoelectric layer thickness on the microphone sensitivity and signal-to-noise ratio, the optimization process of the whole microphone structure with the optimal design of a microphone matching the audio frequency bandwidth is run first. To carry out this, the limits for the structure dimensions are set and static and modal finite-element analyses (FEA) are executed. The obtained results are filtered through the No. 1 condition of the optimization process. This condition is satisfied only by the solutions with the frequency of the first mode in a predefined range, which are related to the maximal frequency of the bandwidth. A number of suitable designs, depending on the range size defined for the first condition, passes back to Ansys for the lumped element extraction based on the static simulation, and then to Matlab for the evaluation of the microphone characteristics. For the final step, the maximal value of the SNR is selected as the No. 2 condition. The frequency response of the optimized geometry is then compared with electromechanical harmonic analysis from Ansys [37]. The thickness of the piezoelectric layer can be fine-tuned in an additional step, which gives the results shown in

Figure 4. In the example shown in Figure 4, the diaphragm diameter was 880 µm and the thickness of the piezoelectric layer was considered in the range from 0.1 µm to 2 µm.

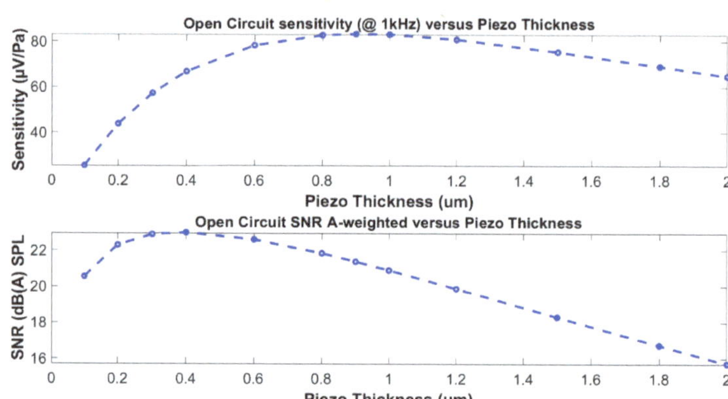

Figure 4. Effect of the piezoelectric layer thickness on microphone sensitivity and SNR.

If both graphs of Figure 4 are compared, it is evident that, in this case study, the maximal microphone sensitivity is achieved when the thickness of the piezoelectric layer is close to 1 µm, whereas the maximal SNR is reached when the piezoelectric layer thickness is 0.4 µm. This difference is due to the dependence of the noise density on the total electrical impedance of the piezoelectric layer and on the thickness at which it reaches its minimal value.

4.2. Optimization of the Piezoelectric Layer Width

The optimization of the piezoelectric layer width is performed on two similar structures illustrated in Figure 5. The only difference between these two cases is the fact that the structure in Figure 5a has the piezoelectric layer deposited on the whole diaphragm surface, while in Figure 5b, the piezoelectric layer is patterned in the same way as the top metallic layer. This difference may become important in view of the fabrication process, as the 'full piezo' layer from Figure 5a will require fewer etching steps than the 'ring piezo' layer from Figure 5b.

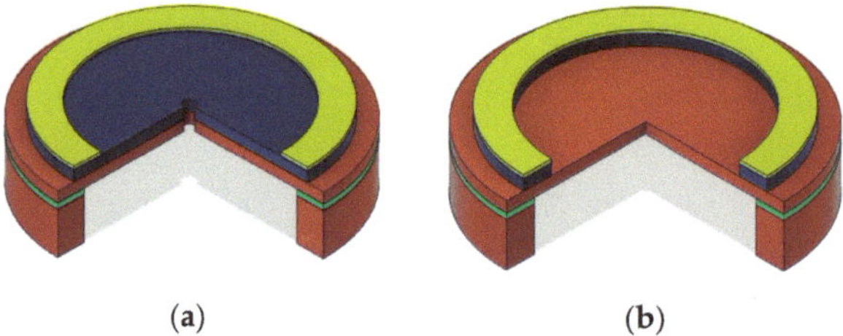

(a)   (b)

Figure 5. Microphone structure with the electrodes at the edge, with the piezoelectric layer deposited (**a**) on the whole diaphragm surface (full piezo); (**b**) under the metallic electrode (ring piezo).

The optimal width of the piezoelectric layer can be found based on a similar process as in the previous paragraph. The primary structure, optimized for the audio frequency bandwidth, is subjected to an additional optimization step in which the width of the piezoelectric layer is fine-tuned. The main layers of this primary structure have a diameter of 880 µm and the thickness of the silicon layer is 3 µm and that of the piezoelectric AlN layer is 1 µm. The obtained results shown in Figure 6 document, again, that the sensitivity and SNR culminate at different values of the electrode width. Nevertheless, it is possible to find a compromise width, for which both parameters maintain reasonably high values.

Moreover, the structure with the piezoelectric layer covering the whole diaphragm surface, which is technologically more suitable, shows substantially better sensitivity and SNR.

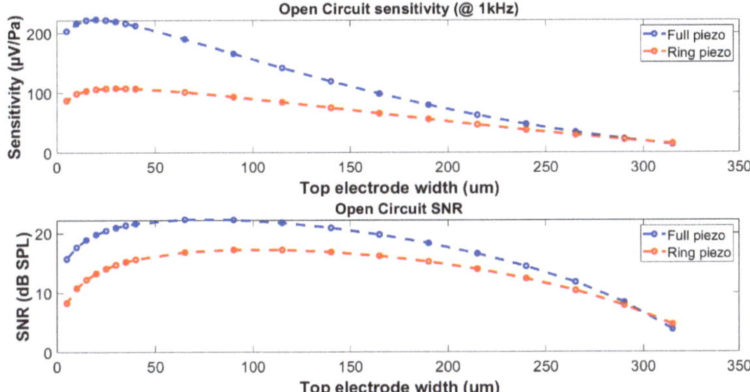

Figure 6. Effect of the electrode width on the microphone sensitivity and SNR with the piezo-electric layer fully covering the diaphragm (full piezo) or localized only under the electrode (ring piezo).

4.3. Effect of the Sensing Electrode Localization

According to the plate bending theory, if an originally stress-free thin circular plate rigidly clamped on its boundary is uniformly charged, two zones with opposite stresses can be distinguished on its surface. Both zones are delimited with a circle of radius r equal to the following [38]:

$$r = a\sqrt{\frac{1+\nu}{3+\nu}},\qquad(17)$$

where a is the plate radius and ν is the Poison's ratio of the plate material. Expression (17), predicting a change in curvature and thus a change in stress polarity located at approximately 60% of the plate radius for currently used materials, can serve only as a rough estimation of electrode placement. Advanced mechanical analysis of the multi-layer circular composite plate including a piezoelectric layer and initial stresses for the range of parameters used in the microphone design was carried out in [39].

In this case study, the same structure dimensions as in Section 4.2 were applied. In Figure 7, the sensitivity and SNR of structures shown in Figure 2a,b are compared. Both configurations clearly demonstrate a sensitivity decrease for dimensions over the limit given by the Expression (17). Even if the sensitivity for the structure in Figure 2b is substantially more important than for the case of the structure in Figure 2a, both structures can reach a similar performance in terms of the SNR.

If the zones defined by Expression (17) are respected, the annular ring electrode and the circular central electrode are located on diaphragm sections exposed to an opposite polarity of stress induced by the acoustic pressure. This fact can be exploited by appropriately connecting both electrodes in series or parallel circuit configurations in order to optimize raw voltage versus capacitance trade-offs [32].

A similar consideration of induced charge distribution of opposite signs in the central and surrounding region of a piezoelectric ultrasonic transducer was exploited in [17]. The respective two charge signals can be summed up effectively by complementary connection of the capacitors corresponding to each region. Moreover, statically deflected diaphragms cause a larger lateral strain in the piezoelectric layer compared to flat diaphragms. Sensors combining these two effects have been designed and they have shown sensitivity over five times higher than that of a conventional sensor.

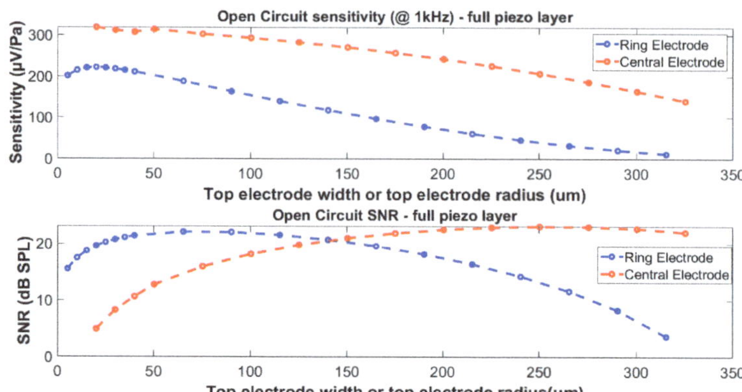

Figure 7. Comparison of piezoelectric microphones with metallic electrodes at the edge (ring electrode), and at the center (central electrode).

4.4. Effect of the Pressure Equalization Hole

An important part of the microphone is a venting system designed to equalize the static pressure on both sides of the diaphragm. The venting system is typically realized by connecting the inside cavity of the microphone to the exterior space by a capillary. As the capillary, representing the acoustic resistance with acoustic mass, together with the inside cavity, representing the acoustic compliance, create a high-pass filter, all dimensions must be carefully tuned in a way that only useless frequencies are cut off, and the required frequency behaviour does not deteriorate. Figure 8 compares the effect of the vent hole diameter on the frequency response of the microphone.

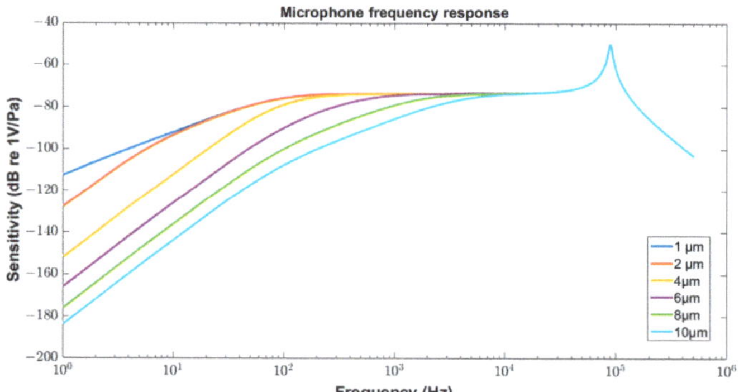

Figure 8. Effect of vent hole diameter on the microphone frequency response.

For the purpose of the example shown here, the standard structure from the previous paragraphs was used. In the simulation process, a minimal capillary length, corresponding to the total thickness of the Si device layer (3 μm), and AlN layer (1 μm), was considered and its diameter size from 1 μm to 10 μm was taken into account. It is worth noting that the curve in Figure 8 corresponding to the smallest capillary diameter is masked by the effect of the electrical resistance R_{ep} of the piezoelectric layer, which results in the change in the frequency response slope in the low frequency range.

4.5. Optimization Towards the Acoustic Overload Point

If a microphone withstanding a high acoustic pressure is designed, it is necessary to set the limits for the structure dimensions and to determine, through the nonlinear static analysis, the maximal diaphragm displacements for a given range of the input static pressures. Based on the deviation between the obtained value of the nonlinear displacement and the corresponding linear displacement, the acoustic overload point is obtained [14,37]. Figure 9 shows that there is an important difference between the linear and the nonlinear displacement responses of the microphone structure. To evaluate the maximal displacement, the optimization condition No. 1 corresponding to the difference between the linear and the nonlinear displacement equal to 3% was set. Each structure fulfilling the No. 1 condition for a predefined range of acoustic pressures passes towards selection No. 2. In this second step, the maximal value of the SNR is used as the No. 2 condition. It can be expected that the microphone with the structure described in the paper, with a diaphragm diameter of 880 μm, can reach the AOP of 163 dB, sensitivity of 212 μV/Pa, and the resonant frequency of 88 kHz.

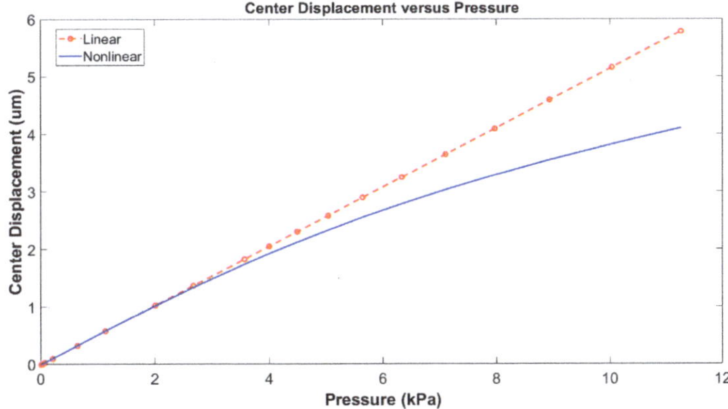

Figure 9. Maximal diaphragm displacement for the microphone structure obtained in linear and nonlinear static FEA.

5. Microphone with Electrically Controlled Sensitivity

In this section, the idea of microphone sensitivity control through a DC bias voltage is outlined. To this aim, the structure shown in Figure 2c, with one electrode at the edge and the other at the centre, is exploited. A similar configuration is typically used in PMUTs, where one electrode is activated with the electrical signal to be emitted and the other one, associated with the sensing layer, provides the electrical signal corresponding to a received stimulus. Unlike PMUTs, for microphone sensitivity control, a DC bias voltage is applied to one of the electrodes to induce stress through the associated piezoelectric layer to control the diaphragm overall stiffness.

5.1. Basic Approach

As discussed in Section 4.3, each electrode is located in a clearly distinguished zone of the diaphragm, separated by the radius given by Expression (17). Two SoI-based structures were considered, one with the AlN, and the other with the PZT piezoelectric layers, optimized for a wide-frequency band, with a resonant frequency in the vicinity of 60 kHz. For a better demonstration of strains variations in the system, the residual stress was not considered in the models of both structures. Nevertheless, as it was already stated earlier, the components of the residual stress must be involved in future models to confirm the hypotheses described in this section.

In the first simulation steps, FEM static analyses were carried out in order to compare the diaphragms' deformation profile due to the injection of a bias voltage to the external and the central electrodes, respectively. Figure 10 shows the displacement profiles of the diaphragms with AlN and PZT as a function of the bias voltage. Only half of the symmetrical diaphragm is shown, with the central point located in *0*. Due to the superior performance of the structure with the PZT layer, confirmed with the deformation profile, only this structure was considered for the next simulation step.

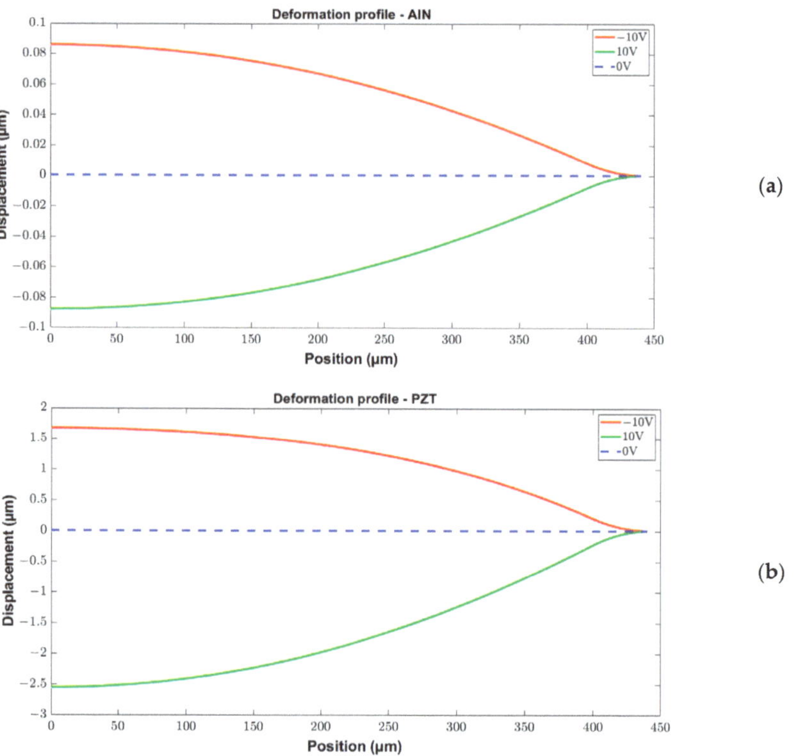

Figure 10. Displacement profile as a function of the static voltage applied on the ring electrode of the microphone diaphragm with the piezoelectric layer formed by (**a**) AlN; (**b**) PZT.

5.2. Simulation Results

The initial FEM static analysis confirmed that the bias voltage of ± 10 V applied on a PZT layer results in a clear change in the diaphragm profile. Such a modification of the deformation curve gives evidence of a modified stress situation in the structure, which can lead to the resonant frequency shift and to the variation in the microphone sensitivity. Both these effects on microphone performance can be demonstrated through harmonic analyses simulated with various bias voltages. The resulting frequency characteristics, shown in Figure 11, allow us to expect that the sensitivity of this microphone structure can be modified in the range reaching 10 dB with the bias voltage up to 10 V. The results are focused on the sensitivity levels, and the effect of acoustic elements is not considered at this stage. The decay on the low-frequency side of the response is due to the dielectric losses in the piezoelectric layer, as was already mentioned in Section 4.4.

The electrically controlled sensitivity described above can be further enhanced by the initial buckling of the diaphragm. The relationship between static deflection of the diaphragm and the acoustic sensitivity was demonstrated by other teams. Diaphragm static deflection due to compressive residual stress can be randomly oriented upward or downward due to the bending moment of the diaphragm at the releasing step. The lateral strain of a diaphragm is caused by bending as well as by expanding due to the large static deflection.

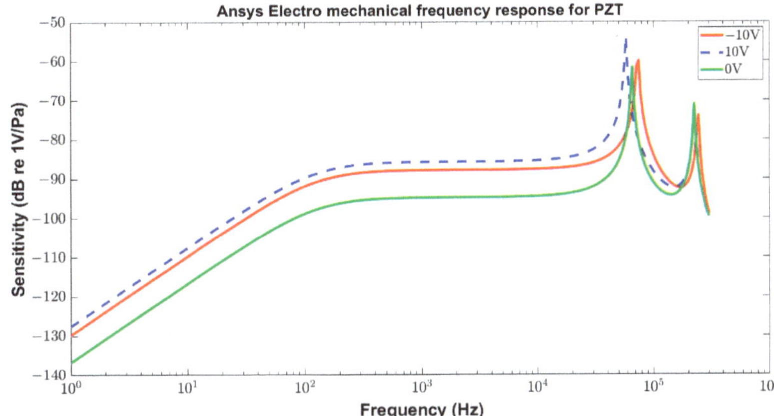

Figure 11. Frequency response of the microphone with the piezoelectric layer formed by PZT as a function of the static voltage applied on the ring electrode.

These two kinds of strains on layers above the neutral plane are summed up in the upward-deflected diaphragm but cancel each other in the downward-deflected diaphragm. Supposing that the piezoelectric layer is located on top of the diaphragm structure, above the neutral plane, it is exposed to higher stress in the upward-deflected diaphragm than in the downward-deflected one. Detailed analysis of the multilayered diaphragm buckling due to its sensitivity and the methods for deflection control using the buckling process were proposed in [40].

6. Conclusions

In this paper, the modeling and optimization approaches applied to the structure of piezoelectric MEMS microphones are presented. The basic aspects to be considered in the micromachined microphone design are discussed on the basis of a typical case study. A realistic estimation of the microphone performance, obtained following the optimization towards the acoustic overload point, is in agreement with the currently published specifications for microphones required for aeroacoustic measurements in application fields where commercialized MEMS microphones are inconvenient. Piezoelectric MEMS microphones can thus satisfy requirements for applications in aircraft design and rocket launching vehicles, as well as for new applications including the detection of gunshots or screams for military and urban security systems, requiring a high sound pressure level detection and harsh environment resistance [41–43].

Finally, a new approach enabling the feasibility of microphone sensitivity control is introduced. This approach is based on the configuration with one circular electrode located in the diaphragm center and one annular electrode placed close to the diaphragm clamped edge. One of the electrodes serves as a sensor providing the microphone output signal, while the other electrode is used as an actuator that can, with a DC bias voltage, mechanically pre-stress the diaphragm and thus modify the microphone sensitivity. Such an electrically controlled sensitivity has strong potential for applications dealing with high acoustic loads that are already enumerated.

Author Contributions: Conceptualization, L.R. and J.E.; methodology, L.R. and J.E.; software, J.E. and D.E.; validation, J.E., D.E. and L.R.; formal analysis, J.E.; investigation, L.R. and J.E.; data curation, J.E.; writing—original draft preparation, L.R.; writing—review and editing, L.R., J.E., D.E. and S.B.; visualization, J.E.; supervision, L.R. and S.B.; project administration, S.B.; funding acquisition, L.R. All authors have read and agreed to the published version of the manuscript.

Funding: This work was supported from 2022 to 2023 by the DGAC (Direction Générale de l'Aviation Civile), by the PNRR (Plan National de Relance et de Résilience Français) and by NextGeneration EU via the MAMBO project (Méthodes Avancées pour la Modélisation du Bruit moteur et aviOn).

Data Availability Statement: The original contributions presented in the study are included in the article, further inquiries can be directed to the corresponding author.

Conflicts of Interest: Libor Rufer was employed by the company ADT MEMS, Didace Ekeom was employed by the company Microsonics. The remaining authors declare that the research was conducted in the absence of any commercial or financial relationships that could be construed as a potential conflict of interest.

References

1. Sessler, G.M.; West, J.E. Self-Biased Condenser Microphone with High Capacitance. *J. Acoust. Soc. Am.* **1962**, *34*, 1787–1788. [CrossRef]
2. Samaun, S.; Wise, K.D.; Nielsen, E.D.; Angel, J.B. An IC Piezoresistive Pressure Sensor for Biomedical Instrumentation. In Proceedings of the IEEE International Solid-State Circuits Conference, Philadelphia, PA, USA, 17–19 February 1971; University of Pennsylvania: Philadelphia, PA, USA, 1971; pp. 104–105.
3. Royer, M.; Holmen, P.; Wurm, M.; Aadland, P.; Glenn, M. ZnO on Si integrated acoustic sensor. *Sens. Actuators A Phys.* **1983**, *4*, 357–362. [CrossRef]
4. Hohm, D.; Sessler, G.M. An integrated silicon-electret-condenser microphone. In Proceedings of the 11th International Congress on Acoustics, Paris, France, 15–16 July 1983; pp. 29–32.
5. Schellin, R.; Hess, G. A silicon subminiature microphone based on piezoresistive polysilicon strain gauges. *Sens. Actuators A Phys.* **1992**, *32*, 555–559. [CrossRef]
6. Kühnel, W. Silicon Condenser Microphone with Integrated Field-effect Transistor. *Sens. Actuators A Phys.* **1991**, *25–27*, 521–525. [CrossRef]
7. Schneider, U.; Schellin, R. A phase-modulating microphone utilizing integrated optics and micromachining in silicon. *Sens. Actuators A Phys.* **1994**, *41–42*, 695–698. [CrossRef]
8. Loeppert, P.V.; Lee, S.B. SiSonic TM—The First Commercialized MEMS Microphone. In Proceedings of the Solid-State Sensors, Actuators, and Microsystems Workshop, Hilton Head Island, SC, USA, 4–8 June 2006; pp. 27–30.
9. Dehé, A.; Wurzer, M.; Füldner, M.; Krumbein, U. Design of a Poly Silicon MEMS Microphone for High Signal-to-Noise Ratio. In Proceedings of the European Solid-State Device Research Conference (ESSDERC), Bucharest, Romania, 24–26 September 2014; pp. 292–295.
10. Oralkan, O.; Ergun, A.S.; Johnson, J.A.; Karaman, M.; Demirci, U.; Kaviani, K.; Lee, T.H.; Khuri-Yakub, B.T. Capacitive micromachined ultrasonic transducers: Next-generation arrays for acoustic imaging? *IEEE Trans. Ultrason. Ferroelectr. Freq. Control* **2002**, *49*, 1596–1610. [CrossRef] [PubMed]
11. Przybyla, R.J.; Shelton, S.E.; Guedes, A.; Izyumin, I.I.; Kline, M.H.; Horsley, D.A.; Boser, B.E. In-Air Rangefinding with an AlN Piezoelectric Micromachined Ultrasound Transducer. *IEEE Sens. J.* **2011**, *11*, 2690–2697. [CrossRef]
12. Lu, Y.; Tang, H.; Fung, S.; Wang, Q.; Tsai, J.; Daneman, M.; Boser, B.; Horsley, D. Ultrasonic fingerprint sensor using a piezoelectric micromachined ultrasonic transducer array integrated with complementary metal oxide semiconductor electronics. *Appl. Phys. Lett.* **2015**, *106*, 5.
13. He, Q.; Liu, J.; Yang, B.; Wang, X.; Chen, X.; Yang, C. MEMS-based ultrasonic transducer as the receiver for wireless power supply of the implantable microdevices. *Sens. Actuators A Phys.* **2014**, *219*, 65–72. [CrossRef]
14. Williams, M.D.; Griffin, B.A.; Reagan, T.N.; Underbrink, J.R.; Sheplak, M. An AlN MEMS Piezoelectric Microphone for Aeroacoustics Applications. *J. Microelectromech. Syst.* **2012**, *21*, 270–283. [CrossRef]
15. Yamashita, K.; Tomiyama, K.; Yoshikawa, K.; Noda, M.; Okuyama, M. Resonant Frequency Tuning of Piezoelectric Ultrasonic Microsensors by Bias Voltage Application to Extra Top-Electrodes on PZT Diaphragms. *Ferroelectrics* **2010**, *408*, 48–54. [CrossRef]
16. Nastro, A.; Ferrari, M.; Rufer, L.; Basrour, S.; Ferrari, V. Piezoelectric MEMS Acoustic Transducer with Electrically-Tunable Resonant Frequency. *Micromachines* **2022**, *13*, 96. [CrossRef]
17. Yamashita, K.; Shimizu, N.; Okuyama, M. Piezoelectric Charge Distribution on Vibrating Diaphragms with Static Deflection and Sensitivity Improvement of Ultrasonic Sensors. *J. Korean Phys. Soc.* **2007**, *51*, 785–789. [CrossRef]
18. Tiersten, H.F. *Linear Piezoelectric Plate Vibrations: Elements of the Linear Theory of Piezoelectricity and the Vibrations Piezoelectric Plates*; eBook; Springer: Berlin/Heidelberg, Germany, 2013; 212p.
19. Safari, A.; Akdogan, E.K. *Piezoelectric and Acoustic Materials for Transducer Applications*; Springer: New York, NY, USA, 2008; 482p.
20. Kanno, I. Piezoelectric MEMS: Ferroelectric Thin Films for MEMS Applications. *Jpn. J. Appl. Phys.* **2018**, *57*, 040101. [CrossRef]
21. Muralt, P. Ferroelectric thin films for micro-sensors and actuators: A review. *J. Micromech. Microeng.* **2000**, *10*, 136–146. [CrossRef]
22. Muralt, P. PZT Thin Films for Microsensors and Actuators: Where Do We Stand? *IEEE Trans. Ultrason. Ferroelectr. Freq. Control* **2000**, *47*, 903–915. [CrossRef]
23. Trolier-McKinstry, S.; Muralt, P. Thin Film Piezoelectrics for MEMS. *J. Electroceram.* **2004**, *12*, 7–17. [CrossRef]

24. Ngoc Thao, P.; Yoshida, S.; Tanaka, S. Fabrication and Characterization of PZT Fibered-Epitaxial Thin Film on Si for Piezoelectric Micromachined Ultrasound Transducer. *Micromachines* **2018**, *9*, 455. [CrossRef]
25. Gangidi, P.; Gupta, N. Optimal selection of dielectric film in piezoelectric MEMS microphone. *Microsyst. Technol.* **2019**, *25*, 4227–4235. [CrossRef]
26. Poudyal, A.; Jackson, N. Characterization of confocal sputtered molybdenum thin films for aluminum nitride growth. *Thin Solid Film.* **2020**, *693*, 137657. [CrossRef]
27. Ali, W.R.; Prasad, M. Piezoelectric MEMS based acoustic sensors: A review. *Sens. Actuators A Phys.* **2020**, *301*, 111756. [CrossRef]
28. Dutta, S.; Pandey, A. Overview of residual stress in MEMS structures: Its origin, measurement, and control. *J. Mat. Sci. Mater. Electron.* **2021**, *32*, 6705–6741. [CrossRef]
29. Yang, H.; Liu, M.; Zhu, Y.; Wang, W.; Qin, X.; He, L.; Jiang, K. Characterization of Residual Stress in SOI Wafers by Using MEMS Cantilever Beams. *Micromachines* **2023**, *14*, 1510. [CrossRef]
30. Pandey, A.; Dutta, S.; Prakash, R.; Raman, R.; Kapoor, A.K.; Kaur, D. Growth and Comparison of Residual Stress of AlN Films on Silicon (100), (110) and (111) Substrates. *J. Electron. Mater.* **2018**, *47*, 1405–1413. [CrossRef]
31. Schlögl, M.; Weißenbach, J.; Schneider, M.; Schmid, U. Stress engineering of polycrystalline aluminum nitride thin films for strain sensing with resonant piezoelectric microbridges. *Sens. Actuators A Phys.* **2023**, *349*, 114067. [CrossRef]
32. Fazzio, R.S.; Lamers, T.; Buccafusca, O.; Goel, A.; Dauksher, W. Design and performance of aluminum nitride piezoelectric microphones. In Proceedings of the 14th International Conference on Solid-State Sensors, Actuators and Microsystems, Lyon, France, 10–14 June 2007; pp. 1255–1258.
33. Merhaut, J. *Theory of Electroacoustics*; McGraw-Hill: New York, NY, USA, 1981; 317p.
34. Reutter, T.; Schrag, G. Reliable Piezoelectric FEM-Simulations of MEMS Microphones: Basis for Intrinsic Stress Reduction. In Proceedings of the IEEE Sensors Conference, Waikoloa, HI, USA, 1–4 November 2010; pp. 193–196.
35. Kohnke, P. *Mechanical APDL Theory Reference*; Ansys, Inc.: Canonsburg, PA, USA, 2021; 934p.
36. Esteves, J.; Rufer, L.; Ekeom, D.; Basrour, S. Lumped-parameters equivalent circuit for condenser microphones modeling. *J. Acoust. Soc. Am.* **2017**, *142*, 2121–2132. [CrossRef]
37. Esteves, J.; Rufer, L.; Ekeom, D.; Defoort, M.; Basrour, S. Approaches to Piezoelectric Micromachined Microphone Design: Comparative Study. In Proceedings of the 10th Convention of the European Acoustics Association, Forum Acusticum, Torino, Italy, 11–15 September 2023; pp. 4909–4916.
38. Timoshenko, P.; Krieger, S.W. *Theory of Plates and Shells*; McGraw-Hill: New York, NY, USA, 1959; 604p.
39. Wang, G.; Sankar, B.V.; Cattafesta, L.N.; Sheplak, M. Analysis of a Composite Piezoelectric Circular Plate with Initial Stresses for MEMS. In Proceedings of the IMECE, New Orleans, LA, USA, 17–22 November 2002; pp. 1–8.
40. Yamashita, K.; Nishimoto, H.; Okuyama, M. Diaphragm deflection control of piezoelectric ultrasonic microsensors for sensitivity improvement. *Sens. Actuators A Phys.* **2007**, *139*, 118–123. [CrossRef]
41. Kumar, A.; Varghese, A.; Sharma, A.; Prasad, M.; Janyani, V.; Yadav, R.P.; Elgaid, K. Recent development and futuristic applications of MEMS based piezoelectric microphones. *Sens. Actuators A Phys.* **2022**, *347*, 113887. [CrossRef]
42. Ali, W.R.; Prasad, M. Design and Fabrication of Piezoelectric MEMS Sensor for Acoustic Measurements. *Silicon* **2022**, *14*, 6737–6747. [CrossRef]
43. Wu, L.; Chen, X.; Ngo, H.D.; Julliard, E.; Spehr, C. Design of dual-frequency piezoelectric MEMS microphones for wind tunnel testing. In Proceedings of the AIAA SciTech Forum, San Diego, CA, USA, 3–4 January 2022; pp. 1–7.

 *micromachines*

Article

Modeling of MEMS Transducers with Perforated Moving Electrodes

Karina Šimonová and Petr Honzík *

Faculty of Transportation Sciences, Czech Technical University in Prague, Konviktská 20,
110 00 Praha, Czech Republic; abramkar@fd.cvut.cz
* Correspondence: honzikp@fd.cvut.cz

Abstract: Microfabricated electroacoustic transducers with perforated moving plates used as microphones or acoustic sources have appeared in the literature in recent years. However, optimization of the parameters of such transducers for use in the audio frequency range requires high-precision theoretical modeling. The main objective of the paper is to provide such an analytical model of a miniature transducer with a moving electrode in the form of a perforated plate (rigid elastically supported or elastic clamped at all boundaries) loaded by an air gap surrounded by a small cavity. The formulation for the acoustic pressure field inside the air gap enables expression of the coupling of this field to the displacement field of the moving plate and to the incident acoustic pressure through the holes in the plate. The damping effects of the thermal and viscous boundary layers originating inside the air gap, the cavity, and the holes in the moving plate are also taken into account. The analytical results, namely, the acoustic pressure sensitivity of the transducer used as a microphone, are presented and compared to the numerical (FEM) results.

Keywords: analytical modeling; electroacoustic transducers; MEMS microphones; perforated plate

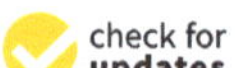
check for
updates

Citation: Šimonová, K.; Honzík, P. Modeling of MEMS Transducers with Perforated Moving Electrodes. *Micromachines* **2023**, *14*, 921. https://doi.org/10.3390/mi14050921

Academic Editor: Libor Rufer

Received: 24 March 2023
Revised: 18 April 2023
Accepted: 20 April 2023
Published: 24 April 2023

1. Introduction

Currently, the vast majority of MEMS microphones production, increasing rapidly in recent years, uses the electrostatic principle of electroacoustic transduction [1] (although piezoelectric types exist [2]). Such devices consist of moving electrodes of circular [3], square [4,5], or other [6] shapes and perforated single [3] or double backplates [7,8]. Note that such MEMS structures can be employed in other domains than audio, such as energy transfer, energy harvesters, and resonators [9–11]. However, new designs presenting technological advances have been proposed recently in the literature, such as a microphone with moving microbeam [12], or transducers (sources and microphones) with perforated moving electrodes. The mean motivation for the work presented herein is the latter case with electrodes in the form of elastic perforated plates clamped at all boundaries [13] or rigid elastically supported perforated plates [14–17]. Although these experimental studies contain approximate theoretical models, mainly based on the lumped elements approach, the precise analytical modeling is still of high interest.

In order to provide high-precision results on sensitivity and bandwidth, the models of electroacoustic transducers (miniaturized or not) should take into account the damping effects of the viscous and thermal boundary layers originating in the narrow regions such as the air gap between the moving and fixed electrodes. The strong coupling between the displacement field of the moving electrode and the acoustic field inside the transducer should be also accounted for when appropriate. In addition to these effects, the model of the transducer with a perforated moving electrode has to deal with the acoustic short circuit between the incident acoustic pressure and the pressure field inside the transducer caused by the perforation. This leads to the sensitivity roll-off at lower frequencies, which has to be calculated correctly when precise theoretical modeling allowing the optimization of the transducer behavior in the audio frequency range is required.

While the precise models of the transducers with perforated moving electrodes are still missing, to our knowledge, several models taking into account the perforation of the fixed electrodes and the acoustic short circuit can be found in the literature. The classical lumped-element models of condenser microphones, such as [18], use the "porosity" approach; the more recent lumped-element model [19] deals with acoustic short circuit through the venting hole. With regard to more advanced models, ref. [20] and, more recently, [21,22] took into account the effects of holes in the fixed electrode accounting for the position of the holes, and ref. [23] employed the impedance approach. Vibration of a very thin perforated backplate of an MEMS transducer was taken into account in [24]. In [25], the effect of the acoustic short circuit through thin slits surrounding the moving electrode in the form of a microbeam was included in the complex wavenumber for the acoustic pressure in the air gap. In the same reference, the acoustic pressure in the air gap was expressed using integral formulation with appropriate Green's function, which was not expressed as a series expansion over the eigenfunctions of the moving electrode. Such a fomulation is also advantageous in the case of rectangular geometries [26,27] and is therefore used herein. It is worth mentioning the numerical methods, namely, the finite element method, which can take into account the thermoviscous losses and the coupling effects without geometry-dependent approximations [28]. However, numerical methods generally suffer from high computational costs, compared to analytical methods, and are usually used as a reference against which the analytical results can be tested.

The present paper deals mainly with the theoretical modeling of the acoustic field inside a miniaturized electroacoustic transducer with a square perforated moving electrode, taking into account its coupling with the vibration of the moving electrode, the acoustic short circuit through the perforation, and the thermoviscous losses originating in the narrow regions inside the transducer. Two types of the perforated moving electrode are considered: (i) the rigid elastically supported square plate, partially inspired by [14] (see Figure 1a), and (ii) the flexible square plate clamped at all boundaries, partially inspired by [13] (see Figure 1b).

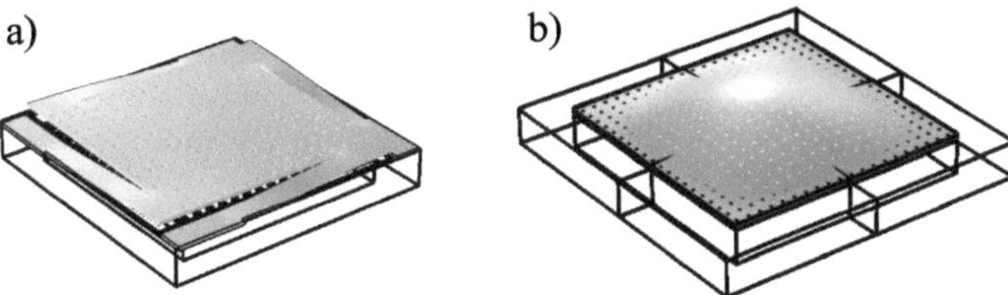

Figure 1. MEMS transducers with perforated moving electrodes in the form of (**a**) the rigid elastically supported square plate and (**b**) the flexible square plate clamped at all boundaries.

Section 2 presents viscous effects in short narrow holes and governing equations for the acoustic pressure field in the thin air gap between the perforated moving electrode and the fixed one (backplate) using the porosity approach. Then, the solutions for the acoustic pressure are expressed for the case of uniform (piston-like) and nonuniform movement of the moving electrode, corresponding to the rigid elastically supported and flexible plate, respectively. The coupling of the acoustic field with vibration of the plates of both types, leading to the expression of their displacements, is finally derived, with the eigenfunctions of the perforated flexible clamped plate being given approximately in Appendix A. In Section 3, the analytically calculated acoustic pressure sensitivities of the transducers used as microphones are depicted and compared with the numerical (FEM) results. The influence of some geometrical parameters is discussed. This section is followed by the conclusion in Section 4.

2. Analytical Solution

In this section, the analytical solution of the problem is expressed in frequency domain (the time dependence being $e^{j\omega t}$; ω is the angular frequency). The acoustic field inside the transducer and the displacement of the moving electrode is searched for as a response to harmonic incident acoustic pressure p_{inc} (assumed to be uniform over the moving electrode surface).

2.1. Description of the Device

The device consists of a moving electrode in the form of a square perforated plate of side $2a$ and thickness h_p with N square holes of side a_h, with the air gap between the plate and the backplate of thickness h_g surrounded by a peripheral cavity described by its volume V_c and acoustic impedance Z_c (see Figure 2). The perforation ratio $\mathcal{R} = Na_h^2/(4a^2)$ is the ratio of total surface occupied by the holes and the area of the plate. In the case of a moving electrode in the form of a rigid elastically supported square plate (Figure 1a), the plate displacement is uniform and the cavity is connected with the incident acoustic pressure p_{inc} through slits of thickness h_s along the arms supporting the plate.

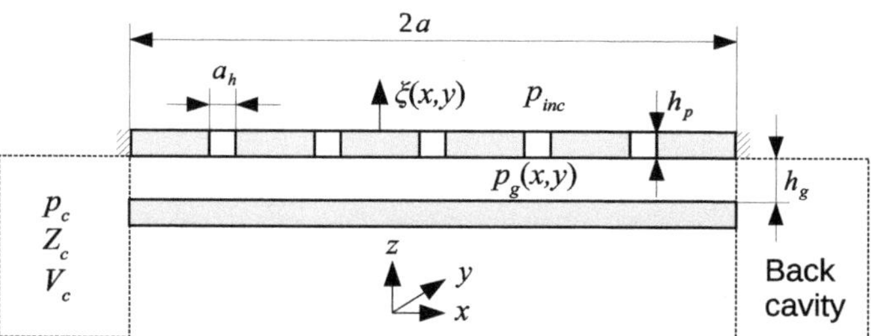

Figure 2. Sketch of the whole system.

2.2. The Acoustic Pressure Field inside the Transducer

The system is supposed to be filled with thermoviscous fluid (air in this case) with the following properties: the density ρ_0, the adiabatic speed of sound c_0, the heat capacity at constant pressure per unit mass C_p, the specific heat ratio γ, the shear viscosity coefficient μ, and the thermal conduction coefficient λ_h. Since the air gap thickness h_g is supposed to be much smaller than other dimensions of the air gap and smaller than the wavelengths considered, even at high frequencies, the acoustic pressure in the air gap is assumed to depend on the x, y spatial coordinates only and denoted $p_g(x, y)$. The particle velocity and temperature variation (that generally depend on the z axis due to the viscous and thermal boundary layers effects) are then replaced by their mean values over the air gap thickness. The acoustic pressure in the cavity volume p_c is supposed to be uniform. The displacement of the moving electrode is denoted ζ.

2.2.1. Viscous Effects Originating in the Holes in the Perforated Moving Electrode

For the sake of simplicity, the holes in the moving electrode are supposed to have circular cross-section instead of the square one, with the radius of the equivalent cylindrical hole being given by $R_h = a_h/\sqrt{\pi}$ (thus $\mathcal{R} = \frac{N\pi R_h^2}{4a^2}$), see Figure 3. The particle velocity $v_z(r, z)$ in such a hole is governed by the diffusion equation [29]

$$\left(\frac{1}{r}\frac{\partial}{\partial r}r\frac{\partial}{\partial r} + k_v^2\right)v_z(r, z) = \frac{1}{\mu}\frac{\partial}{\partial z}p(z), \tag{1}$$

with the diffusion wavenumber

$$k_v = \frac{1 - j}{\sqrt{2}}\sqrt{\frac{\omega\rho_0}{\mu}}, \tag{2}$$

with j being the imaginary unit, and subjected to the nonslip boundary condition at $r = R_h$

$$v_z(R_h, z) = j\omega\xi(x, y). \tag{3}$$

The velocity of the moving electrode $j\omega\xi(x, y)$ at the position of the hole is assumed to be approximately uniform on the whole internal surface of the hole. The solution of the problem (1) and (3) is given by

$$v_z(r, z) = -\frac{1}{j\omega\rho_0}\frac{\partial}{\partial z}p(z)\left[1 - \frac{J_0(k_v r)}{J_0(k_v R_h)}\right] + j\omega\xi(x, y)\frac{J_0(k_v r)}{J_0(k_v R_h)}, \tag{4}$$

where J_n denotes the cylindrical Bessel functions of the first kind of order n. After relying on the approximation of the pressure derivative in a very short hole of length h_p

$$\frac{\partial}{\partial z}p(z) \approx \frac{p_{inc} - p_g(x, y)}{h_p}, \tag{5}$$

the mean value of the particle velocity over the cross-section of the hole $S_h = \pi R_h^2$ is

$$\langle v_z(r, z)\rangle_r = \frac{1}{S_h}\iint_{S_h} v_z(r, z)\mathrm{d}S_h \approx -\frac{1}{j\omega\rho_0}\frac{p_{inc} - p_g(x, y)}{h_p}F_{vh} + j\omega\xi(x, y)K_{vh}, \tag{6}$$

with

$$F_{vh} = 1 - K_{vh},$$
$$K_{vh} = \frac{2}{k_v R_h}\frac{J_1(k_v R_h)}{J_0(k_v R_h)}. \tag{7}$$

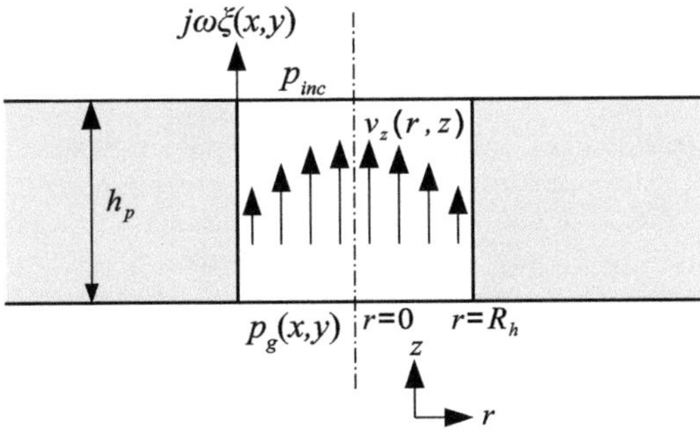

Figure 3. Sketch of the hole in the moving electrode.

The viscous force acting on the interior surface of the hole $2\pi R_h h_p$ is proportional to the normal derivative (here, $\partial/\partial n = -\partial/\partial r$) of the particle velocity (4)

$$F_z = -2\pi R_h h_p \mu\left(\frac{\partial v_z(r, z)}{\partial r}\right)_{r=R_h}, \tag{8}$$

and in using (5) takes the following form

$$F_z(x, y) \approx j\omega\xi(x, y)\Pi_h - \pi R_h^2 K_{vh}\left[p_{inc} - p_g(x, y)\right], \tag{9}$$

with

$$\Pi_h = 2\pi R_h h_p \mu\frac{k_v J_1(k_v R_h)}{J_0(k_v R_h)}. \tag{10}$$

Dividing this force by the area associated with one hole ($4a^2/N$) leads to the equivalent pressure caused by the viscosity effects originating in the hole

$$p_v(x,y) = j\omega\xi(x,y)\Pi_h\frac{N}{4a^2} - \mathcal{R}K_{vh}\left[p_{inc} - p_g(x,y)\right].\tag{11}$$

2.2.2. Wave Equation Governing the Acoustic Pressure in the Air Gap

In order to express the wave equation for the acoustic pressure $p_g(x,y)$ in the air gap, the following contributions of the mass per unit of time in the gap element of dimensions $dx \times dy \times h_g$ have to be taken into account (see velocity contributions in Figure 4):

- Change of the mass per unit of time in both x and y directions $-\frac{\partial}{\partial w}\langle v_{gw}(w,z)\rangle_z\, \rho_0\, dx\, dy\, h_g$, where w designates x and y.
- Contribution from the moving electrode $-j\omega\xi(x,y)\,\rho_0\,(1-\mathcal{R})\,dx\,dy$.
- Contribution form the holes $-\langle v_z(r,z)\rangle_r\,\rho_0\,\mathcal{R}\,dx\,dy$, where $\langle v_z(r,z)\rangle_r$ is given by Equation (6).

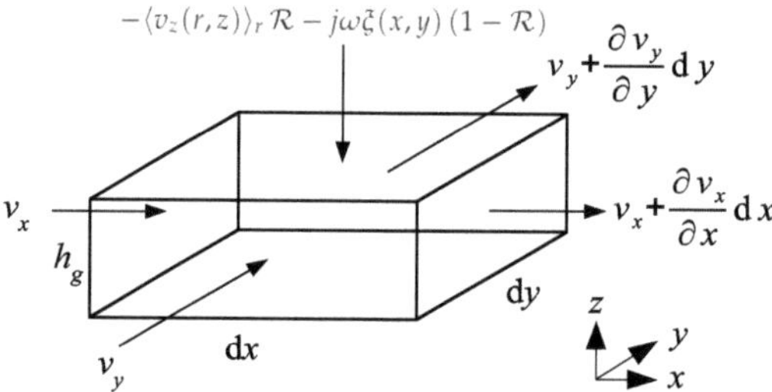

Figure 4. Element of the air gap.

The sum of these terms is equal to $j\omega\,\langle\rho\rangle_z\,dx\,dy\,h_g$ (conservation of mass) where $\langle\rho\rangle_z$ is the time-dependent acoustic density in the gap element averaged over the gap thickness. The classical solutions of linearized Navier–Stokes equation and Fourier equation for the heat conduction, under several approximations [29], give, respectively, the particle velocity and temperature variations profiles in the air gap, leading to the relations (after introducing the latter into the gas state equation) $\langle v_{gw}(w,z)\rangle_z = -\frac{1}{j\omega\rho_0}\frac{\partial p_g(x,y)}{\partial w}F_{vg}$ and $\langle\rho\rangle_z = p_g(x,y)\left[\gamma - (\gamma-1)F_{hg}\right]/c_0^2$ [29], with the mean values of the velocity and temperature variation profiles over the air gap thickness given by

$$F_{vg} = 1 - \frac{\tan\left(k_v h_g/2\right)}{k_v h_g/2},$$

$$F_{hg} = 1 - \frac{\tan\left(k_h h_g/2\right)}{k_h h_g/2},\tag{12}$$

where k_v is given by (2) and $k_h = \frac{1-j}{\sqrt{2}}\sqrt{\frac{\omega\rho_0 C_p}{\lambda_h}}$. Note that these mean values are calculated for nonslip and isothermal boundary conditions at both (nonperforated) electrodes. Alternatively, the relation accounting for more realistic boundary conditions on the perforated plate [21],

$$F_{(v,h)g} = 1 - \frac{2-\mathcal{R}}{2}\frac{\tan\left(k_{(v,h)}h_g/2\right)}{k_{(v,h)}h_g/2},\tag{13}$$

can be used.

The combination of the above mentioned terms leads to the wave equation governing the acoustic pressure $p_g(x,y)$ in the air gap

$$\left[\Delta + \chi^2\right] p_g(x,y) = -U(x,y), \tag{14}$$

where the source term is composed from $U(x,y) = U_1 \xi(x,y) + U_2 p_{inc}$ with

$$U_1 = \frac{\omega^2 \rho_0 (1 - \mathcal{R} F_{vh})}{F_{vg} h_g},$$

$$U_2 = \frac{F_{vh} \mathcal{R}}{F_{vg} h_g h_p}, \tag{15}$$

and the complex wavenumber is given by

$$\chi^2 = \frac{\omega^2}{c_0^2} \frac{\gamma - (\gamma - 1) F_{hg}}{F_{vg}} - \frac{F_{vh} \mathcal{R}}{F_{vg} h_g h_p}. \tag{16}$$

2.2.3. Solution for the Acoustic Pressure in the Air Gap in Case of Piston-like Movement of the Moving Electrode

Since the source term U in (14) does not depend on the spatial coordinates x, y in this case, the solution of (14) takes the classical form

$$p_g(x,y) = A \cos(\kappa_x x) \cos(\kappa_y y) - U/\chi^2, \tag{17}$$

where $\chi^2 = \kappa_x^2 + \kappa_y^2$ (for square geometry $\kappa_x = \kappa_y = \chi/\sqrt{2}$) and A is an integration constant. The boundary condition is given by the acoustic pressure in the peripheral cavity (supposed to be uniform in the whole cavity volume) $p_c = Z_c w_{tot}$, where Z_c is the acoustic impedance of the cavity and w_{tot} is the total volume velocity entering to the cavity. This volume velocity is composed of the volume velocity at the output of the air gap and the volume velocity entering to the cavity through the slits

$$w_{tot} = 8 a h_g \left\langle v_{gw}(w,z) \right\rangle_z - S_s \bar{v}_s, \tag{18}$$

where $8 a h_g$ is the output surface of the air gap, S_s is the total input surface of the slits, and $\bar{v}_s \cong -\frac{F_{vs}}{j\omega\rho_0} \frac{p_{inc} - p_c}{h_p}$ is the velocity in the slit, with $F_{vs} = 1 - \frac{\tan(k_v h_s/2)}{k_v h_s/2}$ being the mean value of the velocity profile through the thickness of the slit h_s (the influence of the plate velocity on the fluid particle velocity in the slits is supposed to be negligible here). This leads directly to the boundary condition for the normal derivative of the acoustic pressure at the output of the air gap

$$\partial_n p_g = -\Lambda_c p_c + \Lambda_2 p_{inc}, \tag{19}$$

with

$$\Lambda_c = \Lambda_1 + \Lambda_2, \quad \Lambda_1 = \frac{j\omega\rho_0}{8 a h_g F_{vg} Z_c}, \quad \Lambda_2 = \frac{S_s F_{vs}}{8 a h_g h_p F_{vg}}.$$

The continuity of the acoustic pressure at the boundary between the air gap and the cavity can be approximately expressed using the value at the middle of the square gap side $p_c = p_g(a,0)$ (alternatively, the value at the corner $p_g(a,a)$ or the mean value over the side of the gap could be used). Replacing $\partial_n p_g$ and p_c in (19) by (for example) $\partial_x p_g(a,0)$ and $p_g(a,0)$, respectively, and substituting from the solution (17) readily gives the integration constant

$$A = (A_1 \xi + A_2 p_{inc})/A_3, \tag{20}$$

with

$$A_1 = \Lambda_c U_1/\chi^2, \quad A_2 = \Lambda_2 + \Lambda_c U_2/\chi^2, \quad A_3 = \Lambda_c \cos(\kappa_x a) - \kappa_x \sin(\kappa_x a). \tag{21}$$

2.2.4. Solution for the Acoustic Pressure in the Air Gap in Case of Nonuniform Movement of the Moving Electrode

Due to the symmetry of the transducer's geometry, the solution of (14) for nonuniform $U(x,y)$ is expressed here in the first quadrant only (namely, $x,y \in (0,a)$). The chosen Green's function used in the integral formulation for the solution of (14) satisfies the same Neumann's condition (the first derivative vanishes) at $x = 0, y = 0$ as the solution for the acoustic pressure, which can be expressed as follows [26,27,29]:

$$
\begin{aligned}
p_g(x,y) = &\int_0^a \int_0^a G(x,x_0;y,y_0)U(x_0,y_0)\mathrm{d}x_0\mathrm{d}y_0 \\
&+ \int_0^a \left[G(x,x_0;y,a)\partial_{y_0}p_g(x_0,a) - \partial_{y_0}G(x,x_0;y,a)p_g(x_0,a) \right]\mathrm{d}x_0 \\
&+ \int_0^a \left[G(x,a;y,y_0)\partial_{x_0}p_g(a,y_0) - \partial_{x_0}G(x,a;y,y_0)p_g(a,y_0) \right]\mathrm{d}y_0,
\end{aligned}
\tag{22}
$$

with the Green's function being given by

$$
G(x,x_0;y,y_0) = g(x,x_0;y,y_0) + g(x,-x_0;y,y_0) + g(x,x_0;y,-y_0) + g(x,-x_0;y,-y_0),
\tag{23}
$$

with

$$
g(x,x_0;y,y_0) = -\frac{j}{4}H_0^- \left(\chi\sqrt{(x-x_0)^2 + (y-y_0)^2} \right),
\tag{24}
$$

where H_0^- denotes the cylindrical Hankel function of the second kind of order "0".

Taking into account the boundary condition (19), here without the slits (Λ_2 vanishes), solution (22) becomes

$$
p_g(x,y) = \int_0^a \int_0^a G(x,x_0;y,y_0)U(x_0,y_0)\mathrm{d}x_0\mathrm{d}y_0 - p_c I_g(x,y),
\tag{25}
$$

where

$$
\begin{aligned}
I_g(x,y) = \Lambda_1 &\left[\int_0^a G(x,x_0;y,a)\mathrm{d}x_0 + \int_0^a G(x,a;y,y_0)\mathrm{d}y_0 \right] \\
&+ \left[\int_0^a \partial_{y_0}G(x,x_0;y,a)\mathrm{d}x_0 + \int_0^a \partial_{x_0}G(x,a;y,y_0)\mathrm{d}y_0 \right].
\end{aligned}
\tag{26}
$$

The acoustic pressure in the cavity, calculated here as the mean value over the edge of the gap, $p_c = \langle p_g(x,a) \rangle_x$, where $\langle f(w) \rangle_w$ denotes $\int_0^a f(w)\mathrm{d}w/a$, can be then expressed from (25) as follows:

$$
p_c = \frac{1}{1 + \langle I_g(x,a) \rangle_x} \int_0^a \int_0^a \langle G(x,x_0;a,y_0) \rangle_x U(x_0,y_0)\mathrm{d}x_0\mathrm{d}y_0.
\tag{27}
$$

2.3. Coupling of the Moving Electrode Displacement Field and the Acoustic Pressure Field

In this section, the strong coupling between the acoustic field inside the transducer, described in previous sections, and the displacement of the moving electrode in the form of an elastically supported rigid perforated plate and a flexible perforated plate clamped at all edges is presented.

2.3.1. Elastically Supported Rigid Perforated Plate

The equation governing the displacement ξ of the elastically supported rigid plate takes the form

$$
\left[-M_p\omega^2 + j\omega R_p + K_p \right]\xi = \int_{-a}^a \int_{-a}^a \left[p_g(x,y) - p_{inc} - p_v(x,y) \right]\mathrm{d}x\mathrm{d}y,
\tag{28}
$$

where M_p is the mass of the plate, K_p is the stiffness of the elastic support, and R_p is the structural damping coefficient which is neglected here (all the damping in the system taken into account here originates in the acoustic fluid-filled parts of the transducer).

Reporting Equations (11) and (17) using (20) to (28) gives, after straightforward calculation, the solution for the displacement of the rigid plate:

$$\xi = \frac{4a^2(1 + \mathcal{R}K_{vh})\left[\frac{\sin(\kappa_x a)\sin(\kappa_y a)A_2}{\kappa_x \kappa_y a^2 A_3} - \left(1 + \frac{U_2}{\chi^2}\right)\right]p_{inc}}{-M_p\omega^2 + j\omega R_p + K_p + j\omega\pi N + 4a^2(1 + \mathcal{R}K_{vh})\left[\frac{U_1}{\chi^2} - \frac{\sin(\kappa_y a)\sin(\kappa_y a)A_1}{\kappa_x \kappa_y a^2 A_3}\right]}. \tag{29}$$

2.3.2. Flexible Perforated Plate Clamped at All Edges

We will depart here from the classical equation governing the displacement of the nonperforated plate [30] with the mass per unit area $M_s = \rho_p h_p$ (ρ_p designates the density of the plate) and the flexural rigidity $D = \frac{Eh_p}{12(1-\nu^2)}$ (E and ν being the Young's modulus and Poisson's ratio, respectively)

$$\left[D\Delta\Delta - M_s\omega^2\right]\xi(x,y) = p_g(x,y) - p_{inc} - p_v(x,y), \tag{30}$$

clamped at all edges

$$\begin{aligned}
\xi(x,y) &= \frac{\partial}{\partial x}\xi(x,y) = 0, x = \pm a, \forall y \in (-a,a), \\
\xi(x,y) &= \frac{\partial}{\partial y}\xi(x,y) = 0, y = \pm a, \forall x \in (-a,a).
\end{aligned} \tag{31}$$

The displacement field can be searched for in the following form of series expansion (with some truncation in practical implementation):

$$\xi(x,y) = \sum_{mn}\xi_{mn}\psi_{mn}(x,y), \tag{32}$$

where the orthonormal eigenfunctions $\psi_{mn}(x,y)$ satisfy the homogeneous equation associated with Equation (30):

$$\left[\Delta\Delta - k_{m,n}^4\right]\psi_{mn}(x,y) = 0, \tag{33}$$

where $k_{m,n}^4 = (k_{xm}^2 + k_{yn}^2)^2$. An approximate form of such eigenfunctions can be obtained as a series expansion over known functions from numerically (FEM) calculated results using the method described in [31] for nonperforated rectangular clamped plates and in [32] for perforated square clamped plates, the latter being used herein (see Appendix A).

Using the properties of the eigenfunctions [29], the modal coefficients ξ_{mn} in (32) can be obtained from the relation (using Equation (11))

$$\xi_{mn}\left[Dk_{m,n}^4 - M_s\omega^2 + j\omega\Pi_h\frac{N}{4a^2}\right] = (1 - \mathcal{R}K_{vh})\int_{-a}^{a}\int_{-a}^{a}[p_g(x,y) - p_{inc}]\psi_{mn}(x,y)dxdy. \tag{34}$$

Using the relation for the acoustic pressure in the air gap $p_g(x,y)$ (25) along with Equations (15) and (32), Equation (34) can be expressed as follows:

$$\left[Dk_{m,n}^4 - M_s\omega^2 + j\omega\Pi_h\frac{N}{4a^2}\right]\xi_{mn} = c_{mn} - \sum_{qr}^{\infty}\xi_{qr}A_{(mn),(qr)}, \tag{35}$$

or in the matrix form

$$[\mathbb{B} - \mathbb{A}]\Xi = \mathbb{C}, \tag{36}$$

where Ξ is the column vector of elements ζ_{mn}, $\mathbb{B}$ is the diagonal matrix of elements $Dk_{m,n}^4 - M_s\omega^2 + j\omega\Pi_h\frac{N}{4a^2}$, $\mathbb{C}$ is the column vector, and $\mathbb{A}$ is the matrix whose elements c_{mn} and $A_{(mn),(qr)}$ are given, respectively, by

$$c_{mn} = p_{inc}\int_{-a}^{a}\int_{-a}^{a}\psi_{mn}(x,y)\left\{U_2\left[\int_0^a\int_0^a G(x,x_0;y,y_0)dx_0dy_0 - MI_g(x,y)\right] - 1\right\}dxdy, \quad (37)$$

and

$$A_{(mn),(qr)} = U_1\int_{-a}^{a}\int_{-a}^{a}\psi_{mn}(x,y)\left[\int_0^a\int_0^a G(x,x_0;y,y_0)\psi_{qr}(x,y)dx_0dy_0 - N_{qr}I_g(x,y)\right]dxdy, \quad (38)$$

with

$$M = \frac{1}{1 + \langle I_g(x,a)\rangle_x}\int_0^a\int_0^a \langle G(x,x_0;a,y_0)\rangle_x\, dx_0dy_0,$$

$$N_{qr} = \frac{1}{1 + \langle I_g(x,a)\rangle_x}\int_0^a\int_0^a \langle G(x,x_0;a,y_0)\rangle_x\, \psi_{qr}(x_0,y_0)\, dx_0dy_0. \quad (39)$$

Solving Equation (36) for Ξ gives the modal coefficients ζ_{mn} and thus the displacement field of the plate $\xi(x,y)$.

3. Analytical Results and Comparisons with the Numerical (FEM) Ones

In this section, the analytical results calculated using the present method are discussed and compared with the numerical (FEM) results provided by the software Comsol Multiphysics, version 6.0. The numerical formulation for the acoustic field in thermoviscous fluid inside the transducer, involving the acoustic particle velocity $\vec{v}$, acoustic temperature variation τ, and acoustic pressure p using the Acoustics Module [33], was coupled with the classical linear mechanical formulation for the plate provided by the Structural Mechanics Module [34]. One quarter of the transducer geometry was used for the simulation (the rest was symmetric), and the mesh consisted of tetrahedral elements combined with layered prism elements (in the boundary layers). The number of degrees of freedom varied between approximately 1 million and 3 million, depending on the dimensions of the holes in the plate (smaller holes lead to finer mesh and thus higher number of degrees of freedom). The properties of the air used in both numerical and analytical calculations are given in Table 1, and the properties of the material of the plate (silicon) are summarized in Table 2.

Table 1. Properties of the air.

Parameter	Value	Unit
Adiabatic sound speed c_0	343.2	m s^{-1}
Air density ρ_0	1.2	kg m^{-3}
Shear dynamic viscosity μ	1.814×10^{-5}	Pa s
Thermal conductivity λ_h	25.77×10^{-3}	$\text{W m}^{-1}\,\text{K}^{-1}$
Specific heat coefficient at constant pressure per unit of mass C_p	1005	$\text{J kg}^{-1}\,\text{K}^{-1}$
Ratio of specific heats γ	1.4	-

Table 2. Material properties of the plate (silicon).

Parameter	Value	Unit
Plate density ρ_p	2330	kg m^{-3}
Young's modulus E	160	G Pa
Poisson's ratio ν	0.27	-

The displacement of the moving electrode given either by Equation (29) for the elastically supported rigid perforated plate or by Equation (32) for the flexible perforated plate clamped at all edges was used to calculate the acoustic pressure sensitivity of the electrostatic receiving transducer $\sigma = U_0\overline{\xi}/(h_g p_{inc})$, where $\overline{\xi} = [\iint_{S_e} \xi(x,y)\mathrm{d}S_e]/S_e$ is the mean displacement of the plate over the surface of the backing electrode $S_e = 4a^2$, and U_0 is the polarization voltage (here, $U_0 = 30$ V).

Figure 5 shows the acoustic pressure sensitivity of the receiving transducer with an elastically supported rigid perforated plate of dimensions 0.3×0.3 mm ($a = 150$ μm) and thickness $h_p = 5$ μm with $N = 256$ square holes of side dimension a_h varying between 0.3 μm and 3 μm, as per the thickness of the slits h_s. The air gap thickness is $h_g = 4$ μm, and the peripheral cavity of thickness 50 μm has the volume of $V_c = 2.72 \times 10^{-12}$ m^3. The mass of the plate is given by $M_p = \rho_p h_p(4a^2 - Na_h^2)$, and the structural damping coefficient is supposed to be negligible $R_p = 0$ Ns/m. The stiffness of the elastic support K_p was calculated from the simple numerical model of the mechanical moving part only (*in vacuo*) at very low frequencies. The dimensions of the arms of 145×30 μm lead to $K_p = 200$ N/m. Very good agreement between the analytical results was obtained using the present method (Equation (29)), and the ones provided by the complete numerical model of the transducer can be observed, especially for small a_h and h_s. When the values of a_h and h_s approach the gap thickness h_g (Figure 5d)), the damping seems to be slightly underestimated. Generally, it seems that the "porosity" approach using the ratio $\mathcal{R}$ works better when the dimensions of the holes are much smaller than the gap thickness.

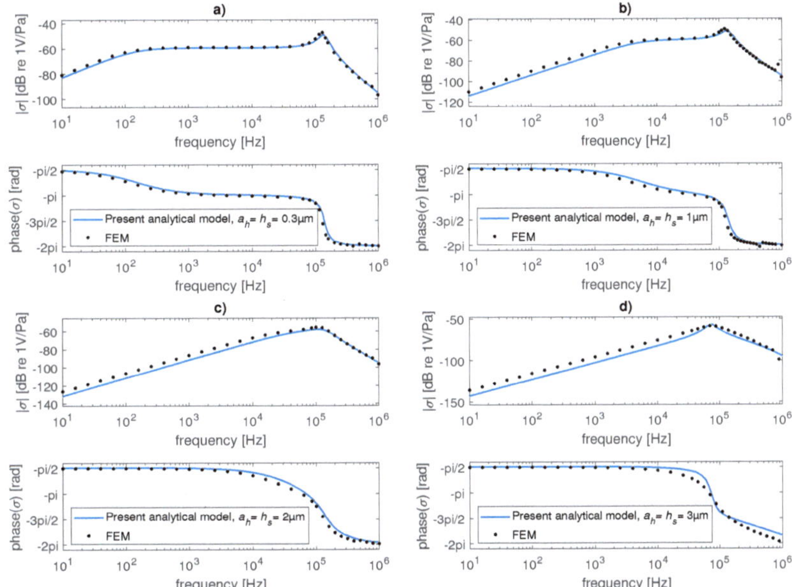

Figure 5. Magnitude (upper curves) and phase (lower curves) of pressure sensitivity of the transducer with elastically supported perforated plate: comparison of the present analytical results (continuous lines) with the numerical (FEM) result (black points) for the side of the holes and the thickness of the slits being equal to (**a**) 0.3 μm, (**b**) 1 μm, (**c**) 2 μm, and (**d**) 3 μm.

The acoustic pressure sensitivities of the receiving transducer with flexible perforated plate clamped at all edges of dimensions $a = 0.5$ mm and thickness $h_p = 10$ μm with $N = 400$ square holes of side dimension a_h varying between 2 μm and 7 μm are shown in Figure 6. Here, the air gap thickness is $h_g = 10$ μm, and the volume of the cavity is $V_c = 10^{-10}$ m^3. The analytical result, calculated using the method described in Section 2.3.2, here takes into account only the first mode of the vibration of the plate ψ_{11} in Equations (37) and (38), which is sufficient in the audio frequency range. Very good agreement between this analytical result and the reference numerical one can be found in the pass band of the transducer. At very high frequencies, the higher modes of the plate vibrations (not contained in the analytical results) appear in the numerical results.

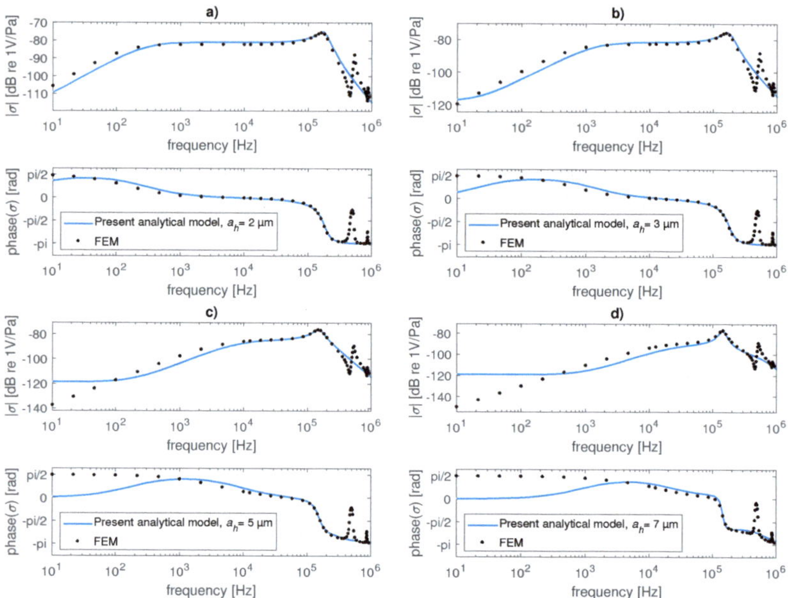

Figure 6. Magnitude (upper curves) and phase (lower curves) of pressure sensitivity of the transducer with flexible perforated clamped plate: comparison of the present analytical results (continuous lines) with the numerical (FEM) result (black points) for the side of the holes being equal to (**a**) 2 μm, (**b**) 3 μm, (**c**) 5 μm, and (**d**) 7 μm.

At very low frequencies, the difference between the acoustic pressure in the gap $p_g(x, y)$ and the incident acoustic pressure p_{inc} is very small due to the acoustic short circuit through the holes (see Figure 7 for $p_{inc} = 1$ Pa and hole side of 2 μm (left) and 5 μm (right) at $f = 100$ Hz). This leads to the numerical noise in the results of the integrals in Equations (25), (37) and (38), especially for larger holes (Figure 7b)). Since this pressure difference is the source for the plate displacement (see Equation (30)), the analytical results are perturbed by the noise in the low-frequency range (see Figure 6c,d). However, when using the transducer in the audio frequency range, the effect of the acoustic short circuit should be reduced, which leads to the use of small holes. For this case, the present analytical model gives correct results (Figure 6a)).

Using the present analytical model, the dimensions of the transducer can be further optimized, as shown in Figure 8. Smaller dimensions of the holes improve the pass band of the transducer in the lower frequency range ($a_h = 1$ μm in Figure 8a). The dependence of the sensitivity σ on the air gap thickness h_g in the pass band of the transducer presents the common sensitivity doubling (+6 dB) when halving the gap thickness for small holes ($a_h = 1$ μm in Figure 8a, $a_h = 2$ μm in Figure 8b), while in the case of larger holes, this effect almost disappears ($a_h = 5$ μm in Figure 8c, $a_h = 7$ μm in Figure 8d). However, the impact of decreasing h_g on increasing damping of the resonance, which is usual in condenser microphones, is preserved in the case of a perforated moving electrode. Note that the thickness of the plate h_p influences the mass and stiffness of the plate, hence the resonance frequency and amplitude of the plate displacement. Higher thickness h_p leads to higher resonance frequency (thus, higher pass band of the transducer) and lower sensitivity.

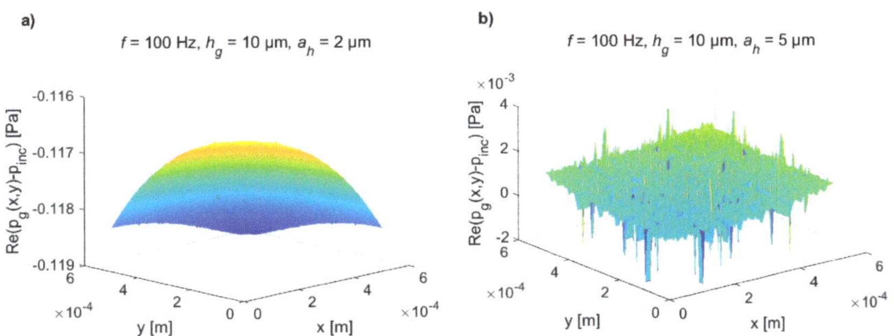

Figure 7. Real part of the difference of the acoustic pressure between both sides of the plate $\Re\left[p_g(x,y) - p_{inc}\right]$ in the first quadrant ($x, y \in (0, a)$) calculated using the present method at $f = 100$ Hz for $p_{inc} = 1$Pa and for the side of the holes being equal to (**a**) $a_h = 2$ μm (left) and (**b**) $a_h = 5$ μm (right).

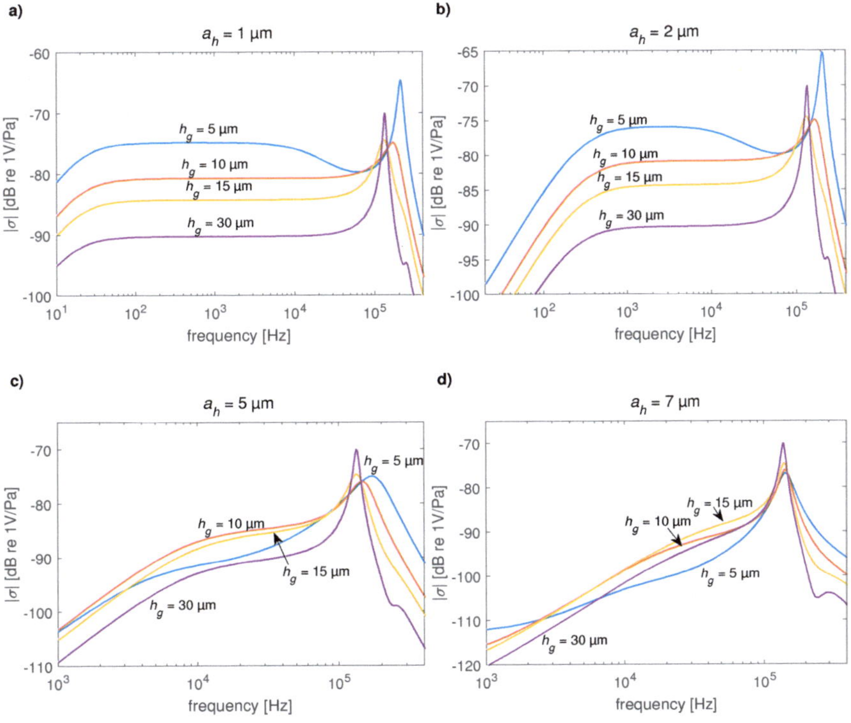

Figure 8. Magnitude of pressure sensitivity of the transducer calculated using the present method: the effect of varying air gap thickness h_g for the side of the holes being equal to (**a**) 1 μm, (**b**) 2 μm, (**c**) 5 μm, and (**d**) 7 μm.

4. Conclusions

The analytical model of an electroacoustic transducer with the moving electrode in the form of a perforated plate (rigid elastically supported or flexible clamped at all boundaries) was developed. The formulation for the acoustic pressure field in the air gap between the moving and the fixed electrode was derived, taking into account the acoustic short circuit through the holes, the thermal and viscous boundary layers effects in the thin fluid film, and the coupling with the displacement of the plate. The displacement of the plate and the acoustic pressure sensitivity of the transducer used as a microphone was calculated, and the latter was compared to the reference numerical (FEM) results. Very good agreement between these models was found in the transducer pass band, and some discrepancies, appearing generally out of the frequency range of interest, were discussed and explained. The influence of some geometrical parameters of the transducer, such as dimensions of the holes in the plate or air gap thickness, was investigated.

Note that only the first mode of the flexible plate vibration was taken into account here using the analytically expressed approximation of its first eigenfunction calculated numerically. This is sufficient in the audio frequency range; however, further research should focus on improved expression of the eigenfunctions, providing better results at the frequencies above the first resonance, where the higher modes of the perforated plate vibration occur.

Author Contributions: Conceptualization, K.Š. and P.H.; methodology, K.Š. and P.H.; software, K.Š. and P.H.; validation, K.Š. and P.H.; formal analysis, K.Š. and P.H.; investigation, K.Š. and P.H.; resources, K.Š. and P.H.; data curation, K.Š. and P.H.; writing—original draft preparation, K.Š. and P.H.; writing—review and editing, K.Š. and P.H.; visualization, K.Š. and P.H.; supervision, P.H. All authors have read and agreed to the published version of the manuscript.

Funding: This research was funded by the Grant Agency of the Czech Technical University in Prague grant number SGS18/200/OHK2/3T/16. The APC was funded by the Future Fund of the Czech Technical University in Prague (2003).

Data Availability Statement: Not applicable.

Acknowledgments: This work was supported by the Grant Agency of the Czech Technical University in Prague, grant No. SGS18/200/OHK2/3T/16. The authors would like to thank Michel Bruneau from Le Mans University for helpfull discussions.

Conflicts of Interest: The authors declare no conflicts of interest.

Appendix A. Approximate Eigenfunctions of the Perforated Flexible Square Plate Clamped at All Edges [32]

Since no exact analytical expression for the eigenfunctions of perforated plates is known, to our knowledge, the eigenfunctions used in Section 2.3.2 are approximated using the series expansion [32]

$$\psi_{mn}(x,y) = \sum_{qr} c_{(qr),(mn)} \phi_q(x)\phi_r(y), \tag{A1}$$

where the basis functions $\phi_q(x), \phi_r(y)$ are the symmetrical eigenfunctions of 1D beam clamped at both ends, given by [27,35]

$$\phi_q^s(x) = \frac{1}{\sqrt{2a}}\left[\frac{\cos\left(\alpha_q^s x\right)}{\cos\left(\alpha_q^s a\right)} - \frac{\cosh\left(\alpha_q^s x\right)}{\cosh\left(\alpha_q^s a\right)}\right], \quad \text{with} \quad \tan\left(\alpha_q^s a\right) = -\tanh\left(\alpha_q^s a\right), \tag{A2}$$

and similarly for $\phi_r^s(y)$. Such a form of the basis functions ensures that the approximated eigenfunctions $\psi_{mn}(x,y)$ verify the boundary conditions (31). A simple numerical (FEM) simulation of the perforated plate only (not loaded by the acoustic parts of the transducer) was performed in order to obtain the numerically calculated eigenfunctions ${}^n\psi_{mn}(x,y)$. The coefficients $c_{(qr),(mn)}$ in (A1) were then calculated from these numerical eigenfunctions as follows:

$$c_{(qr),(mn)} = \frac{1}{2a^2}\int_{-a}^{a}\int_{-a}^{a} {}^n\psi_{mn}(x,y)\phi_q(x)\phi_r(y)\mathrm{d}x\mathrm{d}y. \tag{A3}$$

Other important information provided by the simple numerical model of the perforated plate are the eigenfrequencies $^{n}f_{mn}$ associated with each eigenmode m, n. The eigenvalues k_{xm}, k_{yn} are then expressed from these numerically calculated eigenfrequencies $^{n}f_{mn}$ as follows:

$$k_{xm} = \sqrt{\frac{2\pi\,^{n}f_{mm}}{2c_p}},\qquad\text{(A4)}$$

where $c_p = \sqrt{D/M_s}$ is the wave speed on the plate. Note that k_{yn} has the same values as k_{xm} in the case of a square plate.

References

1. Malcovati, P.; Baschirotti, A. The Evolution of Integrated Interfaces for MEMS Microphones. *Micromachines* **2018**, *9*, 323. [CrossRef] [PubMed]
2. Ali, W.R.; Prasad, M. Piezoelectric MEMS based acoustic sensors: A review. *Sens. Actuator A Phys.* **2020**, *301*, 111756. [CrossRef]
3. Dehé, A. Silicon microphone development and application. *Sens. Actuator A Phys.* **2007**, *133*, 283–287. [CrossRef]
4. Bergqvist, J.; Rudolf, F. A silicon condenser microphone using bond and etch-back technology. *Sens. Actuator A Phys.* **1994**, *45*, 115–124. [CrossRef]
5. Iguchi, Y.; Goto, M.; Iwaki, M.; Ando, A.; Tanioka, K.; Tajima, T.; Takeshi, F.; Matsunaga, S.; Yasuno, Y. Silicon microphone with wide frequency range and high linearity. *Sens. Actuator A Phys.* **2007**, *135*, 420–425. [CrossRef]
6. Scheeper, P.R.; Nordstrand, B.; Gulløv, J.O.; Liu, B.; Clausen, T.; Midjord, L.; Storgaard-Larsen, T. A new measurement microphone based on MEMS technology. *J. Microelectromech. Syst.* **2003**, *12*, 880–891. [CrossRef]
7. Füldner, M.; Dehé, A. Dual Back Plate Silicon MEMS microphone: Balancing High Performance! In Proceedings of the DAGA 2015, Nürnberg, Germany, 16–19 March 2015; pp. 41–43.
8. Peña-García, N.N.; Aguilera-Cortés, L.A.; Gonzáles-Palacios, M.A.; Raskin, J.-P.; Herrera-May, A.L. Design and Modeling of a MEMS Dual-Backplate Capacitive Microphone with Spring-Supported Diaphragm for Mobile Device Applications. *Sensors* **2018**, *18*, 3545. [CrossRef]
9. Rong, Z.; Zhang, M.; Ning, Y.; Pang, W. An ultrasound-induced wireless power supply based on AlN piezoelectric micromachined ultrasonic transducers. *Sci. Rep.* **2022**, *12*, 16174. [CrossRef]
10. Pinto, R.M.R.; Gund, V.; Dias, R.A.; Nagaraja, K.K.; Vinayakumar, K.B. CMOS-Integrated Aluminum Nitride MEMS: A Review. *J. Microelectromech. Syst.* **2022**, *31*, 500–523. [CrossRef]
11. Lynes, D.D.; Chandrahalim, H. Influence of a Tailored Oxide Interface on the Quality Factor of Microelectromechanical Resonators. *Adv. Mater. Interfaces* **2023**, *10*, 2202446. [CrossRef]
12. Verdot, T.; Redon, E.; Ege, K.; Czarny, J.; Guianvarc'h, C.; Guyader, J.-L. Microphone with planar nano-gauge detection: Fluid-structure coupling including thermo-viscous effects. *Acta Acust. United Acust.* **2016**, *102*, 517–529. [CrossRef]
13. Rufer, L.; De Pasquale, G.; Esteves, J.; Randazzo, F.; Basrour, S.; Somà, A. Micro-acoustic source for hearing applications fabricated with 0.35 μm CMOS-MEMS process. *Procedia Eng.* **2015**, *120*, 944–947. [CrossRef]
14. Ganji, B.A.; Sedaghat, S.B.; Roncaglia, A.; Belsito, L. Design and fabrication of very small MEMS microphone with silicon diaphragm supported by Z-shape arms using SOI wafer. *Solid State Electron.* **2018**, *148*, 27–34. [CrossRef]
15. Ganji, B.A.; Majlis, B.Y. Design and fabrication of a new MEMS capacitive microphone using a perforated aluminum diaphragm. *Sens. Actuator A Phys.* **2009**, *149*, 29–37. [CrossRef]
16. Ganji, B.A.; Sedaghat, S.B.; Roncaglia, A.; Belsito, L. Design and fabrication of high performance condenser microphone using C-slotted diaphragm. *Microsyst. Technol.* **2018**, *24*, 3133–3140. [CrossRef]
17. Sedaghat, S.B.; Ganji, B.A.; Ansari, R. Design and modeling of a frog-shape MEMS capacitive microphone using SOI technology. *Microsyst. Technol.* **2018**, *24*, 1061–1070. [CrossRef]
18. Škvor, Z. On the Acoustical Resistance due to Viscous Losses in the Air Gap of Electrostatic Transducers. *Acustica* **1967**, *19*, 295–299.
19. Estèves, J.; Rufer, L.; Ekeom, D.; Basrour, S. Lumped-parameters equivalent circuit for condenser microphones modeling. *J. Acoust. Soc. Am.* **2017**, *142*, 2121–2132. [CrossRef]
20. Zuckerwar, A.J. Theoretical response of condenser microphones. *J. Acoust. Soc. Am.* **1978**, *64*, 1278–1285. [CrossRef]
21. Lavergne, T.; Durand, S.; Bruneau, M.; Joly, N. Dynamic Behavior of Circular Membrane and An Electrostatic Microphone: Effect of Holes In The Backing Electrode. *J. Acoust. Soc. Am.* **2010**, *128*, 3459–3477. [CrossRef]
22. Lavergne, T.; Durand, S.; Bruneau, M.; Joly, N. Analytical Modeling of Electrostatic Transducers in Gases: Behavior of Their Membrane and Sensitivity. *Acta Acust. United Acust.* **2014**, *100*, 440–447. [CrossRef]
23. Naderyan, V.; Raspet, R.; Hickey, C. Thermo-viscous acoustic modeling of perforated micro-electro-mechanical systems (MEMS). *J. Acoust. Soc. Am.* **2020**, *148*, 2376–2385. [CrossRef] [PubMed]
24. Pedersen, M.; Olthuis, W.; Bergveld, P. On the electromechanical behaviour of thin perforated backplates in silicon condenser microphones. In Proceedings of the 8th International Conference on Solid-state Sensors and Actuators, and Eurosensors IX, Stockholm, Sweden, 25–29 June 1995; p. 234 A7.

25. Novak, A.; Honzík, P.; Bruneau, M. Dynamic behaviour of a planar micro-beam loaded by a fluid-gap: Analytical and numerical approach in a high frequency range, benchmark solutions. *J. Sound Vib.* **2017**, *401*, 36–53. [CrossRef]

26. Honzík, P.; Bruneau, M. Acoustic fields in thin fluid layers between vibrating walls and rigid boundaries: Integral method. *Acta Acust. United Acust.* **2015**, *101*, 859–862. [CrossRef]

27. Šimonová, K.; Honzík, P.; Bruneau, M.; Gatignol, P. Modelling approach for MEMS transducers with rectangular clamped plate loaded by a thin fluid layer. *J. Sound Vib.* **2020**, *473*, 115246. [CrossRef]

28. Herring Jensen, M.J.; Sandermann Olsen, E. Virtual prototyping of condenser microphone using the finite element method for detailed electric, mechanic, and acoustic characterisation. *Proc. Meet. Acoust.* **2013**, *19*, 030039.

29. Bruneau, M.; Scelo, T. *Fundamentals of Acoustics*; ISTE: London, UK, 2006.

30. Leissa, A.W. *Vibration of Plates*; Scientific and Technical Information Division, National Aeronautics and Space Administration: Washington, DC, USA, 1969.

31. Šimonová, K.; Honzík, P.; Joly, N.; Durand, S.; Bruneau, M. Modelling of a MEMS transducer using approximate eigenfunctions of a square clamped plate. In Proceedings of the 23rd International Congress on Acoustics, Aachen, Germany, 9–13 September 2019; pp. 7361–7368.

32. Šimonová, K.; Honzík, P.; Joly, N.; Durand, S.; Bruneau, M. Modelling of a MEMS Transducer with a Moving Electrode in Form of Perforated Square Plate. In Proceedings of Forum Acusticum 2020, Lyon, France, 7–11 December 2020; pp. 2539–2542.

33. COMSOL Multiphysics. Acoustics Module User's Guide. 2022. Available online: https://doc.comsol.com/6.1/doc/com.comsol.help.aco/AcousticsModuleUsersGuide.pdf (accessed on 23 March 2023).

34. COMSOL Multiphysics. Structural Mechanics Module User's Guide. 2022. Available online: https://doc.comsol.com/6.1/doc/com.comsol.help.sme/StructuralMechanicsModuleUsersGuide.pdf (accessed on 23 March 2023).

35. Le Van Suu, T.; Durand, S.; Bruneau, M. On the modelling of a clamped plate loaded by a squeeze fluid film: Application to miniaturized sensors. *Acta Acust. United Acust.* **2010**, *96*, 923–935. [CrossRef]

micromachines

Review

Review of Recent Development of MEMS Speakers

Haoran Wang [1], Yifei Ma [2], Qincheng Zheng [2], Ke Cao [2], Yao Lu [2] and Huikai Xie [2,3,*]

1   Department of Electrical and Computer Engineering, University of Florida, Gainesville, FL 32611, USA; wanghaoran@ufl.edu
2   School of Information and Electronics, Beijing Institute of Technology, Beijing 100081, China; 3120200697@bit.edu.cn (Y.M.); 1120171582@bit.edu.cn (Q.Z.); 3220210564@bit.edu.cn (K.C.); y.lu@bit.edu.cn (Y.L.)
3   BIT Chongqing Center for Microelectronics and Microsystems, Chongqing 400030, China
*   Correspondence: hk.xie@ieee.org

Abstract: Facilitated by microelectromechanical systems (MEMS) technology, MEMS speakers or microspeakers have been rapidly developed during the past decade to meet the requirements of the flourishing audio market. With advantages of a small footprint, low cost, and easy assembly, MEMS speakers are drawing extensive attention for potential applications in hearing instruments, portable electronics, and the Internet of Things (IoT). MEMS speakers based on different transduction mechanisms, including piezoelectric, electrodynamic, electrostatic, and thermoacoustic actuation, have been developed and significant progresses have been made in commercialization in the last few years. In this article, the principle and modeling of each MEMS speaker type is briefly introduced first. Then, the development of MEMS speakers is reviewed with key specifications of state-of-the-art MEMS speakers summarized. The advantages and challenges of all four types of MEMS speakers are compared and discussed. New approaches to improve sound pressure levels (SPLs) of MEMS speakers are also proposed. Finally, the remaining challenges and outlook of MEMS speakers are given.

Keywords: microelectromechanical systems; MEMS; loudspeakers; microspeakers

Citation: Wang, H.; Ma, Y.; Zheng, Q.; Cao, K.; Lu, Y.; Xie, H. Review of Recent Development of MEMS Speakers. *Micromachines* **2021**, *12*, 1257. http://doi.org/10.3390/mi12101257

Academic Editor: Libor Rufer

Received: 8 September 2021
Accepted: 13 October 2021
Published: 16 October 2021

Publisher's Note: MDPI stays neutral with regard to jurisdictional claims in published maps and institutional affiliations.

1. Introduction

With the rapid advancement of consumer electronics, the worldwide audio market has been seeing a growing trend towards smaller devices with lower power consumption and better performance in the last decade. Speakers, as one of the core components in mobile electronic devices such as laptops, smartphones, wireless earbuds, and human-machine interfaces, are highly demanded to be smaller, lighter, and more power efficient. Currently, speakers in those mobile electronic devices are dominated by conventional speakers with bulky moving coils, which are still challenging to be batch fabricated since voice coils and permanent magnets must be assembled [1]. The miniaturization of these conventional speakers also has a negative impact on the sound quality and reaches some limits due to the employed materials and the fabrication approaches [2]. For example, the plastic or polymer diaphragms of conventional speakers are too soft to be used as high-quality radiator surfaces [3]. The simplification of the mechanical suspensions and electromagnetic parts in the miniaturization would lead to reduced bandwidths and increased nonlinearities, thus deteriorating the sound quality [2,4]. It is also difficult for conventional manufacturing technologies to achieve high dimensional precision and good reproducibility in the miniaturization of speakers.

By contrast, microelectromechanical systems (MEMS) speakers, or microspeakers, have been drawing more and more attention due to their inherent advantages, e.g., small form factors, low power consumption, batch fabrication, and potential on-chip integration with electronic circuits. Many researchers have developed MEMS speakers based on various transduction mechanisms and achieved promising results, including electrodynamic MEMS speakers [5–7], electrostatic MEMS speakers [8,9], piezoelectric MEMS speakers [10–12], and thermoacoustic MEMS speakers [13,14]. Various materials and fabrication approaches have also been explored for developing MEMS speakers [15–17]. The performances of MEMS speakers have been evaluated and compared with conventional speakers in terms of several key specifications, such as device footprint, output sound pressure level (SPL), power consumption, bandwidth, and total harmonic distortion (THD) [2,9,10,17]. Among them, the SPL and bandwidth are two widely used parameters to evaluate the acoustic performance of MEMS speakers. THD, defined as the sum of all power radiated in frequencies other than the fundamental frequency relative to the total emitted sound power, is an important parameter to evaluate the sound quality of MEMS speakers [9].

To date, MEMS speakers have been developed mainly for in-ear applications (e.g., hearing aids) and headphones [15,18]. It is challenging for MEMS speakers and conventional electrodynamic microspeakers as well to achieve both high SPL output and flat audio frequency response due to the vibration mode complexity of the diaphragm and the limited space for actuation. Thus, the actuation method, structure, and electrode pattern design of the diaphragm as well as the enclosure design are crucial to the overall response and performance of a MEMS speaker. Both finite element analysis (FEA) and lumped element modelling (LEM) are typically employed to study the effects of various design parameters and to optimize the overall performance of MEMS speakers [19,20]. Several approaches in terms of material selection [12], special structural design [21,22], and electrode configuration [23,24] have also been demonstrated to achieve the better acoustic performance of MEMS speakers. Extensive research efforts have been devoted to developing better MEMS speakers with promising results demonstrated, which is evidenced by a large amount of literature produced.

With so many research efforts paid to the development of MEMS speakers, significant progress has been made in their commercialization. For example, piezoelectric MEMS speakers developed by Usound have reached the market. With a chip size of 6.7 mm × 4.7 mm × 1.58 mm, the developed piezoelectric MEMS speaker can generate a high SPL of around 116 dB in an acoustic coupler, under a driving voltage of 15 V [25]. The TDK Corporation has developed a series of piezoelectric speakers called PiezoListen. With a thickness of as small as 0.49 mm and footprints ranging from 20 mm × 10 mm to 66 mm × 30 mm, the developed speakers can be installed on almost any kinds of displays or surfaces to generate sound over a wide frequency range from 400 Hz to 20 kHz [26]. In addition, Audio Pixels has successfully implemented a digital sound reconstruction (DSR) technique in a commercially feasible manner and developed MEMS speaker arrays to generate high quality sounds [27]. Furthermore, by using moving beams with electrostatic actuation to generate sound inside silicon chips, Arioso Systems has developed MEMS speakers with high-fidelity sound and CMOS-compatible process for in-ear applications [28].

In order to better leverage the existing achievements, it is necessary to sort out the recent development of MEMS speakers, understand the barriers, compare different types of MEMS speakers, and point out the future perspectives with respect to these challenges. Thus, the main purpose of this article is to provide a state-of-the-art review of MEMS speakers and a future outlook as well.

This review article is organized as follows. In Section 2, we introduce the theories and modeling of MEMS speakers, including device concepts, LEMs, and several transduction mechanisms. In Section 3, we review different types of MEMS speakers, including their fabrication technologies, characterization results, and approaches to improve the SPLs of MEMS speakers with regard to structures, materials, and actuation methods. The focus is on the piezoelectric MEMS speaker. In Section 4, we compare and discuss the performances of different MEMS speakers. In Section 5, we summarize the review and discuss future perspectives of MEMS speakers.

2. Theory and Modeling of MEMS Speakers

2.1. Basic Structure

In general, the main structure of a MEMS speaker consists of an acoustic diaphragm, an actuation mechanism, and an air chamber. When an AC voltage is applied to drive the MEMS speaker, a bending moment will be generated by the actuation mechanism, forcing the diaphragm to vibrate and thus generating a sound pressure output. Considering a circular vibrating diaphragm, as shown in Figure 1, the pressure amplitude can be calculated based on the Helmholtz equation and the Rayleigh integral and is readily given by [29]:

$$P(z) = \rho(2\pi f)^2 \int_0^a \frac{w(r)}{\sqrt{z^2 + r^2}} r\, dr \tag{1}$$

where ρ is the air density, f is the vibration frequency, a is the radius of the diaphragm, $w(r)$ is the vibration amplitude at the radial distance of r, and z is the distance from the diaphragm to the listener.

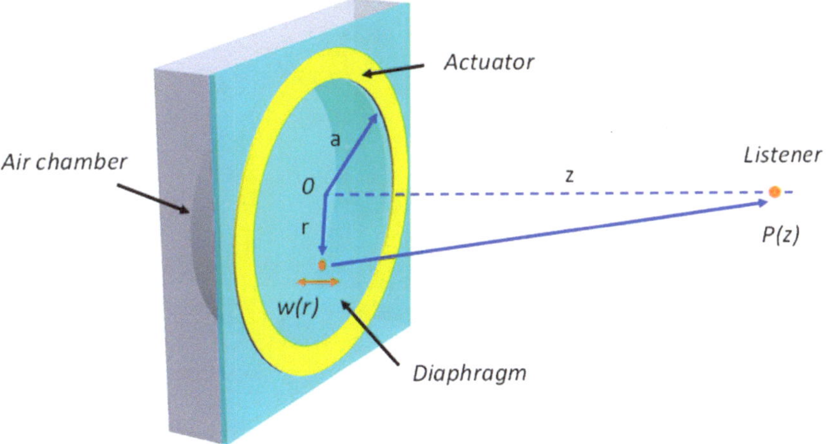

Figure 1. Schematic of a MEMS speaker with a piston-move diaphragm and the geometries for sound pressure calculation.

When the vibration of the acoustic diaphragm is simplified as a piston on an infinite baffle, the effective sound pressure output $P_e(z)$ and the sound pressure level (SPL) in decibels (dB) can be further simplified as:

$$P_e(z) = \frac{P(z)}{\sqrt{2}} = \frac{\sqrt{2}\pi\rho S w f^2}{z} \tag{2}$$

$$SPL = 20\lg\left(\frac{P_e(z)}{P_{ref}}\right) \tag{3}$$

where S and w are the surface area and vibration amplitude of the diaphragm, respectively. The reference effective sound pressure value P_{ref} is 20 µPa [30]. Typically, the SPLs of MEMS speakers are measured by microphones placed at 1 cm away from the MEMS speakers in open air. For MEMS speakers specifically developed for in-ear or hearing-aid applications, their SPLs are measured in a 2cc coupler (a coupler with a volume of 2 cm³ that conforms to the ANSI S3.7 and IEC 60318-5 standards) [31].

The acoustic diaphragm is important in MEMS speaker designs. According to Equation (2), the sound pressure output generated by the acoustic diaphragm is proportional to its surface area and vibration amplitude, and the square of the working frequency. Thus, generating high sound pressure output at lower frequencies is more challenging, which requires larger deflections under the same diaphragm size constraint, as indicated in Figure 2. Figure 2 plots the required deflection amplitudes for circular diaphragms with different frequencies and different diameters to achieve a 90 dB SPL at 1 cm. This plot shows the decreasing trend of the required deflection amplitudes with the increasing frequencies and the diaphragm sizes and gives a general indication of the design values. As can be seen, for a circular diaphragm with a diameter of 4 mm, achieving a 90 dB SPL at 1 cm requires a diaphragm deflection of 5.9 μm, 94.4 μm, and 1.05 mm at frequencies of 4 kHz, 1 kHz, and 300 Hz, respectively.

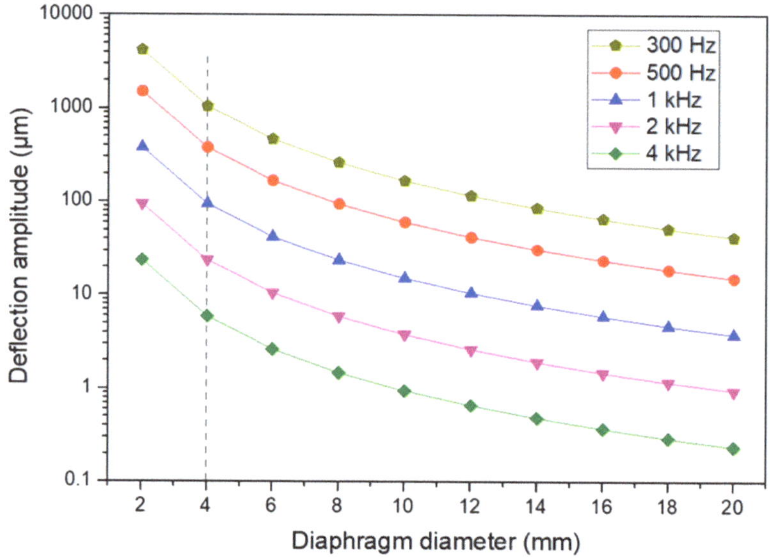

Figure 2. Required deflection amplitudes for different diaphragm diameters to achieve a 90 dB SPL at 1 cm at frequencies of 300 Hz, 500 Hz, 1 kHz, 2 kHz, and 4 kHz.

MEMS speakers are usually designed to work in a frequency range from 20 Hz to 20 kHz, which is consistent with the hearing range of humans. Since the frequencies of audible sounds for humans in daily life typically varies from 100 Hz to 10 kHz, including speeches in low frequencies (300 Hz–3.4 kHz) and musical harmonics in high frequencies (>6 kHz) [32], MEMS speakers are normally evaluated in both of these low-frequency and high-frequency bands. In general, piezoelectric, electrodynamic, and electrostatic actuation are the three most commonly used approaches to excite acoustic diaphragms. Details of these transduction mechanisms will be introduced in Section 2.2.

In addition to the deflection, resonant frequency is another important design parameter of acoustic diaphragms. Most of MEMS speakers presented in literatures are developed based on deformable diaphragms with edges clamped on the substrate. Their fundamental resonant frequencies are dependent on the dimensions and material properties of the diaphragms. For a circular clamped vibrating diaphragm, the fundamental resonant frequency f_0 is given by [33]:

$$f_0 = 0.47 \frac{t}{a^2} \sqrt{\frac{E}{\rho_m(1 - v^2)}} \tag{4}$$

where t, a, E, ρ_m, and v are the thickness, radius, effective Young's modulus, mass density, and Poisson's ratio of the circular diaphragm, respectively.

The fundamental vibration mode of the clamped diaphragm is the so-called drum mode, whose deflection profile peaks at the center of the diaphragm and decreases from the center to the edge. When designing the fundamental resonant frequency, there are two considerations. On one hand, to achieve high SPL at low frequencies and thus improve the acoustic performance over a wide frequency range, the fundamental drum mode frequencies are typically designed at around 2 kHz to 3 kHz [16,34,35]. On the other hand, from the acoustic point of view, the drum mode vibration with the deformed emissive surface and higher harmonics stimulation due to nonlinearities will distort the acoustic wavefront, therefore causing sound distortions and deteriorating the sound quality [2]. Thus, some special diaphragm designs other than edge clamped diaphragms have been developed, such as rigid diaphragms with radial rib structures supported by suspension beams [2] and circular diaphragms supported by four flexible dual-curve actuators [21], in which piston mode vibrations at low frequencies can be employed to generate the sound while the drum mode vibrations can be shifted to high frequencies to avoid the sound distortion of MEMS speakers.

2.2. Transduction Mechanisms

MEMS speakers have been developed based on various transduction mechanisms, including the piezoelectric transduction [36], electrodynamic transduction [37], electrostatic transduction [38], and thermoacoustic transduction [13]. Figure 3 shows the schematics of MEMS speakers with different transduction mechanisms. Among them, MEMS speakers developed based on the first three types of transduction mechanisms rely on the mechanical vibration of the acoustic diaphragm to generate the sound. By contrast, thermoacoustic MEMS speakers produce the sound by the periodic contraction and expansion of the medium around the diaphragm due to the heat exchange between the diaphragm and the surrounding medium.

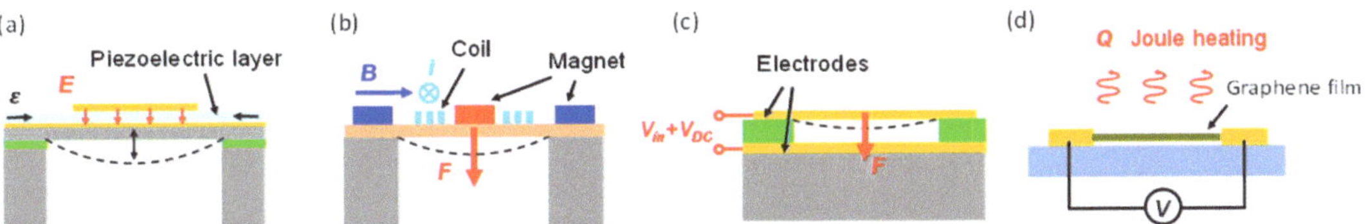

Figure 3. Schematic of MEMS speakers based on (**a**) piezoelectric, (**b**) electrodynamic, (**c**) electrostatic, and (**d**) thermoacoustic transduction mechanisms.

As shown in Figure 3a, piezoelectric MEMS speakers work on the flexural vibration of the piezoelectric diaphragm. When an AC voltage is applied across the piezoelectric film sandwiched by two metal electrodes, an in-plane strain will be generated based on the converse piezoelectric effect, thus causing the out-of-plane vibration of the diaphragm. The relation between the in-plane strain ε and the applied electric field E can be expressed by [21]:

$$\varepsilon = d_{31} E \tag{5}$$

where d_{31} is the piezoelectric constant of the employed piezoelectric film.

For electrodynamic MEMS speakers, the acoustic diaphragm is actuated by electromagnetic (Lorentz) force. As shown in Figure 3b, when the current flows through coils, Lorentz force will be generated due to the interaction between the external magnetic field and the electric current, thus bending the acoustic diaphragm. For a planar concentric coil with N turns carrying an electric current I, the Lorentz force $F_{Lorentz}$ generated by a magnetic field with a flux density B can be expressed as [2]:

$$F_{Lorentz} = I \int_0^l \vec{B} d\vec{l} = \sum_{i=1}^{N} 2\pi I R_i B_i \tag{6}$$

where l is the length of the coil, R_i is the radius of the ith turn, and B_i is the radial component of the magnetic flux density on the coil plane corresponding to the ith turn.

Electrostatic MEMS speakers are driven by the electrostatic force between two conductive plates. As shown in Figure 3c, the acoustic diaphragm is suspended over the substrate by a small gap d. Considering this structure as a parallel-plate capacitor with flat and rigid electrodes for simplification, the electrostatic force exerted on the diaphragm under an AC driving voltage V_{in} and a DC bias V_{DC} is given by [39]:

$$F_E = \frac{1}{2}\epsilon A\left(\frac{V_{in}+V_{DC}}{d}\right)^2 \tag{7}$$

where ϵ is the electric permittivity of air and A is the area of the diaphragm. Advanced models considering the bending of the plate and pull-in limitations are presented in [40,41].

Different from the mechanical vibration sound generators described above, thermoacoustic MEMS speakers emit sound by the thermoacoustic effect, which converts the Joule heat into sound. As shown in Figure 3d, when an AC current is applied to a conductive film, the film will be heated and exchange the thermal energy with the surrounding air, causing the periodic contraction and expansion of the air, thus generating sounds. The root-mean-square sound pressure amplitude produced by a thermoacoustic thin film speakers can be derived as [42]:

$$p_{rms} = \frac{\sqrt{\alpha}\rho_0}{2\sqrt{\pi}T_0}\cdot\frac{1}{r}\cdot P_{in}\cdot\frac{\sqrt{f}}{C_s}\cdot M \tag{8}$$

where ρ_0, α, and T_0 are the mass density, thermal diffusivity, and temperature of the ambient gas, respectively, r is the distance between the thin film conductor and the listener, P_{in} is the input power, f is the frequency of the sound, C_s is the heat capacity per unit area of the thin film conductor, and M is a frequency-related factor.

2.3. Modeling

The acoustic performance of MEMS speakers is dependent on many design parameters, including material properties, device structures, and acoustic enclosure designs. Lumped element modeling (LEM) and finite element analysis (FEA) can be used to effectively predict the acoustic performance of MEMS speakers and optimize the designs. For example, Neumann Jr. et al. presented CMOS-MEMS diaphragms for acoustic actuation based on electrostatic force, and developed a simplified acoustic model to investigate the effects of the dimensional parameters of the diaphragms [43]. Huang et al. studied the sound pressure response of miniaturized moving-coil loudspeakers using an equivalent circuit method (ECM), which can simulate the electrical, mechanical, and acoustical responses and optimize the device designs [44]. These methods are also called electro-mechanoacoustical modeling, or lumped element modeling (LEM), which can be applied to study the effects of different acoustic enclosures and model the performances of MEMS speakers based on different transduction mechanisms [16,45]. LEM is a simple and efficient tool for designing and analyzing multiphysics systems as well as for predicting their responses. In this method, the representation of spatially distributed physical systems is simplified by using a set of lumped elements when the length scale of the device is much smaller than the wavelength of the governing physical phenomenon. Since the acoustic wavelengths (34.3–343 mm for 1–10 kHz) for MEMS speakers are much greater than their sizes (1–10 mm), LEM is applicable.

Typically, MEMS speakers are packaged in enclosures with a front cover, a back chamber, and vent holes. Figure 4a illustrates a simplified structure of a MEMS speaker in a package. The LEM of this device is shown in Figure 4b, representing a multiphysics system consisting of electrical, mechanical, and acoustical energy domains. In the electrical domain, the effort and flow are voltage (in V) and current (in A), respectively. The electrical and mechanical domains are coupled by a transformer or a gyrator that models the transduction mechanism of the MEMS speaker. In the mechanical domain, the effort represents the force (in N) that actuates the vibrating diaphragm while the flow represents the velocity of the diaphragm (in m/s). The acoustical domain is coupled to the mechanical domain by the effective area of the diaphragm. Thus, the effort and flow in the acoustical domain correspond to the pressure (in Pa) and volume velocity (in m^3/s), respectively. The lumped elements sharing the same effort are connected in parallel, while those sharing the same flow are connected in series.

In the electrical domain, the electrical input impedance of the MEMS speaker is modeled as Z_e, which can be resistance and inductance from the wires and coils for electrodynamic MEMS speakers, or capacitance and resistance for electrostatic MEMS speakers and piezoelectric MEMS speakers. The electrical domain is coupled to the mechanical domain by a transformer (or a gyrator), representing the energy transformation from the electromagnetic force, electrostatic force, or the piezoelectric force.

In the mechanical domain, the vibrating diaphragm is modeled as a mass-spring-damper system, governed by the following equation:

$$M_d \frac{d^2 w}{dt^2} + R_d \frac{dw}{dt} + \frac{w}{C_d} = F_t \tag{9}$$

where w is the vibration amplitude of the diaphragm, F_t is the total force applied on the diaphragm, and M_d, C_d, and R_d are the equivalent mass, compliance, and damping of the diaphragm, respectively. The mechanical and acoustical domains are coupled with the effective area of the diaphragm, which converts the actuation force to the acoustic pressure. Two separate transformers are used to account for the front and the back sides of the diaphragm [44].

In the acoustical domain, the air in an acoustic chamber with a volume V_a can be modeled as an acoustic compliance C_a that is readily given by [44]:

$$C_a = \frac{V_a}{\rho_a c^2} \tag{10}$$

where ρ_a and c are the air density and the sound speed, respectively. Therefore, the front volume and the back chamber can be modeled as acoustic compliances $C_{f,v}$ and $C_{b,c}$, respectively. The air flow inside narrow spaces can be modeled as acoustic resistances and masses, such as $R_{f,h}$ and $M_{f,h}$ of the acoustic holes in the front side and $R_{b,v}$ and $M_{b,v}$ of the backside vent. The acoustic radiation impedance of the diaphragm is also approximated as the acoustic resistance $R_{f,rad}$ ($R_{b,rad}$) and mass $M_{f,rad}$ ($M_{b,rad}$). Details of the calculation of these lumped elements are described in [44].

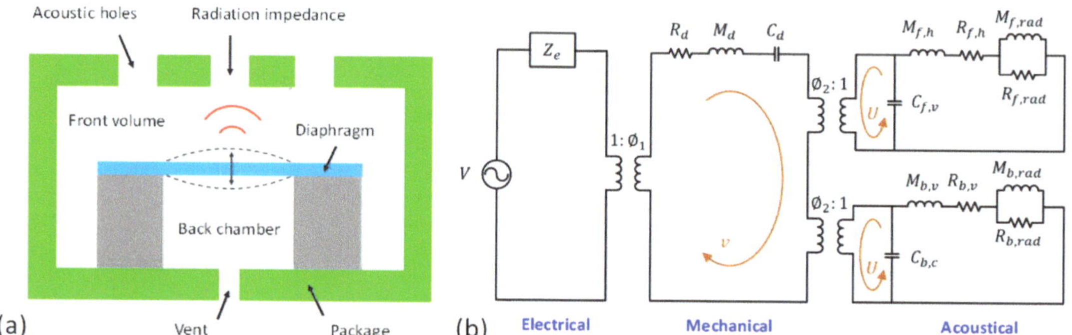

Figure 4. Lumped element model (LEM) of a packaged MEMS speaker: (**a**) illustration of the structures and (**b**) the equivalent circuit in multiple domains.

By solving the equivalent circuit in the LEM, the volume velocity U generated by the MEMS speaker in the acoustical domain can be obtained. Thus, by assuming the MEMS speaker as a point source at a far-field distance r (much larger than the Rayleigh distance), the sound pressure output P can be calculated as [45]:

$$P = j\frac{k\rho_a c}{2\pi r} \cdot U \cdot e^{j(\omega t - kr)} \tag{11}$$

where k and ω are the wave number and angular frequency of the acoustic wave, respectively. Here, it is worthy of note that the MEMS speaker is considered as a monopole mounted on a baffle plate for far-field calculation. Since the plate restricts the acoustic radiation only to the forward hemisphere, the pressure is twice that of a free radiation without a baffle plate [46].

The LEM has been widely applied to predict the dynamic responses of MEMS speakers, especially at low-frequency regions or in the neighborhood of the fundamental resonant frequency due to its simplicity [16,47]. However, the LEM is not sufficient to model higher order resonant modes and incapable to well predict the high frequency responses of MEMS speakers. Therefore, LEM is often used together with FEA to calculate the dynamic responses [5,44,48], analyze the enclosure designs [45,49], and optimize the diaphragm structural designs.

3. Development of MEMS Speakers

The study of MEMS speakers started in the late 1990s. Since then, significant progress has been made to develop MEMS speakers based on different transduction mechanisms, especially on piezoelectric, electrodynamic, and electrostatic transduction. To achieve a small size, high output sound pressure, and flat frequency response, various materials, structure designs, and fabrication techniques have been employed. In this section, the development of MEMS speakers will be reviewed based on their transduction mechanisms.

3.1. Piezoelectric MEMS Speakers

3.1.1. Design and Fabrication of Piezoelectric MEMS Speakers

Piezoelectric actuation, with the advantages of small driving voltage and large actuation force, has been widely used in many MEMS devices, including ink-jet printer heads [50], MEMS scanning mirrors [51], ultrasonic motors [52], RF resonators [53], and acoustic generators [54]. Among them, piezoelectric MEMS speakers are important applications and are attracting more and more interest. Piezoelectric MEMS speakers based on different piezoelectric materials, such as zinc oxide (ZnO), aluminum nitride (AlN), and lead zirconate titanate (PZT), have been presented for hearing aid or earphone applications [35,55,56]. Piezoelectric MEMS speakers mainly consist of a piezoelectric vibration diaphragm and an acoustic cavity. Typical vibration diaphragms can be designed as beam-like piezoelectric actuators [57] (Figure 5a), fully clamped diaphragms with piezoelectric layers embedded [12] (Figure 5b), or partially clamped diaphragms surrounded by piezoelectric actuators [21] (Figure 5c). Various piezoelectric MEMS speakers based on different designs have been demonstrated [12,58].

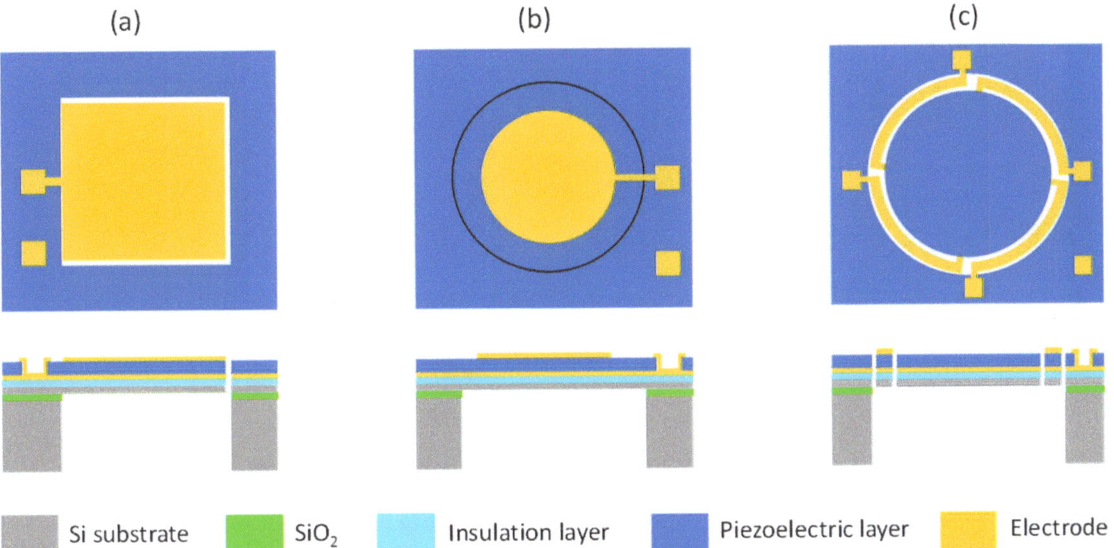

Figure 5. Schematics of typical structures of piezoelectric MEMS speakers in top view (**top**) and cross-sectional view (**bottom**). (**a**) Beam-like piezoelectric actuator. (**b**) Fully clamped diaphragm with piezoelectric layer embedded. (**c**) Partially clamped diaphragm surrounded by piezoelectric actuators.

The fabrication process of piezoelectric MEMS speakers with various structures can be different, depending on whether the diaphragm needs to be released from both sides (Figure 5a,c) or the backside only (Figure 5b), but their general steps are similar. Here, an example for the design of MEMS speakers with a partially clamped diaphragm (Figure 5c) is presented to illustrate the typical fabrication process. As shown in Figure 6, firstly, an insulation layer (Si_xN_y or SiO_2), a bottom electrode layer, and a piezoelectric layer are deposited in sequence on a silicon-on-insulator (SOI) substrate (Figure 6a). After that, the piezoelectric layer is patterned by wet etching or reactive ion etching (RIE) to expose the bottom electrode [59,60] (Figure 6b). Next, a top electrode is deposited and patterned (Figure 6c). After that, RIE is used to define a diaphragm and a set of piezoelectric actuators on the front side (Figure 6d). Subsequently, the acoustic cavity is defined on the backside with a two-sided photolithography and formed by the deep reactive ion etching (DRIE) of silicon or wet etching with KOH (Figure 6e). The buried oxide layer is used as the etch stop and finally removed by RIE or vapor hydrofluoric acid to release the moveable structures (Figure 6f). For the fabrication of fully clamped diaphragms in Figure 5b, the process step shown in Figure 6d can be skipped.

In the design and fabrication of piezoelectric MEMS speakers, the material of the piezoelectric layer is important as it will affect the selection of the fabrication method and the performance of the fabricated devices. Next, the piezoelectric materials for making MEMS speakers will be discussed.

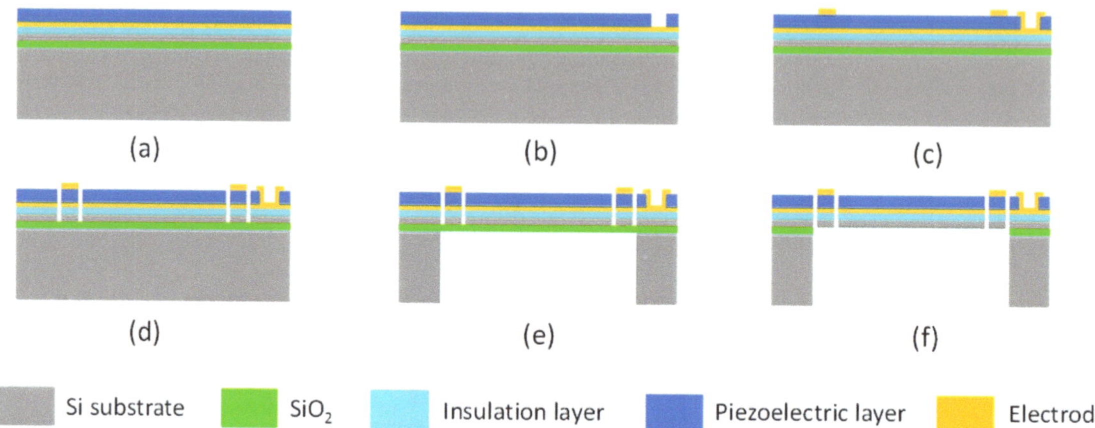

Figure 6. Typical fabrication process flow of a piezoelectric MEMS speaker. (**a**) Deposit insulation, electrode, and piezoelectric layers. (**b**) Pattern the piezoelectric layer. (**c**) Deposit and pattern the top electrode layer. (**d**) Reactive ion etching to define the diaphragm and piezoelectric actuators. (**e**) Etch the acoustic cavity. (**f**) Release the moveable structures.

3.1.2. Piezoelectric Materials

Lead zirconate titanate (PZT) ceramics, single-crystal lithium niobate (LiNbO$_3$), and single-crystal lead magnesium niobate-lead titanate (PMN-PT) are widely used bulk piezoelectric materials with high piezoelectric coefficients and electromechanical coupling factors for piezoelectric transducers [61]. However, how to thin down these materials remains an issue in fabricating piezoelectric MEMS devices. With the advancement of thin film deposition technologies, piezoelectric thin films including ZnO, AlN, and PZT can be fabricated by sputtering or sol-gel methods, which have been applied to fabricate piezoelectric MEMS devices, such as microspeakers [62,63]. Among these materials, ZnO is one of the most commonly used for making piezoelectric thin film devices such as film bulk acoustic wave resonators (FBAR), surface acoustic wave (SAW) resonators, piezoelectric micromachined ultrasonic transducers (pMUTs), and microspeakers in early years. ZnO-based piezoelectric MEMS speakers have been developed as early as in 1996, when Lee et al. fabricated a piezoelectric cantilever transducer that worked both as a microphone and a microspeaker [58]. In their system, the $2000 \times 2000 \times 4.5$ μm^3 piezoelectric cantilever was fabricated based on a 0.5 μm-thick ZnO layer with the magnetron sputtering method. In 2003, Ko et al. presented a piezoelectric microspeaker based on a clamped $3000 \times 3000 \times 3$ μm^3 diaphragm. This micromachined transducer also has a thin ZnO film as the piezoelectric layer, which is deposited on a membrane of low-stress silicon nitride of 1.5 μm [64].

Another type of piezoelectric material, AlN, has also been well studied and characterized in the past few decades. A thin film of AlN is normally deposited by the reactive magnetron sputtering method. Sputtered AlN thin films have better chemical and thermal stability than ZnO. The lower conductivity of AlN compared to ZnO also results in lower power loss [65]. With these advantages, AlN has also been a good candidate for fabricating the piezoelectric layer of MEMS speakers. In 2007, Seo et al. presented piezoelectric microspeakers with circular-type and cross-type electrode configurations based on a 0.5-μm-thick AlN film [36]. With a diaphragm size of 4×4 mm^2, the AlN-based microspeakers achieved good acoustic performance with a high sound pressure level (SPL).

However, it is challenging to sputter ZnO and AlN with controlled properties. Their morphology and crystalline quality will highly affect the piezoelectric constants of materials. In a fabrication process, the sputtering rate and residual stress are dependent on the sputtering condition and film thicknesses [66,67]. Sputtering with heated substrates (above 300 °C) have been reported with large residual stresses [35,68], which will wrinkle the diaphragm of fabricated piezoelectric MEMS speakers and affect the sound pressure output. It is possible to deal with such residual stress problem by adding a stress compensation layer or fabricating dome-shaped diaphragms to reduce the effect of the residual stress. For example, in 2000, Han et al. reported dome-shaped piezoelectric MEMS speakers built on 1.5-μm-thick Parylene diaphragms, which can easily release the residual stress through volumetric shrinkage or expansion [69]. In 2009, Yi et al. reported piezoelectric AlN MEMS speakers with improved performance by controlling the residual stress of the compressively stressed diaphragm using Si_xN_y films [35]. The results revealed that the SPLs of the piezoelectric AlN microspeakers were increased by more than 10 dB when the residual stresses became more compressive, especially at the low frequency region.

Other limitations of sputtering ZnO and AlN thin films include low deposition rates (tens of nm/min), small film thicknesses, and small piezoelectric constants [67,70]. The lower value of piezoelectric constants will directly limit the vibration amplitude of a piezoelectric diaphragm and lead to poor acoustic performance. By contrast, PZT thin films have greater piezoelectric constants and are favorable for the applications of piezoelectric actuation. The sputtering and sol-gel methods have also been employed to deposit PZT thin films with typical thicknesses of 0.5–2 μm, which can be applied to a wide range of applications [63]. For example, in 2009, Cho et al. fabricated a piezoelectric MEMS speaker based on a sol-gel PZT thin film with a thickness of 700 nm [11]. The fabricated MEMS speaker had a circular diaphragm with a diameter of 2 mm, which achieved SPLs of 79 dB at 1 kHz, 87 dB at 5 kHz, and 90 dB at 10 kHz under a driving voltage of 13 V. However, sputtered and sol-gel PZT films also suffer from residual stresses and limited thicknesses. Thicker sol-gel PZT films require multiple coatings and high temperature annealing, which will cause serious stress issues. Moreover, since the piezoelectric properties of deposited thin films are largely dependent on the crystal orientation and substrate condition, proper buffer layers are required to prevent the material interdiffusion and oxidation and help to obtain good piezoelectric properties with lower residual stress.

The material properties of these commonly used piezoelectric thin films and the commercial ceramic PZT are summarized in Table 1. Since most of piezoelectric MEMS speakers work on the d_{31} mode of the piezoelectric layer, only the d_{31} piezoelectric constant is listed in the table for comparison. Among these materials, AlN thin films have the smallest piezoelectric constant, while PZT thin films exhibit the highest piezoelectric constant, which is about 10 to 20 times greater than that of ZnO thin films. However, the piezoelectric constant of PZT films also vary in a wide range, dependent on the film thickness, deposition, and poling conditions. In particular, the piezoelectric coefficient of the commercial ceramic PZT (e.g., PZT-5H) can reach 300 pm/V [71], which makes it a promising candidate for the construction of piezoelectric transducers.

Table 1. Material properties of commonly used piezoelectric thin films and the commercial ceramic PZT [21,71–76].

Property	ZnO	AlN	Sol-Gel PZT	Sputtered PZT	Ceramic PZT-5H
Density (kg/m^3)	5700	3260	7700	7700	7800
Young's modulus (GPa)	98.6	283	96	96	50
Dielectric constant	8.8	8.5–10.7	650–1470	400–980	3400
Piezoelectric constant d_{31} (pm/V)	3.9–5.5	2–2.6	23–76	45–102	270–300

3.1.3. Approaches to Improve SPLs

Although a large number of piezoelectric MEMS speakers have been demonstrated based on various piezoelectric thin films with promising results, inadequate sound pressure level (SPL) outputs and non-flat frequency responses are common challenges of these devices. High SPLs of over 90 dB were achieved in a few piezoelectric MEMS speakers, but they were measured either in canals or ear simulators or at high-frequency resonances. Piezoelectric MEMS speakers with high SPLs (90 dB or above) over wide frequency ranges, especially in open air and low-frequency range, are needed for broader applications such as mobile phones, laptops, wearable electronics, and Internet of Things (IoT) devices. Therefore, several approaches have been proposed to improve the SPLs of piezoelectric MEMS speakers in terms of materials and fabrication processes and structure designs, which will be reviewed in the following.

Materials and Fabrication Processes

As discussed in Section 3.1.1, the commonly used piezoelectric thin films of ZnO and AlN deposited by sputtering or sol-gel methods suffer from large residual stresses and limited thickness. For sputtered or sol-gel PZT, their obtained piezoelectric constants are also not comparable with those of bulk piezoelectric crystals or ceramics. As illustrated in Table 1, the piezoelectric constant of ceramic PZT is over four times greater than that of sputtered or sol-gel PZT. Thus, ceramic PZT was gradually employed in fabricating the piezoelectric layer of MEMS speakers with particular fabrication process to thin down this material. In 2009, Kim et al. thinned ceramic PZT down to around 40 μm and fabricated piezoelectric MEMS speakers based on it, and they measured an SPL of 90 dB ($\pm$5 dB) in the audible frequency range under a 32-V_{pp} drive at 1 cm away from the MEMS speaker in an anechoic box [17]. The fabricated MEMS speaker also exhibited a total harmonic distortion (THD) of less than 15% from 400 Hz to 8 kHz. However, the acoustic diaphragm was as large as 20 mm × 18 mm.

Since the resonant frequency of a diaphragm is affected by its area and thickness, scaling down the diaphragm size requires a thinner piezoelectric layer to maintain a proper resonant frequency. In 2020, Wang et al. presented a piezoelectric MEMS speaker based on thin ceramic PZT [16]. By using wafer bonding and chemical mechanical polishing techniques, ceramic PZT was thinned down to only 5 μm and applied to fabricate MEMS speakers. An optical image of the fabricated MEMS speaker and a cross-section SEM image of the device layers are shown in Figure 7a1,a2. Thin ceramic PZT not only exhibits much greater piezoelectric constants than sol-gel or sputtered PZT thin films but also has a wider range of thicknesses, thus allowing the scaling of diaphragms within size restrictions for different applications. With a 6 mm diameter diaphragm, the fabricated MEMS speaker achieved a maximum SPL of 119 dB measured at 1 cm under a 10-V_{pp} drive, as shown in Figure 9a [16].

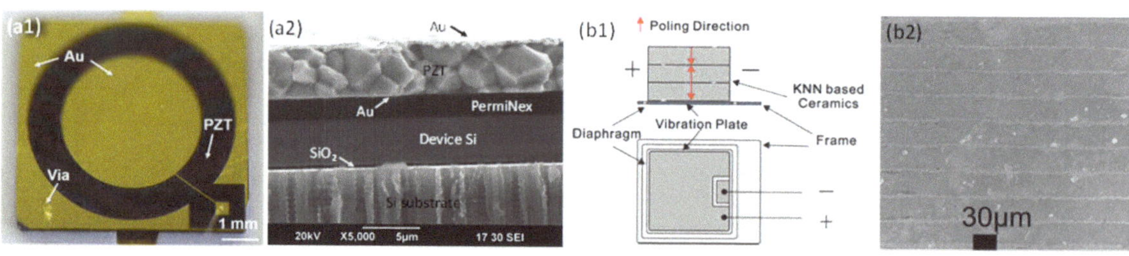

Figure 7. Piezoelectric MEMS speakers based on new materials: (**a1**) optical image and (**a2**) cross-section SEM image of a thin ceramic PZT-based MEMS speaker (Reproduced with permission from Elsevier [16]); (**b1**) schematic and (**b2**) cross-section SEM image of multilayer ceramics of a KNN ceramics-based MEMS speaker (Reproduced with permission from IOP [77]).

Furthermore, lead-free piezoelectric ceramics with high piezoelectric constants have also been explored for fabricating piezoelectric MEMS speakers. For example, in 2014, Gao et al. fabricated piezoelectric MEMS speakers using potassium sodium niobate ((K,Na)NbO$_3$, KNN)-based multilayer piezoelectric ceramics [77]. They employed a tape casting and cofiring process and used Ag–Pd alloys as an inner electrode. A schematic of the multilayer ceramics based piezoelectric MEMS speaker and a cross-section SEM image of the multilayer KNN-based ceramics are shown in Figure 7b1,b2, respectively. With a form factor of 23 × 27 × 0.6 mm^3, using three layers of 30-μm-thick KNN-based ceramics, the fabricated MEMS speakers showed an average SPL of 87 dB from 1 kHz to 20 kHz measured at 3.16 cm under a 5-V_{rms} drive.

Structure Designs

As illustrated in Section 2.1, the output SPL of a MEMS speaker is directly determined by the frequency, area, and displacement of its diaphragm. Increasing the out-of-plane displacement of piezoelectric diaphragms is an effective approach to improve SPLs, especially at low frequency, as a much larger displacement is required at low frequency to achieve the same SPL at high frequency. Therefore, various designs of piezoelectric MEMS speakers have been proposed to improve their SPLs by changing the diaphragm structures, electrode configurations, or using an array form to enhance their acoustic performance.

Diaphragm Structures

In 2018, Stoppel et al. demonstrated a piezoelectric MEMS speaker based on a 2-μm-thick sputtered PZT with two open cuts on a square diaphragm (4 × 4 mm^2) for in-ear applications, as shown in Figure 8a [18]. Without a closed diaphragm, four individual actuators are mechanically decoupled from each other and thus can achieve larger out-of-plane displacements. The measurement in an ear simulator showed a high SPL of above 81 dB from 20 Hz and above 100 dB from 4.7 kHz to 15.8 kHz under a 2-V_{pp} drive, as shown in Figure 9b. The measured THD was less than 2% at most frequencies, except for the subharmonics of the resonance frequency, where the THD was increased to 7%.

In 2020, Cheng et al. presented a piezoelectric MEMS speaker with enhanced SPL by designing suspension-spring actuators with a dual-electrode driving [21]. As shown in Figure 8b, the designed MEMS speaker consisted of a circular moveable diaphragm and four flexible spring actuators. Dual-curve spring actuators with dual-electrode driving were utilized to achieve larger displacements than single-curve spring actuators under the same form factor. Measurements in a 3-cm-long tube showed a maximum SPL of 90.1 dB at the resonance of 1.85 kHz under a 2-V_{pp} drive, which was 28 dB higher than the SPL of a fully clamped diaphragm speaker at the same frequency (Figure 9c). The measured THD of the dual-curve spring device was also lower than those of the clamped diaphragm devices, which was less than 2% at most frequencies and low than 8% at the resonant or harmonic frequencies.

In addition to employing unsealed vibration diaphragms with large displacements, Wang et al. proposed a rigid–flexible vibration coupling mechanism in 2021. By depositing a Parylene film on a pre-etched diaphragm, the fabricated MEMS speaker can maintain large displacements of the unsealed diaphragms without acoustic loss. Measurement in an ear simulator under a 2-V drive showed SPLs can exceed 59 dB from 250 Hz to 20 kHz, with the maximum value of 101.2 dB obtained at the resonance of 6.7 kHz [78].

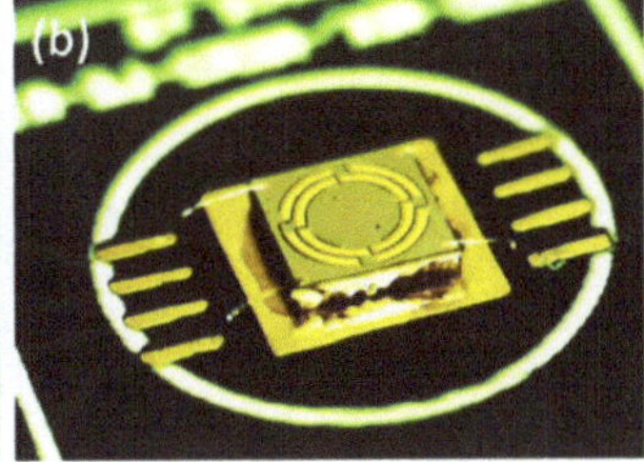

Figure 8. Optical images of piezoelectric MEMS speakers with novel structural designs. (**a**) A diaphragm with two open cuts (Reproduced with permission from IEEE [18]). (**b**) A diaphragm with suspension-spring actuators (Reproduced with permission from Elsevier [21]). (**c**) A diaphragm formed by four piezoelectric cantilevers with different dimensions (Reproduced with permission from IEEE [79]).

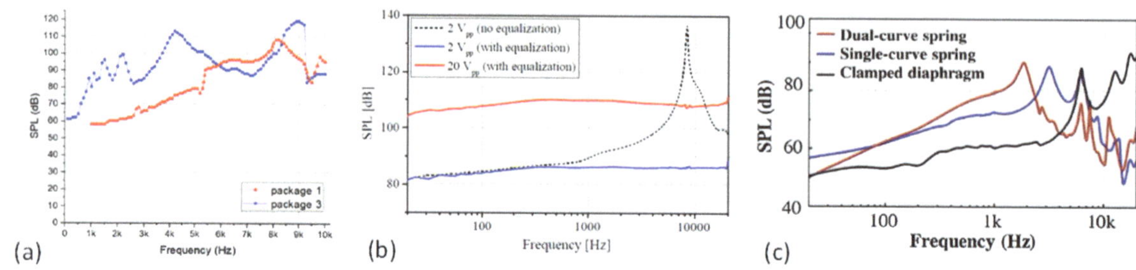

Figure 9. Measured frequency response of piezoelectric MEMS speakers with (**a**) a thin ceramic PZT-based diaphragm in free field at 1 cm (Reproduced with permission from Elsevier [16]), (**b**) a square diaphragm with open cuts in an ear simulator (Reproduced with permission from IEEE [18]), and (**c**) dual-curve spring actuators in a 3-cm-long tube (Reproduced with permission from Elsevier [21]).

To improve SPLs over a broad frequency range, in 2021, Wang et al. proposed a cantilever array design with an in-phase/out-of-phase hybrid driving method to realize a broadband piezoelectric MEMS speaker [79]. As shown in Figure 8c, the device consisted of four piezoelectric cantilevers with different dimensions, the four resonance frequencies of which contribute to the broadband performance of the MEMS speaker. In this device, in order to avoid the sound pressure cancellation due to the large phase shifts around the resonances of the cantilevers, a hybrid drive voltage with a combination of both in-phase and out-of-phase signals was applied to ensure that the cantilevers vibrate in the same direction. Measurements showed a broadband frequency response from 100 Hz to 10 kHz with an SPL of 70 dB or higher and a maximum SPL of 110 dB at 1.54 kHz in an ear simulator under a 2-V_{pp} drive.

Electrode Configurations

Efforts have also been devoted to improving the SPLs of MEMS speakers by the special design of electrode configurations. Electrode configurations on piezoelectric diaphragms are important as they largely determine the excitation mode, vibration displacement, and electromechanical coupling efficiency. As introduced in Section 2.1, most piezoelectric MEMS speakers work on the d_{31} flexural vibration mode of piezoelectric diaphragms with the electrical field applied in the thickness direction and the strain generated in the lateral directions. In addition to the d_{31} vibration mode, piezoelectric materials can also be excited in the d_{33} mode with the applied electrical field and the generated stain in the same direction, typically in the thickness direction. Typically, the magnitude of the d_{33} constant of a piezoelectric material is roughly two times larger than that of the d_{31} constant. Therefore, by proper electrode configurations, the d_{33} mode of piezoelectric diaphragms can be excited with larger out-of-plane displacements than the d_{31} mode. In 2015, Kim et al. presented a piezoelectric MEMS speaker based on the d_{33} mode PMN-PT single crystal diaphragm with a circular inter-digitated electrode (IDE) configuration and studied the effects of the patterned electrodes on the acoustic characteristics of the MEMS speaker [23]. A single crystal PMN-PT was thinned down to 10 μm to form an 8.5 mm diameter diaphragm by grinding, polishing, and inductively-coupled-plasma (ICP) etching, followed by metallization with circular IDE patterns on the top, as shown in Figure 10a. Measurements showed improved SPL with increasing area of the patterned IDE. With an 8 mm diameter IDE, the MEMS speaker showed an average SPL of above 70 dB from 1 kHz to 10 kHz and a maximum SPL of around 100 dB at 1 cm under a 5-V_{rms} drive.

In addition to the IDE configuration that can excite the piezoelectric d_{33} mode for SPL improvement, dual-electrode configuration has been investigated to improve the SPLs of piezoelectric MEMS speakers working on the d_{31} mode. In 2020, Tseng et al. presented a piezoelectric MEMS speaker with the SPL improved by dual-electrode driving [56]. The schematic of the designed MEMS speaker is shown in Figure 10b, where the square diaphragm consists of four triangular plates whose vibrations are synchronized by a connection mass. The low frequency response can be enhanced by reducing the size of the gaps between the triangular plates. Each triangular plate can be driven by an inner electrode and an outer electrode with a 180° phase difference to actuate the piston mode of the diaphragm to increase the SPL. Measurements showed a SPL enhancement of 9.5 dB under the dual-electrode driving in comparison with the single (inner or outer) electrode driving.

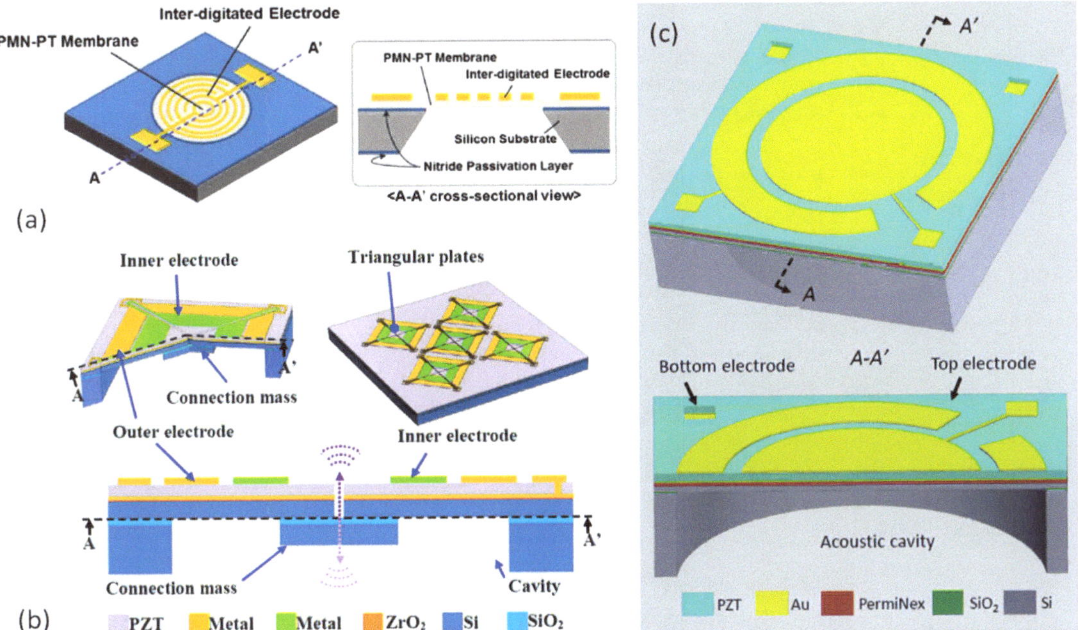

Figure 10. Schematics of piezoelectric MEMS speakers with special electrode configurations. (**a**) Circular inter-digitated electrode (Reproduced with permission from Springer [23]). (**b**) Triangular plates with dual electrode (Reproduced with permission from IEEE [56]). (**c**) Circular diaphragm with dual electrode (Reproduced with permission from IEEE [24]).

In addition to the 180° out-of-phase, other phase differences in dual-electrode driving and their influences on the SPL improvement of piezoelectric MEMS speakers have been studied. In 2021, Wang et al. presented a ceramic PZT-based piezoelectric MEMS speaker with the SPL improved by dual-electrode driving and studied the effects of the phase difference at different frequencies [24]. As shown in Figure 10c, the reported MEMS speaker consists of an inner circular electrode and an outer ring-shaped electrode. By applying sine waves on these two electrodes with a phase difference tuned from 0° to 360° in the experiments, the measurement results revealed that the SPL changed significantly with the phase difference and was frequency dependent, peaking at different phase differences for different frequencies. With the optimal phase differences, a 2–10 dB SPL improvement can be achieved in the frequency band spanning from 600 Hz to 10 kHz, compared with the single-electrode driving method.

Array Structures

Another approach to improve the SPLs of the piezoelectric MEMS speakers is using digital sound reconstruction or speaker arrays. Different from traditional sound generation techniques that rely on the vibration amplitudes and frequencies of a single or a few diaphragms to achieve high SPL at specific frequencies, digital sound reconstruction generates loud sound by adding the outputs of a large number of speaker pixels that can be excited individually by signals with different frequency compositions [80]. Typically, a speaker array containing 2^n speaker pixels is used in digital sound reconstruction, where *n* is the bit number, and each pixel contributes a small amount of sound pressure in the system. In 2015, Casset et al. implemented digital sound reconstruction with piezoelectric MEMS speaker arrays [81]. Figure 11a shows the fabricated speaker array packaged on an electronic board. With a chip size of 4×4 cm^2, the speaker array contains 256 piezoelectric diaphragms based on a 2-µm sol-gel PZT film. The output SPL of the speaker array reached over 100 dB at 13 cm. In 2016, Arevalo et al. increased the bit number and presented a 10-bit (1024 elements) piezoelectric MEMS speaker array with a chip size of 2.3×2.3 cm^2 [82]. An optical image of part of the speaker array is shown in Figure 11b. The characterization results demonstrated the potential of piezoelectric MEMS loudspeaker arrays for digital sound reconstruction, but more efforts are still needed to optimize the design for better acoustic performances.

Figure 11. Schematic and optical images of piezoelectric MEMS speaker arrays: (**a**) a 256-speaker array packaged on an electronic board (Reproduced with permission from Elsevier [81]) and (**b**) part of a 1024-speaker array (Reproduced with permission from IEEE [82]).

3.1.4. Summary of Piezoelectric MEMS Speakers

Piezoelectric MEMS speakers are reviewed above from piezoelectric materials, fabrication techniques, and approaches to improve SPLs. Table 2 summarizes the key results of these piezoelectric MEMS speakers. As shown in the table, sol-gel and sputtered PZT films are popular piezoelectric materials for fabricating piezoelectric MEMS speakers due to their higher piezoelectric constants than those of ZnO or AlN films. Piezoelectric MEMS speakers based on sol-gel or sputtered PZT films with thicknesses of 1–2 µm typically have diaphragm sizes of no more than 4 mm and can generate high SPLs over 90 dB in tubes or ear simulators for in-ear applications. With optimized structure designs, their SPLs can be significantly improved to reach maximum values over 110 dB under small driving voltages. Moreover, piezoelectric MEMS speakers based on ceramic PZT or single-crystal PMN-PT can generate high SPLs in open air, which have potential applications in consumer electronics such as cell phones or laptops. Bulk ceramic PZT or PMN-PT with superior piezoelectric properties can be thinned down to 5–40 µm for fabricating piezoelectric MEMS speakers, which enables larger diaphragm designs ranging from 6 mm to 2 cm and high SPLs of over 100 dB at 1 cm in open air.

Table 2. Key results of different piezoelectric MEMS speakers.

	Ref.	Piezoelectric Layer	Diaphragm Size	1st Resonant Frequency	Maximum SPL	Driving Voltage	Note
In-coupler measurement	[58]	0.5 µm ZnO	2 mm length (square)	890 Hz	100 dB at 4.8 kHz	12 V_{PP}	Measured into a 2 cm^3 coupler
	[21]	1 µm sputtered PZT	1.13 mm diameter (central part)	1.85 kHz	90.1 dB at 1.85 kHz	2 V_{PP}	Measured in a 3-cm-long tube
	[78]	1 µm sputtered PZT	2 mm side length (hexagon)	6.7 kHz	101.2 dB at 6.7 kHz	2V (unspecified)	Measured in an ear simulator
	[79]	2 µm sputtered PZT	4 mm^2 (rectangle)	1.54 kHz	110 dB at 1.54 kHz	2 V_{PP}	Measured in an ear simulator
	[56]	2 µm sputtered PZT	3.24 mm^2 (four triangles)	~6 kHz	118.1 dB at 11.9 kHz	2 V_{PP}	5-speaker array, measured in an ear simulator
	[18]	2 µm sputtered PZT	4 mm length (square)	8.3 kHz	138 dB at 8.3 kHz	2 V_{PP}	Measured in an ear simulator
Free-field measurement	[64]	0.5 µm ZnO	3 mm length (square)	7.3 kHz	83.1 dB at 13.3 kHz	30 V_{PP}	Measured at 1 cm
	[10]	0.5 µm ZnO	5 mm length (square)	2.92 kHz	92.4 dB at 2.92 kHz	6 V_{PP}	Measured at 2 mm
	[36]	0.5 µm AlN	4 mm length (square)	–	100 dB at 10 kHz	20 V_{PP}	Measured at 3 mm
	[35]	0.5 µm AlN	–	–	104 dB at 3 kHz	20 V_{PP}	Device in a 4 cm^3 package, measured at 1 cm
	[11]	0.7 µm sol-gel PZT	2 mm diameter	–	90 dB at 10 kHz	13 V (unspecified)	Measured at 1 cm
	[81]	2 µm sol-gel PZT	2.6 mm diameter	18 kHz	~110 dB	8 V (unspecified)	256-speaker array, measured at 13 cm
	[16]	5 µm ceramic PZT	6 mm diameter	4.3 kHz	119 dB at 9 kHz	10 V_{PP}	Measured at 1 cm
	[17]	40 µm ceramic PZT	18 mm × 20 mm	0.49 kHz	~106 dB at 5.5 kHz	32 V_{PP}	Measured at 1 cm
	[23]	10 µm PMN-PT	8.5 mm diameter	1.4–1.84 kHz	~100 dB at 6.5 kHz	$10\sqrt{2}$ V_{PP}	Measured at 1 cm

3.2. Electrodynamic MEMS Speakers

Electrodynamic MEMS speakers have been developed based on electromagnetic actuation, which is the most widely used actuation mechanism in classical speakers. Electrodynamic MEMS speakers have advantages of high power density, low driving voltage, and linear responses. Efforts have been devoted to the development of electrodynamic MEMS speakers with integrated magnetic materials and small form factors at low cost, while improving their sound performances. However, the full integration of magnetic materials to realize electrodynamic MEMS speakers is still challenging.

In 2004, Cheng et al. presented an electrodynamic MEMS speaker for hearing instruments. The device was fabricated with a low temperature process using an electroplated Ni/Fe soft magnet, which was suitable for post-CMOS processing and potential integration with electronic circuits [5]. A schematic of the designed MEMS speaker is shown in Figure 12a, which has a chip size of 5 mm × 5 mm and consists of a micromachined polymer diaphragm on a silicon wafer, a single-curve Cu coil, an electroplated Ni/Fe soft magnet, and a permanent magnet mounted on the backside. The frequency responses of the fabricated device are measured in air and in a 2-cc coupler with results shown in Figure 13a. At a low driving voltage of 1.5 V, the MEMS speaker generated a maximum SPL of 93 dB at 5 kHz in a 2-cc volume. This work provided a concept and process for micromachining electrodynamic MEMS speakers. Following that, several electrodynamic MEMS speakers have been reported for lower power consumption, high-level integration process, and improved SPL and sound quality.

In 2009, Chen et al. presented an electrodynamic MEMS speaker with improved power efficiency through incorporating Ni nano-composites into Cu to make the voice coil [83]. A cross-sectional view of the MEMS speaker structure is illustrated in Figure 12b, where the coil is made of a Cu–Ni composite by mixing Ni nano-powders with alkaline non-cyanide-based copper-plating solution in a colloidal bath. The frequency responses of the fabricated MEMS speakers driven by the Cu–Ni nanocomposite and pure Cu coils are measured and compared, as shown in Figure 13b. The experimental results showed that the MEMS speaker with a Cu–Ni composite coil can averagely provide about 40% power savings than the one with a Cu coil for the same SPL output at 70 dB.

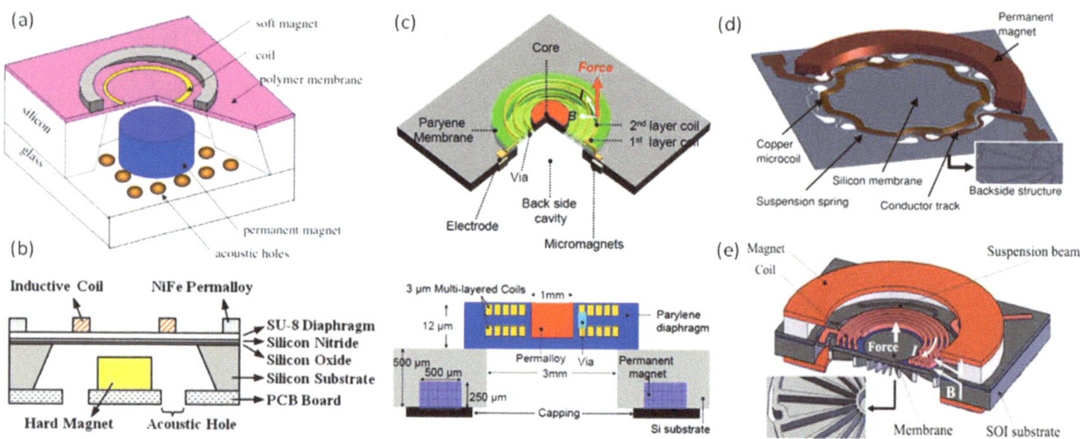

Figure 12. Electrodynamic MEMS speakers: (**a**) typical structure of an electrodynamic MEMS speaker (Reproduced with permission from IOP [5]), (**b**) cross-sectional view of a low-power electrodynamic MEMS speaker with Cu–Ni nanocomposite coil synthesized (Reproduced with permission from Journal of IEEE [83]), (**c**) schematic (top) and cross-sectional view (bottom) of a fully integrated electrodynamic MEMS speaker (Reproduced with permission from IEEE [15]), (**d**) schematic of an electrodynamic MEMS speaker with a rigid silicon diaphragm (Reproduced with permission from Springer [84]), (**e**) schematic of an electrodynamic MEMS speaker showing a rigid silicon diaphragm and the optimized configuration of coil and two face-to-face magnets (Reproduced with permission from IEEE [2]).

As shown in Figure 12a,b, most electrodynamic MEMS speakers require the assembly of a bulky permanent magnet, which will not only increase the overall footprint of the device but also add challenge to the batch fabrication process and precise alignment of the magnet to the diaphragm coil. In order to address this issue, in 2009, Je et al. presented a fully-integrated electrodynamic MEMS speaker with an IC process-compatible micromachined permanent magnets for hearing aid applications [15]. A schematic and cross-sectional view of the presented MEMS speaker is shown in Figure 12c, where a Parylene diaphragm containing embedded multi-turn coils and a soft magnet core is suspended over an acoustic cavity. A rare earth Nd–Fe–B magnetic powder was mixed into a wax binder and dispensed into pre-etched trenches to form the permanent ring micromagnet. The fabricated MEMS speaker produced a 0.64 µm peak displacement at 1 kHz with a 46-mW power consumption. Referring to Figure 2, the achieved displacement is too small to generate sufficient SPLs by a diaphragm with a diameter of 3 mm. Although this work demonstrated the feasibility of fabricating fully integrated electrodynamic MEMS speakers, further design optimization is required to improve the displacement and acoustic performance.

Most MEMS speakers use clamped polymer diaphragms, such as polyimide, Parylene, and SU-8, for flexural vibration and sound generation, whose small mass is in favor of power efficiency and large deflection. However, the flexible nature of polymer diaphragms will lead to dynamic deformations and numerous structural modes within the audio frequency band, thus inducing sound distortion and non-flat frequency response. From 2012 to 2013, to improve the sound performance of electrodynamic MEMS speakers, Shahosseini et al. proposed novel electrodynamic MEMS speakers based on rigid silicon diaphragms and optimized structural designs [2,6,37,84]. The rigid silicon diaphragms were designed with radial ribbed structures for increased stiffnesses and reduced masses, thus enhancing the piston mode vibration for the sound generation and shifting other modes out of the audio frequency band. Figure 12d shows the structure of such an electrodynamic MEMS speaker, where the silicon diaphragm was connected to the substrate by a set of flexible springs to provide out-of-plane displacements [84]. A 14-turn Cu coil was shaped in a special geometry to prevent the damage near the springs' clamp areas, and it was located as close as possible to the permanent magnet to maximize the electromagnetic force. The same research group also investigated the distribution of the magnetic flux density under different configurations of the permanent magnets. In 2013, another electrodynamic MEMS speaker with an optimized microcoil configuration and two face-to-face magnets has been developed [2], as shown in Figure 12e. The fabricated MEMS speaker had a circular diaphragm with a diameter of 15 mm and generated a SPL of around 80 dB at 10 cm starting from 300 Hz to over 20 kHz, as shown in Figure 13c.

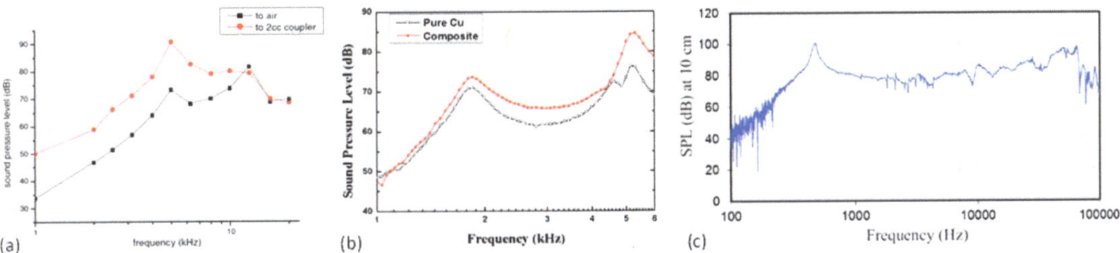

Figure 13. Typical frequency responses of electrodynamic MEMS speakers based on (**a**) a polymer diaphragm and a Cu coil and measured at 2 cm in air and in a 2-cc coupler (Reproduced with permission from IOP [5]), (**b**) pure Cu and Cu–Ni composite coils and measured in a 2-cc coupler (Reproduced with permission from IEEE [83]), (**c**) a rigid silicon diaphragm and measured at 10 cm in air (Reproduced with permission from IEEE [2]).

Table 3 summarizes the key results of these electrodynamic MEMS speakers. As shown in the table, electrodynamic MEMS speakers based on polymer diaphragms typically have small size and low power consumption but limited SPLs. Their maximum SPLs are around 100 dB or less measured in 2-cc couplers or ear simulators. By contrast, electrodynamic MEMS speakers with rigid silicon diaphragms can generate loud sound in open air at large distance but suffer from large diaphragm size and high power consumption.

Table 3. Key results of different electrodynamic MEMS speakers.

Ref	Diaphragm Material	Diaphragm Size	Maximum SPL	Power Consumption	Note
[5]	Polyimide	3.5 mm diameter	93 dB at 5 kHz	320 mW	Measured in a 2 cm^3 volume
[7]	Polyimide	3 mm diameter	106 dB at 1 kHz	0.13 mW	Calculated based on the displacement
[85]	Polyimide	2.5 mm diameter	90 dB at 1,5,10 kHz	–	Measured in a sealed 1500 mm^3 silicone tube
[83]	SU-8	-	Around 85 dB at 5.2 kHz	–	Measured in a 2 cm^3 volume
[86]	PDMS	3.5 mm diameter	106 dB at 1 kHz	1.76 mW	Measured in a 2 cm^3 volume
[2]	Silicon	15 mm diameter	80 dB at 0. 33 kHz	0.5 W	Measured at 10 cm

3.3. Electrostatic MEMS Speakers

MEMS speakers based on electrostatic actuation have also been proposed, which typically consist of parallel or lateral plate actuators. The advantages of such speakers include easy fabrication, high electromechanical efficiency, and relatively flat frequency response. In this section, the recent designs of electrostatic MEMS speakers based on different diaphragm materials will be introduced first. Then, the approaches to improve SPLs of electrostatic MEMS speakers while balancing the design constraints will be reviewed in detail.

3.3.1. Devices with Different Diaphragm Materials

Electrostatic MEMS speakers have been demonstrated based on different diaphragm materials [8,39,87,88]. In 2005, Kim et al. reported an electrostatic MEMS speaker based on a Parylene thin diaphragm. As the cross-sectional SEM image shown in Figure 14a1, the speaker contains two separated chambers on the top and bottom, respectively, which enables bi-directional actuation by electrostatic forces [88]. Figure 14a2 shows the measured frequency response of the speaker. With a diaphragm size of 2×2 mm^2, the fabricated device generated high SPLs of 113.4 dB at 7.68 kHz and 98.8 dB at 13.81 kHz, which were measured at a distance of 1 cm under a driving voltage of 150 V. In 2007, Roberts et al. presented an electrostatically driven touch-mode MEMS speaker based on poly-SiC diaphragms with a diameter of 800 µm, which was robust and operable in harsh environments [8]. Figure 14b1 shows the SEM image of the suspended poly-SiC diaphragm of the fabricated device. At a distance of 1 cm, a maximum SPL of 73 dB was obtained at 16.59 kHz under a driving voltage of 200 V (Figure 14b2). Another material, graphene, has also been explored for fabricating high-quality broad-band audio speakers due to its extremely low mass density and high mechanical strength. In 2013, Zhou et al. presented a miniaturized electrostatic speaker based on a 30 nm thin graphene diaphragm and demonstrated a broad frequency response from 20 Hz to 20 kHz with the performance matching or surpassing a commercial magnetic coil speaker [39].

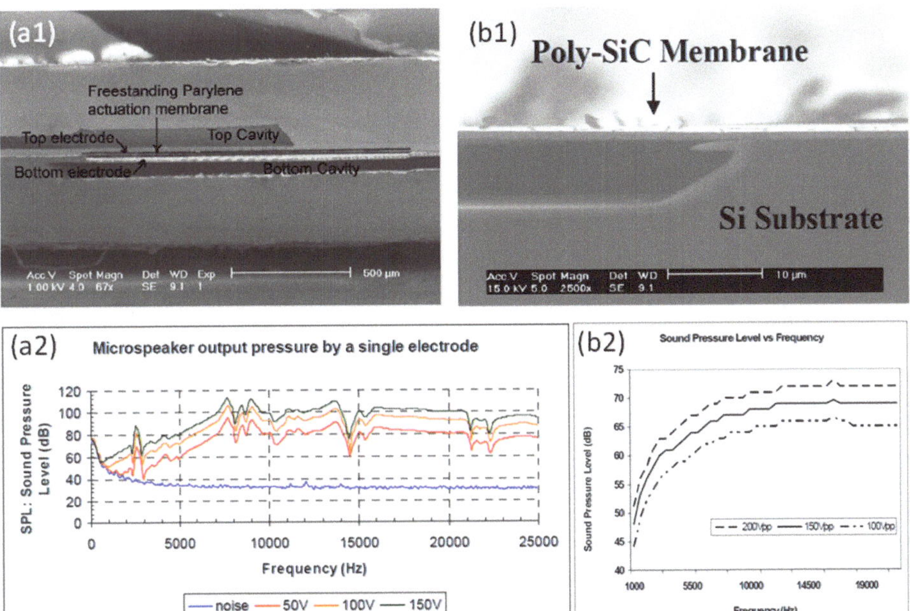

Figure 14. Electrostatic MEMS speakers based on different diaphragm materials: (**a1**) SEM image and (**a2**) measured frequency response of a bi-directional MEMS speaker with a Parylene diaphragm (Reproduced with permission from IEEE [88]). (**b1**) SEM image and (**b2**) measured frequency response of a touch-mode MEMS speaker with a poly-SiC diaphragm (Reproduced with permission from IEEE [8]).

3.3.2. Approaches to Improve SPLs

Most electrostatic MEMS speakers are based on the conventional parallel plate structures and have low SPLs due to the small deflections of their diaphragms, which is a direct result of the balance between the electrostatic force and the mechanical restoring force. In order to overcome the limitation of low SPLs, large electrostatic forces need to be generated. As introduced in Section 2.2 and shown in Equation (7), large electrostatic forces require high driving voltages and small separation gaps. However, the small separation gap will limit the deflection range of the diaphragm and generate large squeeze film air damping [89]. Moreover, the driving voltage must be reasonably lower than the pull-in voltage of the parallel plates to ensure a good reliability. Therefore, tradeoffs have to be made among the electrostatic force, the separation gap between the parallel plates, and the driving voltage to increase the SPLs of electrostatic MEMS speakers.

To generate considerable SPLs and balance the above-mentioned constraints, several approaches in terms of device structure and driving voltage have been applied in the development of electrostatic MEMS speakers [9,38,90–92]. One approach to improve SPL is to use multiple speakers, i.e., array structures. In 2016, Arevalo et al. presented an electrostatic MEMS speaker array for digital sound generation, where each of the individual MEMS speakers had a hexagonal diaphragm connected to an outer hexagonal ring by tethers (Figure 15a) [91]. This work demonstrated the feasibility of generating sounds with electrostatic MEMS speaker arrays but lacked acoustical characterization results.

Different from conventional MEMS speakers that work on the out-of-plane deflection of a diaphragm, Kaiser et al. proposed a novel structure design in 2019, which consisted of in-plane bending electrostatic actuators working in air chambers based on the so-called nanoscopic electrostatic drive (NED) technology, as shown in Figure 15b [9,48]. Utilizing the curvy geometric shape of the moving beams, electrostatic forces are translated into lateral forces and cause the bending of the beams. Therefore, high SPLs can be reached by the large deflection of the beams in the air chambers and a large number of beams in one chip, without the limitation of small separation gaps between electrodes [93]. This novel structure utilized the chip's bulk volume rather than the surface to generate sound pressures. Figure 15c shows an optical image of such a fabricated electrostatic MEMS speaker with in-plane actuators. The acoustic measurement in an ear simulator showed a SPL of 69 dB at 500 Hz with a THD of 4.4%. The maximum SPL reached 104 dB at 11.4 kHz.

In 2020, Garud et al. designed and fabricated a MEMS speaker with peripheral electrostatic actuation [38]. Figure 15d shows the schematic of the designed electrostatic MEMS speaker, where the clamped circular diaphragm has a peripheral electrode configuration that can mitigate the squeeze film damping effect and increase the pull-in voltage. The simulation results showed that as the peripheral electrode width was reduced from 100% (full electrode coverage) to 10%, the pull-in voltage and the vibration amplitude of the diaphragm could be increased by a factor up to 40 and 80, respectively.

To reduce or eliminate the DC bias of electrostatic MEMS speakers, electrets embedded with quasi-permanent electrical charges have been integrated within the electrode structures. In 2020, Sano et al. presented an electret-augmented electrostatic MEMS speaker and demonstrated its sound generation under low driving voltages [92]. The schematic and an SEM image of the fabricated MEMS speaker are shown in Figure 15e,f, respectively. By integrating the electrets into the MEMS speaker, the built-in electrical potential is equivalent to an external DC bias, thus resulting in an increased displacement or a reduced bias voltage. The characterization result showed that a −10 V electret-augmented electrostatic MEMS speaker reached a maximum SPL of 50 dB at 1.5 cm under a 5-V_{pp} AC driving voltage.

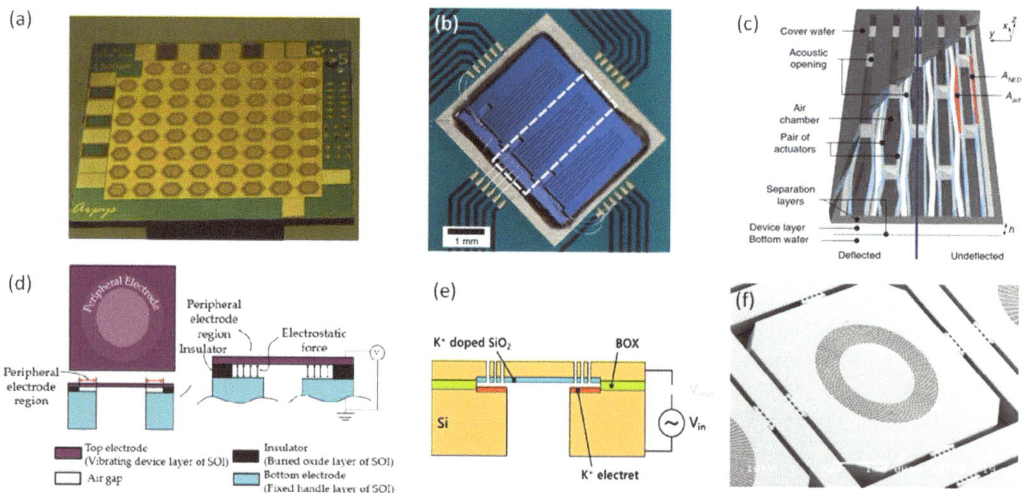

Figure 15. Electrostatic MEMS speakers with special designs: (**a**) optical image of a MEMS speaker array (Reproduced with permission from IEEE [91]), (**b**) schematic and (**c**) optical image of a MEMS speaker with in-plane bending electrostatic actuators working in air chambers (Reproduced with permission from Nature Portfolio [9]), (**d**) schematic of a peripheral electrode configuration (Reproduced with permission from IEEE [38]), (**e**) schematic and (**f**) the corresponding SEM image of an electret-augmented MEMS speaker (Reproduced with permission from MDPI AG [92]).

Table 4 summarizes the representative electrostatic MEMS speakers reported in the literature. It can be seen that electrostatic MEMS speakers typically require high driving voltage and large DC bias to generate considerable diaphragm deflection. Most of electrostatic MEMS speakers have small separation gaps (1–8 μm) and limited sound pressure output. High SPLs are generally obtained only at the high-frequency range.

Table 4. Key results of different electrostatic MEMS speakers.

Ref	Diaphragm Size	Electrode Separation	Maximum SPL	Driving Voltage	Note
[92]	2 mm diameter	2 µm, peripheral electrode	50 dB at around 35 kHz	AC 5 V_{PP}	Measured at 1.5 cm
[8]	0.8 mm diameter	8 µm, touch mode	73 dB at 16.59 kHz	AC 200 V_{PP}	Measured at 1 cm
[38]	3.1 mm diameter	1 µm, peripheral electrode	75–78 dB at above 10 kHz	DC 30 V + AC 30 V	Measured at 1 cm
[88]	2 mm length (square)	7.5 µm	113.4 dB at 7.68 kHz	AC 150 V	Measured at 1 cm
[9]	-	-	104 dB at 11.4 kHz	DC 40 V + AC 10 V_{PP}	Measured in an ear simulator

3.4. Thermoacoustic MEMS Speakers

In addition to the above reviewed three major types of transduction mechanisms, thermoacoustic transduction also has potential to be applied for making MEMS speakers. Several thermoacoustic loudspeakers have been developed based on carbon nanotube or graphene with research efforts focused on increasing the sound pressure output and reducing the power consumption [13,14,42,94,95].

In 2008, Xiao et al. found that thin carbon nanotube films emitted sound when a current in audio frequency was applied, which could be attributed to the thermoacoustic effect [42]. Based on this finding, they successfully fabricated thermoacoustic speakers with A4 paper sizes and cylindrical shapes (9 cm diameter and 8.5 cm height) based on one-layer or four-layer carbon nanotube thin films, which could generate over 70 dB SPLs at 5 cm starting from 1 kHz, with an input power of 3 Watts. Figure 16a shows the photograph of a fabricated thermoacoustic speaker with an A4 paper size. This work demonstrated the feasibility of developing thermoacoustic speakers using carbon nanotube films. However, it required a large device size and a high power consumption to generate high SPLs.

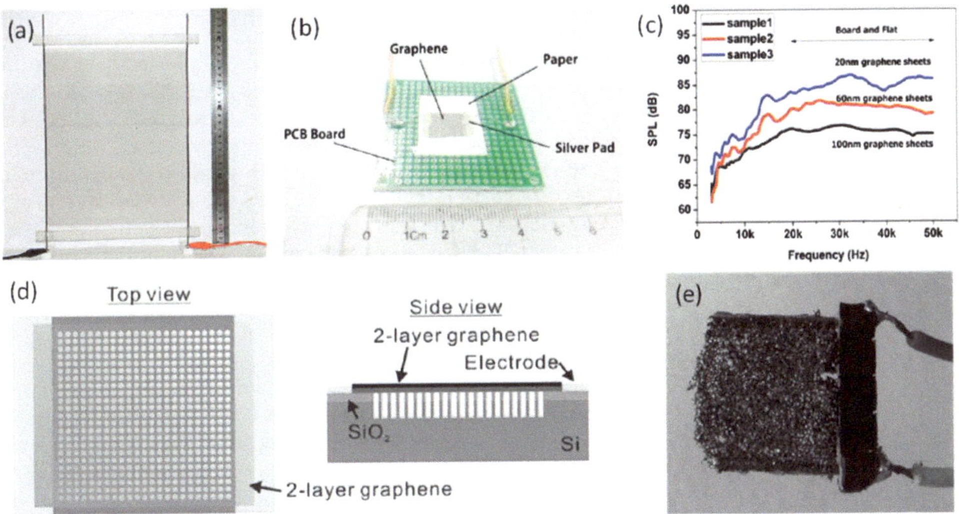

Figure 16. Thermoacoustic speakers: (**a**) photograph of a A4 paper size carbon nanotube thin film thermoacoustic speaker (Reproduced with permission from ACS [42]), (**b**) photograph and (**c**) measured SPL of a graphene-on-paper speaker (Reproduced with permission from ACS [94]), (**d**) schematic of a graphene speaker on a patterned substrate (Reproduced with permission from Wiley [14]), (**e**) photograph of a graphene foam speaker (Reproduced with permission from Wiley [13]).

In 2011, Tian et al. observed thermoacoustic effect on graphene and demonstrated graphene-on-paper speakers [94]. As shown in Figure 16b, the fabricated thermoacoustic speaker had a 1 cm × 1 cm graphene sheet, which was placed on a piece of paper and connected to a printed circuit board (PCB) using silver ink. Graphene sheets with thicknesses of 20 nm, 60 nm, and 100 nm were used to fabricate speakers. Figure 16c shows the SPL curves with the input power density normalized to 1 W/cm^2, which indicated that thinner graphene sheets produced higher SPLs and the SPL of 20 nm graphene sheets reached 85 dB at 5 cm with the frequency increased to over 15 kHz.

The sound performance of thermoacoustic speakers has been further studied and optimized in terms of substrate material and structure design. For example, in 2012, Suk et al. studied thermoacoustic sound generation with graphene on different substrates, including glass, polyethylene terephthalate (PET), and polydimethylsiloxane (PDMS) [14]. The substrate effect was also investigated by transferring graphene onto patterned substrates with different porosities, as shown in Figure 16d. The experiments revealed that graphene on the substrates with lower thermal effusivity and higher porosity exhibited better sound performances. In 2015, Fei et al. presented a low-voltage driven thermoacoustic speaker based on graphene foam synthesized by the nickel-template chemical vapor deposition (CVD) method [13]. A photograph of the fabricated free-standing graphene foam speaker is shown in Figure 16e. Benefited from high thermal conductivity and low in-plane resistance of the 3D graphene foam, the speaker generated a SPL of around 50 dB at 3 cm and 10 kHz with a power consumption of only 0.1 W.

In summary, thermoacoustic speakers made of carbon nanotubes or graphene films have advantages of simple structure, light weight, and easy fabrication. The transparent and stretchable nature of carbon nanotube or graphene films also makes it possible to fabricate them into any shape and size, freestanding or on any insulating surfaces, showing great potentials to be applied for developing thermoacoustic MEMS speakers. However, current thermoacoustic speakers require large size (1–4 cm) and high power consumption (0.1–3 W) to generate adequate sound pressure output.

4. Comparison of Different MEMS Speakers

As reviewed in Sections 2 and 3, MEMS speakers have been demonstrated based on piezoelectric, electrodynamic, electrostatic, and thermoacoustic transduction mechanisms, showing great potentials for various applications including hearing instruments and portable electronic devices. Among them, piezoelectric MEMS speakers and electrodynamic MEMS speakers are the dominant types of MEMS speakers, which have been extensively studied and reported in a vast amount of the literature. Piezoelectric MEMS speakers have advantages of relatively large driving force and high sound pressure output over other MEMS speakers. High SPLs of over 90 dB have been achieved by several piezoelectric MEMS speakers either in ear simulators or in open air. Piezoelectric thin films including ZnO, AlN, PZT, and PMN-PT have been fabricated either by deposition or thinning down bulk materials and applied for fabricating piezoelectric MEMS speakers. However, most current piezoelectric MEMS speakers suffer from non-flat frequency responses due to the resonance behavior of diaphragms. The nonlinearity and hysteresis of piezoelectric materials are also drawbacks of piezoelectric MEMS speakers.

By contrast, electrodynamic MEMS speakers with quasi-linear behaviors are favorable for high-fidelity sound reconstruction. Low power consumption and large mechanical displacements are also advantages of electrodynamic MEMS speakers. Several electrodynamic MEMS speakers have been developed based on polymer diaphragms or rigid silicon diaphragms, with SPLs of 60–100 dB obtained in 2-cc couplers for in-ear applications. However, the requirement of permanent magnets for electrodynamic MEMS speakers not only increases the overall size of devices but also makes the full integration complicated and challenging.

In comparison, electrostatic MEMS speakers do not require complicated fabrication processes but suffer from small displacements, very high driving voltages, and pull-in limitations. Several approaches, such as nanoscopic electrostatic drive (NED) technology, have been proposed to balance the driving voltage, pull-in limitation, and displacement of the diaphragm. Improved SPLs and low THDs have been obtained on these electrostatic MEMS speakers.

Compared with piezoelectric, electrodynamic, and electrostatic MEMS speakers, thermoacoustic MEMS speakers are special acoustic devices that do not rely on mechanical vibration of diaphragms to generate sounds. Therefore, there are no resonant peaks in the frequency response of thermoacoustic MEMS speakers. High transparency, high stretchability, and easy fabrication into any sizes and shapes are the advantages of thermoacoustic MEMS speakers. However, current thermoacoustic speakers all require much larger sizes to achieve comparable SPLs of piezoelectric or electrodynamic MEMS speakers. Large power consumption is another concern of thermoacoustic MEMS speakers.

In common, all these MEMS speakers are required to improve their SPLs at specific frequencies to satisfy a wider range of applications. Approaches have been proposed and demonstrated on these MEMS speakers with improved SPLs, including applying new materials and fabrication processes, designing novel structures and special electrode configurations, and using large speaker arrays.

5. Summary and Outlook

In summary, MEMS speakers have been reviewed in terms of the theory, modeling, transduction mechanisms, and development history in this article. Four types of MEMS speakers, working on piezoelectric, electrodynamic, and electrostatic actuation and the thermoacoustic effect have been introduced; their respective development milestones, performances, advantages, and limitations are also discussed. Approaches to improve the SPLs of MEMS speakers including special structures, new materials, electrode configurations, and speaker arrays are highlighted and discussed, especially for piezoelectric MEMS speakers.

In the future, the SPLs of all types of MEMS speakers will continue to be improved by the incorporation of new materials, novel fabrication techniques, and optimized device and enclosure designs, as well as with deeper understandings of their modeling. In addition to SPLs, fabrication challenges, frequency response, sound quality, and power consumption will also be taken into account. Particularly, piezoelectric MEMS speakers will be extensively investigated to obtain flat frequency responses. Electrodynamic MEMS speakers will be further studied with electroacoustic efficiency improved and permanent magnets fully integrated in batch processes. Electrostatic MEMS speakers, with efforts in reducing driving voltages, and high-level integration with electronic circuits, may find broader applications, especially in digital sound reconstruction. Finally, thermoacoustic MEMS speakers will continue to be explored with efforts to reduce the device size and power consumption. Thereby, MEMS speakers are expected to become a promising candidate not only in the in-ear applications but also in a wide range of consumer electronics.

Author Contributions: Writing—original draft preparation and figures preparation, H.W., Y.M., Q.Z., and K.C.; writing—review and editing, H.W., Y.L., and H.X.; methodology, conceptualization, and editing, H.W. and H.X.; supervision and project administration, H.X.; funding acquisition, H.X. All authors have read and agreed to the published version of the manuscript.

Funding: This work was supported by the National Science and Technology Major Project of China (2018YFF01010904), Beijing Institute of Technology Startup Funds, and the Foshan Science and Technology Innovation Team Project (2018IT100252).

Data Availability Statement: Not applicable.

Conflicts of Interest: The authors declare no conflict of interest.

References

1. Hwang, S.-M.; Lee, H.-J.; Hong, K.-S.; Kang, B.-S.; Hwang, G.-Y. New development of combined permanent-magnet type microspeakers used for cellular phones. *IEEE Trans. Magn.* **2005**, *41*, 2000–2003. [CrossRef]
2. Shahosseini, I.; Lefeuvre, E.; Moulin, J.; Martincic, E.; Woytasik, M.; Lemarquand, G. Optimization and microfabrication of high performance silicon-based MEMS microspeaker. *IEEE Sens. J.* **2013**, *13*, 273–284. [CrossRef]
3. Shahosseini, I.; Lefeuve, E.; Mrtincic, E.; Woytasik, M.; Moulin, J.; Megherbi, S.; Ravaud, R.; Lemarquand, G. Microstructured silicon membrane with soft suspension beams for a high performance MEMS microspeaker. *Microsyst. Technol.* **2012**, *18*, 1791–1799. [CrossRef]
4. Lee, C.-M.; Kwon, J.-H.; Kim, K.-S.; Park, J.-H.; Hwang, S.-M. Design and analysis of microspeakers to improve sound characteristics in a low frequency range. *IEEE Trans. Magn.* **2010**, *46*, 2048–2051. [CrossRef]
5. Cheng, M.C.; Huang, W.S.; Huang, S.R.S. A silicon microspeaker for hearing instruments. *J. Micromech. Microeng.* **2004**, *14*, 859–866. [CrossRef]
6. Lemarquand, G.; Ravaud, R.; Shahosseini, I.; Lemarquand, V.; Moulin, J.; Lefeuvre, E. MEMS electrodynamic loudspeakers for mobile phones. *Appl. Acoust.* **2012**, *73*, 379–385. [CrossRef]
7. Je, S.S.; Rivas, F.; Diaz, R.E.; Kwon, J.; Kim, J.; Bakkaloglu, B.; Kiaei, S.; Fellow; Chae, J. A compact and low-cost MEMS loudspeaker for digital hearing aids. *IEEE Trans. Biomed. Circuits Syst.* **2009**, *3*, 348–358. [CrossRef]
8. Roberts, R.C.; Du, J.; Ong, A.K.; Li, D.; Zorman, C.A.; Tien, N.C. Electrostatically driven touch-mode poly-SiC micro speaker. In Proceedings of the SENSORS, 2007 IEEE, Atlanta, GA, USA, 28–31 October 2007; pp. 284–287. [CrossRef]
9. Kaiser, B.; Langa, S.; Ehrig, L.; Stolz, M.; Schenk, H.; Conrad, H.; Schenk, H.; Schimmanz, K.; Schuffenhauer, D. Concept and proof for an all-silicon MEMS micro speaker utilizing air chambers. *Microsyst. Nanoeng.* **2019**, *5*, 1–11. [CrossRef] [PubMed]
10. Yi, S.H.; Kim, E.S. Micromachined piezoelectric microspeaker. *Jpn. J. Appl. Phys. Part. 1 Regul. Pap. Short Notes Rev. Pap.* **2005**, *44*, 3836–3841. [CrossRef]
11. Cho, I.J.; Jang, S.; Nam, H.J. A Piezoelectrically actuated mems speaker with polyimide membrane and thin film Pb(Zr,Ti)O$_3$(PZT) actuator. *Integr. Ferroelectr.* **2009**, *105*, 27–36. [CrossRef]
12. Wang, H.; Li, M.; Yu, Y.; Chen, Z.; Ding, Y.; Jiang, H.; Xie, H. A piezoelectric MEMS loud speaker based on ceramic PZT. In Proceedings of the 2019 20th International Conference on Solid-State Sensors, Actuators and Microsystems & Eurosensors XXXIII (TRANSDUCERS & EUROSENSORS XXXIII), Berlin, Germany, 23–27 June 2019; IEEE: Piscataway, NJ, USA, 2019; pp. 857–860.
13. Fei, W.; Zhou, J.; Guo, W. Low-voltage driven graphene foam thermoacoustic speaker. *Small* **2015**, *11*, 2252–2256. [CrossRef]
14. Suk, J.W.; Kirk, K.; Hao, Y.; Hall, N.A.; Ruoff, R.S. Thermoacoustic sound generation from monolayer graphene for transparent and flexible sound sources. *Adv. Mater.* **2012**, *24*, 6342–6347. [CrossRef] [PubMed]
15. Je, S.-S.; Wang, N.; Brown, H.C.; Arnold, D.P.; Chae, J. An electromagnetically actuated microspeaker with fully-integrated wax-bonded Nd-Fe-B micromagnets for hearing aid applications. In Proceedings of the TRANSDUCERS 2009—2009 International Solid-State Sensors, Actuators and Microsystems Conference, Denver, CO, USA, 21–25 June 2009; IEEE: Piscataway, NJ, USA, 2009; pp. 885–888.
16. Wang, H.; Chen, Z.; Xie, H. A high-SPL piezoelectric MEMS loud speaker based on thin ceramic PZT. *Sens. Actuators A Phys.* **2020**, *309*, 112018. [CrossRef]
17. Kim, H.J.; Koo, K.; Lee, S.Q.; Park, K.H.; Kim, J. High performance piezoelectric microspeakers and thin speaker array system. *ETRI J.* **2009**, *31*, 680–687. [CrossRef]
18. Stoppel, F.; Mannchen, A.; Niekiel, F.; Beer, D.; Giese, T.; Wagner, B. New integrated full-range MEMS speaker for in-ear applications. In Proceedings of the 2018 IEEE Micro Electro Mechanical Systems (MEMS), Belfast, UK, 21–25 January 2018; IEEE: Piscataway, NJ, USA, 2018; pp. 1068–1071.
19. Park, K.-H.; Jiang, Z.-X.; Hwang, S.-M. Design and Analysis of a Novel Microspeaker with Enhanced Low-Frequency SPL and Size Reduction. *Appl. Sci.* **2020**, *10*, 8902. [CrossRef]
20. Liu, W.; Huang, J.; Shen, Y.; Cheng, J. Theoretical Modeling of Piezoelectric Cantilever MEMS Loudspeakers. *Appl. Sci.* **2021**, *11*, 6323. [CrossRef]
21. Cheng, H.H.; Lo, S.C.; Huang, Z.R.; Wang, Y.J.; Wu, M.; Fang, W. On the design of piezoelectric MEMS microspeaker for the sound pressure level enhancement. *Sens. Actuators A Phys.* **2020**, *306*, 111960. [CrossRef]
22. Stoppel, F.; Eisermann, C.; Gu-Stoppel, S.; Kaden, D.; Giese, T.; Wagner, B. Novel membrane-less two-way MEMS loudspeaker based on piezoelectric dual-concentric actuators. In Proceedings of the TRANSDUCERS 2017—19th International Conference on Solid-State Sensors, Actuators and Microsystems, Kaohsiung, Taiwan, 18–22 June 2017; IEEE: Piscataway, NJ, USA, 2017; pp. 2047–2050.
23. Kim, H.J.; Yang, W.S. The effects of electrodes patterned onto the piezoelectric thin film on frequency response characteristics of PMN-PT MEMS acoustic actuators. *J. Electroceramics* **2015**, *35*, 45–52. [CrossRef]
24. Wang, H.; Feng, P.X.L.; Xie, H. A dual-electrode MEMS speaker based on ceramic PZT with improved sound pressure level by phase tuning. In Proceedings of the 2021 IEEE 34th International Conference on Micro Electro Mechanical Systems (MEMS), Gainesville, FL, USA, 25–29 January 2021; IEEE: Piscataway, NJ, USA, 2021; Volume 2021, pp. 701–704.
25. Usound GmbH. Available online: https://www.usound.com/home/ (accessed on 13 October 2021).
26. TDK Corporation. Available online: https://www.tdk.com/en/news_center/press/20190521_01.html (accessed on 13 October 2021).

27. Audio Pixels Limited. Available online: https://www.audiopixels.com.au/index.cfm/technology/ (accessed on 13 October 2021).
28. Arioso Systems GmbH. Available online: https://arioso-systems.com/ (accessed on 13 October 2021).
29. Chiang, H.-Y.; Huang, Y.-H. Vibration and sound radiation of an electrostatic speaker based on circular diaphragm. *J. Acoust. Soc. Am.* **2015**, *137*, 1714–1721. [CrossRef]
30. Švec, J.G.; Granqvist, S. Tutorial and guidelines on measurement of sound pressure level in voice and speech. *J. Speech Lang. Hear. Res.* **2018**, *61*, 441–461. [CrossRef]
31. Fedtke, T.; Grason, L. Sound level calibration: Microphones, ear simulators, couplers, and sound level meters. *Semin. Hear.* **2014**, *35*, 295–311. [CrossRef]
32. Jax, P.; Vary, P. Bandwidth extension of speech signals: A catalyst for the introduction of wideband speech coding? *IEEE Commun. Mag.* **2006**, *44*, 106–111. [CrossRef]
33. Blevins, R.D. *Formulas for Natural Frequency and Mode Shape*; Van Nostrand Reinhold Company: New York, NY, USA, 1979; ISBN 9780442207106.
34. Cheng, H.-H.; Huang, Z.-R.; Wu, M.; Fang, W. Low frequency sound pressure level improvement of piezoelectric MEMS microspeaker using novel spiral spring with dual electrode. In Proceedings of the 2019 20th International Conference on Solid-State Sensors, Actuators and Microsystems & Eurosensors XXXIII (TRANSDUCERS & EUROSENSORS XXXIII), Berlin, Germany, 23–27 June 2019; IEEE: Piscataway, NJ, USA, 2019; pp. 2013–2016.
35. Yi, S.; Ur, S.C.; Kim, E.S. Performance of packaged piezoelectric microspeakers depending on the material properties. In Proceedings of the 2009 IEEE 22nd International Conference on Micro Electro Mechanical Systems, Sorrento, Italy, 25–29 January 2009; IEEE: Piscataway, NJ, USA, 2009; pp. 765–768.
36. Seo, K.; Park, J.; Kim, H.; Kim, D.; Ur, S.; Yi, S. Micromachined piezoelectric microspeakers fabricated with high quality AlN thin film. *Integr. Ferroelectr.* **2007**, *95*, 74–82. [CrossRef]
37. Sturtzer, E.; Shahosseini, I.; Pillonnet, G.; Lefeuvre, E.; Lemarquand, G. High fidelity microelectromechanical system electrodynamic micro-speaker characterization. *J. Appl. Phys.* **2013**, *113*, 214905. [CrossRef]
38. Garud, M.V.; Pratap, R. A novel MEMS speaker with peripheral electrostatic actuation. *J. Microelectromechanical Syst.* **2020**, *29*, 592–599. [CrossRef]
39. Zhou, Q.; Zettl, A. Electrostatic graphene loudspeaker. *Appl. Phys. Lett.* **2013**, *102*, 223109. [CrossRef]
40. Melnikov, A.; Schenk, H.A.G.; Monsalve, J.M.; Wall, F.; Stolz, M.; Mrosk, A.; Langa, S.; Kaiser, B. Coulomb-actuated microbeams revisited: Experimental and numerical modal decomposition of the saddle-node bifurcation. *Microsyst. Nanoeng.* **2021**, *7*. [CrossRef]
41. Younis, M.I.; Abdel-Rahman, E.M.; Nayfeh, A. A reduced-order model for electrically actuated microbeam-based MEMS. *J. Microelectromechanical Syst.* **2003**, *12*, 672–680. [CrossRef]
42. Xiao, L.; Chen, Z.; Feng, C.; Liu, L.; Bai, Z.-Q.; Wang, Y.; Qian, L.; Zhang, Y.; Li, Q.; Jiang, K.; et al. Flexible, Stretchable, Transparent Carbon Nanotube Thin Film Loudspeakers. *Nano Lett.* **2008**, *8*, 4539–4545. [CrossRef]
43. Neumann, J.J.; Gabriel, K.J. CMOS-MEMS membrane for audio-frequency acoustic actuation. *Sens. Actuators A Phys.* **2002**, *95*, 175–182. [CrossRef]
44. Huang, J.H.; Her, H.-C.; Shiah, Y.C.; Shin, S.-J. Electroacoustic simulation and experiment on a miniature loudspeaker for cellular phones. *J. Appl. Phys.* **2008**, *103*, 033502. [CrossRef]
45. Chang, J.R.; Wang, C.N. Acoustical analysis of enclosure design parameters for microspeaker system. *J. Mech.* **2019**, *35*, 1–12. [CrossRef]
46. Blackstock, D.T. *Fundamentals of Physical Acoustics*; Wiley: New York, NY, USA, 2000.
47. Chiang, H.-Y.; Huang, Y.-H. Experimental modeling and application of push-pull electrostatic speakers. *J. Acoust. Soc. Am.* **2019**, *146*, 2619–2631. [CrossRef] [PubMed]
48. Spitz, B.; Wall, F.; Schenk, H.; Melnikov, A.; Pufe, W. Audio-transducer for in-ear-applications based on CMOS compatible electrostatic actuators. In Proceedings of the MikroSystemTechnik Kongress, Berlin, Germany, 28–30 October 2019.
49. Sun, P.; Xu, D.P.; Hwang, S.M. Design of microspeaker module considering added stiffness. *J. Mech. Sci. Technol.* **2014**, *28*, 1623–1628. [CrossRef]
50. Kim, B.H.; Lee, H.S.; Kim, S.W.; Kang, P.; Park, Y.S. Hydrodynamic responses of a piezoelectric driven MEMS inkjet print-head. *Sens. Actuators A Phys.* **2014**, *210*, 131–140. [CrossRef]
51. Zhu, Y.; Liu, W.; Jia, K.; Liao, W.; Xie, H. A piezoelectric unimorph actuator based tip-tilt-piston micromirror with high fill factor and small tilt and lateral shift. *Sens. Actuators A Phys.* **2011**, *167*, 495–501. [CrossRef]
52. Smith, G.L.; Rudy, R.Q.; Polcawich, R.G.; DeVoe, D.L. Integrated thin-film piezoelectric traveling wave ultrasonic motors. *Sens. Actuators A Phys.* **2012**, *188*, 305–311. [CrossRef]
53. Rinaldi, M.; Zuniga, C.; Chengjie, Z.; Piazza, G. Super-high-frequency two-port AlN contour-mode resonators for RF applications. *IEEE Trans. Ultrason. Ferroelectr. Freq. Control* **2010**, *57*, 38–45. [CrossRef]
54. Wang, H.; Yu, Y.; Chen, Z.; Yang, H.; Jiang, H.; Xie, H. Design and fabrication of a piezoelectric micromachined ultrasonic transducer array based on ceramic PZT. In Proceedings of the 2018 IEEE SENSORS, New Delhi, India, 28–31 October 2018; IEEE: Piscataway, NJ, USA, 2018; Volume 2018, pp. 1–4.

55. Yi, S.; Yoon, M.; Ur, S. Piezoelectric microspeakers with high compressive ZnO film and floating electrode. *J. Electroceramics* **2009**, *23*, 295–300. [CrossRef]
56. Tseng, S.-H.; Lo, S.-C.; Wang, Y.-J.; Lin, S.-W.; Wu, M.; Fang, W. Sound pressure and low frequency enhancement using novel PZT MEMS microspeaker design. In Proceedings of the 2020 IEEE 33rd International Conference on Micro Electro Mechanical Systems (MEMS), Vancouver, BC, Canada, 18–22 January 2020; IEEE: Piscataway, NJ, USA, 2020; Volume 2020, pp. 546–549.
57. Ren, T.; Zhang, L.; Liu, L.; Li, Z. Design optimization of beam-like ferroelectrics-silicon microphone and microspeaker. *IEEE Trans. Ultrason. Ferroelectr. Freq. Control* **2002**, *49*, 266–270. [CrossRef]
58. Lee, S.S.; Ried, R.P.; White, R.M. Piezoelectric cantilever microphone and microspeaker. *J. Microelectromechanical Syst.* **1996**, *5*, 238–242. [CrossRef]
59. Wang, H.; Godara, M.; Chen, Z.; Xie, H. A one-step residue-free wet etching process of ceramic PZT for piezoelectric transducers. *Sens. Actuators A Phys.* **2019**, *290*, 130–136. [CrossRef] [PubMed]
60. Jung, J.K.; Lee, W.J. Dry etching characteristics of Pb(Zr,Ti)O$_3$ films in CF$_4$ and Cl$_2$/CF$_4$ inductively coupled plasmas. *Jpn. J. Appl. Phys. Part. 1 Regul. Pap. Short Notes Rev. Pap.* **2001**, *40*, 1408–1419. [CrossRef]
61. Shung, K.K.; Cannata, J.M.; Zhou, Q.F. Piezoelectric materials for high frequency medical imaging applications: A review. *J. Electroceramics* **2007**, *19*, 139–145. [CrossRef]
62. Muralt, P. Recent progress in materials issues for piezoelectric MEMS. *J. Am. Ceram. Soc.* **2008**, *91*, 1385–1396. [CrossRef]
63. Muralt, P. PZT thin films for microsensors and actuators: Where do we stand? *IEEE Trans. Ultrason. Ferroelectr. Freq. Control* **2000**, *47*, 903–915. [CrossRef]
64. Ko, S.C.; Kim, Y.C.; Lee, S.S.; Choi, S.H.; Kim, S.R. Micromachined piezoelectric membrane acoustic device. *Sens. Actuators A Phys.* **2003**, *103*, 130–134. [CrossRef]
65. Shelton, S.; Chan, M.L.; Park, H.; Horsley, D.; Boser, B.; Izyumin, I.; Przybyla, R.; Frey, T.; Judy, M.; Nunan, K.; et al. CMOS-compatible AlN piezoelectric micromachined ultrasonic transducers. In Proceedings of the IEEE International Ultrasonics Symposium, Rome, Italy, 20–23 September 2009; pp. 402–405. [CrossRef]
66. Ait Aissa, K.; Achour, A.; Camus, J.; Le Brizoual, L.; Jouan, P.-Y.; Djouadi, M.-A. Comparison of the structural properties and residual stress of AlN films deposited by dc magnetron sputtering and high power impulse magnetron sputtering at different working pressures. *Thin Solid Films* **2014**, *550*, 264–267. [CrossRef]
67. Ababneh, A.; Schmid, U.; Hernando, J.; Sánchez-Rojas, J.L.; Seidel, H. The influence of sputter deposition parameters on piezoelectric and mechanical properties of AlN thin films. *Mater. Sci. Eng. B* **2010**, *172*, 253–258. [CrossRef]
68. Lim, W.T.; Lee, C.H. Highly oriented ZnO thin films deposited on Ru/Si substrates. *Thin Solid Film.* **1999**, *353*, 12–15. [CrossRef]
69. Han, C.-H.; Kim, E.S. Parylene-diaphragm piezoelectric acoustic transducers. In Proceedings of the Proceedings IEEE Thirteenth Annual International Conference on Micro Electro Mechanical Systems, Miyazaki, Japan, 23–27 January 2000; IEEE: Piscataway, NJ, USA, 2000; pp. 148–152.
70. Zhang, Y.; Du, G.; Liu, D.; Wang, X.; Ma, Y.; Wang, J.; Yin, J.; Yang, X.; Hou, X.; Yang, S. Crystal growth of undoped ZnO films on Si substrates under different sputtering conditions. *J. Cryst. Growth* **2002**, *243*, 439–443. [CrossRef]
71. CTS Incorporation. Available online: https://www.ctscorp.com/products/piezoelectric-move-products/speakers/ (accessed on 13 October 2021).
72. Li, J.; Wang, C.; Ren, W.; Ma, J. ZnO thin film piezoelectric micromachined microphone with symmetric composite vibrating diaphragm. *Smart Mater. Struct.* **2017**, *26*, 55033. [CrossRef]
73. Watanabe, S.; Fujiu, T.; Fujii, T. Effect of poling on piezoelectric properties of lead zirconate titanate thin films formed by sputtering. *Appl. Phys. Lett.* **1995**, *66*, 1481–1483. [CrossRef]
74. Cheng, J.-R.; Zhu, W.; Li, N.; Cross, L.E. Electrical properties of sol-gel-derived Pb(Zr$_{0.52}$Ti$_{0.48}$)O$_3$ thin films on a PbTiO$_3$-coated stainless steel substrate. *Appl. Phys. Lett.* **2002**, *81*, 4805–4807. [CrossRef]
75. Tsaur, J.; Wang, Z.J.; Zhang, L.; Ichiki, M.; Wan, J.W.; Maeda, R. Preparation and application of lead zirconate titanate (PZT) films deposited by hybrid process: Sol-gel method and laser ablation. *Jpn. J. Appl. Phys. Part. 1 Regul. Pap. Short Notes Rev. Pap.* **2002**, *41*, 6664–6668. [CrossRef]
76. Moriyama, M.; Totsu, K.; Tanaka, S. Sol–gel deposition and characterization of lead zirconate titanate thin film using different commercial sols. *Sens. Mater.* **2019**, *31*, 2497–2509. [CrossRef]
77. Gao, R.; Chu, X.; Huan, Y.; Sun, Y.; Liu, J.; Wang, X.; Li, L. A study on (K, Na) NbO$_3$ based multilayer piezoelectric ceramics micro speaker. *Smart Mater. Struct.* **2014**, *23*, 105018. [CrossRef]
78. Wang, Q.; Yi, Z.; Ruan, T.; Xu, Q.; Yang, B.; Liu, J. Obtaining high SPL piezoelectric MEMS speaker via a rigid-flexible vibration coupling mechanism. *J. Microelectromechanical Syst.* **2021**, *30*, 725–732. [CrossRef]
79. Wang, Y.-J.; Lo, S.; Hsieh, M.; Wang, S.; Chen, Y.; Wu, M.; Fang, W. Multi-way in-phase/out-of-phase driving cantilever array for performance enhancement of PZT MEMS microspeaker. In Proceedings of the 2021 IEEE 34th International Conference on Micro Electro Mechanical Systems (MEMS), Gainesville, FL, USA, 25–29 January 2021; IEEE: Piscataway, NJ, USA, 2021; pp. 83–84.
80. Diamond, B.M.; Neumann, J.J.; Gabriel, K.J. Digital sound reconstruction using arrays of CMOS-MEMS microspeakers. In Proceedings of the Technical Digest. MEMS 2002 IEEE International Conference on Micro Electro Mechanical Systems, Las Vegas, NV, USA, 24 January 2002; IEEE: Piscataway, NJ, USA, 2002; Volume 1, pp. 292–295.
81. Casset, F.; Dejaeger, R.; Laroche, B.; Desloges, B.; Leclere, Q.; Morisson, R.; Bohard, Y.; Goglio, J.P.; Escato, J.; Fanget, S. A 256 MEMS Membrane Digital Loudspeaker Array Based on PZT Actuators. *Procedia Eng.* **2015**, *120*, 49–52. [CrossRef]

82. Arevalo, A.; Conchouso, D.; Castro, D.; Kosel, J.; Foulds, I.G. Piezoelectric transducer array microspeaker. In Proceedings of the IEEE 11th Annual International Conference on Nano/Micro Engineered and Molecular Systems (NEMS), Sendai, Japan, 17–20 April 2016; pp. 180–183. [CrossRef]

83. Chen, Y.C.; Liu, W.T.; Chao, T.Y.; Cheng, Y.T. An optimized Cu-Ni nanocomposite coil for low-power electromagnetic microspeaker fabrication. In Proceedings of the TRANSDUCERS 2009—2009 International Solid-State Sensors, Actuators and Microsystems Conference, Denver, CO, USA, 21–25 June 2009; pp. 25–28. [CrossRef]

84. Shahosseini, I.; Lefeuvre, E.; Moulin, J.; Woytasik, M.; Martincic, E.; Pillonnet, G.; Lemarquand, G. Electromagnetic MEMS microspeaker for portable electronic devices. *Microsyst. Technol.* **2013**, *19*, 879–886. [CrossRef]

85. Majlis, B.Y.; Sugandi, G.; Noor, M.M. Compact electrodynamics MEMS-speaker. In Proceedings of the 2017 China Semiconductor Technology International Conference (CSTIC), Shanghai, China, 12–13 March 2017; IEEE: Piscataway, NJ, USA, 2017; pp. 1–3.

86. Chen, Y.C.; Cheng, Y.T. A low-power milliwatt electromagnetic microspeaker using a PDMS membrane for hearing aids application. In Proceedings of the IEEE 24th International Conference on Micro Electro Mechanical Systems, Cancun, Mexico, 23–27 January 2011; pp. 1213–1216. [CrossRef]

87. Murarka, A.; Lang, J.H.; Bulovic, V. Printed membrane electrostatic MEMS microspeakers. In Proceedings of the IEEE 29th International Conference on Micro Electro Mechanical Systems (MEMS), Shanghai, China, 24–28 January 2016; pp. 1118–1121. [CrossRef]

88. Kim, H.; Astle, A.A.; Najafi, K.; Bernal, L.P.; Washabaugh, P.D.; Cheng, F. Bi-directional electrostatic micro speaker with two large-deflection flexible membranes actuated by single/dual electrodes. In Proceedings of the SENSORS, 2005 IEEE, Irvine, CA, USA, 30 October–3 November 2005; pp. 89–92. [CrossRef]

89. Bao, M.; Yang, H. Squeeze film air damping in MEMS. *Sens. Actuators A Phys.* **2007**, *136*, 3–27. [CrossRef]

90. Glacer, C.; Dehé, A.; Tumpold, D.; Laur, R. Silicon microspeaker with out-of-plane displacement. In Proceedings of the IEEE International Conference on Nano/micro Engineered & Molecular Systems, Waikiki Beach, HI, USA, 13–16 April 2014.

91. Arevalo, A.; Castro, D.; Conchouso, D.; Kosel, J.; Foulds, I.G. Digital electrostatic acoustic transducer array. In Proceedings of the IEEE 11th Annual International Conference on Nano/Micro Engineered and Molecular Systems (NEMS), Sendai, Japan, 17–20 April 2016; pp. 225–228. [CrossRef]

92. Sano, C.; Ataka, M.; Hashiguchi, G.; Toshiyoshi, H. An electret-augmented low-voltage MEMS electrostatic out-of-plane actuator for acoustic transducer applications. *Micromachines* **2020**, *11*, 267. [CrossRef] [PubMed]

93. Conrad, H.; Schenk, H.; Kaiser, B.; Langa, S.; Gaudet, M.; Schimmanz, K.; Stolz, M.; Lenz, M. A small-gap electrostatic micro-actuator for large deflections. *Nat. Commun.* **2015**, *6*, 10078. [CrossRef] [PubMed]

94. Tian, H.; Ren, T.-L.; Xie, D.; Wang, Y.-F.; Zhou, C.-J.; Feng, T.-T.; Fu, D.; Yang, Y.; Peng, P.-G.; Wang, L.-G.; et al. Graphene-on-paper sound source devices. *ACS Nano* **2011**, *5*, 4878–4885. [CrossRef] [PubMed]

95. Wang, D.; He, X.; Zhao, J.; Jin, L.; Ji, X. Research on the electrical-thermal-acoustic conversion behavior of thermoacoustic speakers based on multilayer graphene film. *IEEE Sens. J.* **2020**, *20*, 14646–14654. [CrossRef]

 *micromachines*

Article

Temperature Stable Piezoelectric Imprint of Epitaxial Grown PZT for Zero-Bias Driving MEMS Actuator Operation

Marco Teuschel, Paul Heyes , Samu Horvath, Christian Novotny and Andrea Rusconi Clerici *

USound GmbH, 1100 Vienna, Austria
* Correspondence: andrea.rusconi@usound.com

Abstract: In piezoelectric transducer applications, it is common to use a unipolar operation signal to avoid switching of the polarisation and the resulting nonlinearities of micro-electromechanical systems. However, semi-bipolar or bipolar operation signals have the advantages of less leakage current, lower power consumption and no additional need of a DC−DC converter for low AC driving voltages. This study investigates the potential of using piezoelectric layers with an imprint for stable bipolar operation on the basis of epitaxially grown lead zirconate titanate cantilevers with electrodes made of a metal and metal oxide stack. Due to the manufacturing process, the samples exhibit high crystallinity, rectangular shaped hysteresis and a high piezoelectric response. Furthermore, the piezoelectric layers have an imprint, indicating a strong built-in field, which shifts the polarisation versus electric field hysteresis. To obtain the stability of the imprint, laser doppler vibrometry and switching current measurements were performed at different temperatures, yielding a stable imprinted electric field of −1.83 MV/m up to at least 100 °C. The deflection of the cantilevers was measured with a constant AC driving voltage while varying the DC bias voltage to examine the influence of the imprint under operation, revealing that the same high deflection and low nonlinearities, quantified by the total harmonic distortion, can be maintained down to low bias voltages compared to unipolar operation. These findings demonstrate that a piezoelectric layer with a strong imprint makes it possible to operate with low DC or even zero DC bias, while still providing strong piezoelectric response and linear behaviour.

Keywords: MEMS; speaker; PZT; imprint; bipolar driving

Citation: Teuschel, M.; Heyes, P.; Horvath, S.; Novotny, C.; Rusconi Clerici, A. Temperature Stable Piezoelectric Imprint of Epitaxial Grown PZT for Zero-Bias Driving MEMS Actuator Operation. *Micromachines* **2022**, *13*, 1705. https://doi.org/10.3390/mi13101705

Academic Editor: Huikai Xie

Received: 1 September 2022
Accepted: 6 October 2022
Published: 10 October 2022

Publisher's Note: MDPI stays neutral with regard to jurisdictional claims in published maps and institutional affiliations.

1. Introduction

Piezoelectric transducers have been used for decades for a wide variety of applications, such as energy harvesting [1], micropumps [2], electro-optical modulators [3], sensors [4] and actuators [5]. The expectations of micro-electromechanical systems (MEMS) in terms of cost, linearity and performance are constantly increasing. Due to its superior piezoelectric properties, in particular the strong electromechanical coupling, $Pb(Zr_\chi Ti_{1-\chi})O_3$ (PZT) is a popular choice for the active layer material [6]. Thin film technologies are needed for submicron-thick PZT layers to enable low operation voltages. The most commonly used deposition technique is the sol-gel process due to its simplicity and low manufacturing costs [7]. However, by depositing PZT layers via a sputtering or pulsed laser process, high crystallinity, low impurity content and defined interfaces can be achieved [8,9]. These two processes allow controlling the stoichiometric properties of the PZT layer by its underlaying layers. The seed layer and the electrode stack on which the piezoelectric layer is grown play an important role for the physical properties such as crystallinity, piezoelectric constant and imprint [10,11]. The imprint is the property that one polarisation state is more likely than the other one; thus, electric fields of different strengths are needed to switch the polarisation from one state to the other. This asymmetric effect is reflected by a shifted polarisation versus electric field (P/E) hysteresis loop. While the mechanism of the imprint is not yet fully understood, it has been attributed to interface effects between the piezoelectric layer and the electrode stack which lead to self-polarisation [12,13]. The literature offers various explanations for the underlying mechanisms of self-polarisation, such as flexoelectricity where a strain gradient induces an electric field [14,15], charge defects between the bottom electrode and the PZT layer [16,17] and asymmetric Schottky barriers created by using different electrode materials or stacks which could lead to an internal electric field [18,19].

The property of an imprint can be useful for MEMS applications, since complicated poling procedures can be avoided [19]. Furthermore, it is usual for MEMS actuators to use a unipolar operation signal by applying a DC bias to circumvent switching of polarisation and hysteresis behaviour, which leads to increased nonlinear behaviour and energy loss [20]. The effect of the imprint allows overcoming these problems due to a shifted hysteresis loop, making semi-bipolar or bipolar operation possible, which is shown in this work. In addition, the temperature stability and benefit of the self-polarisation regarding the operation signal of a MEMS device is explained. For this purpose, the strength of the built-in electric field was determined by the shift of the P/E hysteresis loop. To understand the stability of the imprint, cantilever deflection measurements with a laser doppler vibrometer and switching current measurements were performed at different temperatures. To substantiate the benefit of the imprint, different bias voltages and a constant AC driving signal were applied to the cantilever to see the difference regarding performance.

2. Materials and Methods

For the experiments, MEMS devices with cantilever structures were used. Cantilever structures often form the foundation of MEMS applications, and due to their simplicity, analytical and numerical equations can be used to compare different designs [21]. PZT films of 2 μm thickness were grown epitaxially by sputtering on an electrode stack using silicon as the wafer material. As seed layer yttria-stabilized zirconia was used underneath the bottom electrode stack, consisting of one layer each of platinum and $SrRuO_3$ with a total stack thickness of 150 nm. The 200 nm thick top electrode was made of $SrRuO_3$. On top of the electrode is an insulation and protective layer stack made of metal oxides, metals and a 45 μm thick polymer layer. Each of the manufactured devices consist of six trapezoidal cantilevers with a total area of 4 mm^2, which are electrically connected in parallel. Mechanically, they are connected at their tips with a piston (Figure S1 in Supplementary Materials). This design enables the central part of the device to move vertically in Z direction, compared to the bending motion of just a cantilever. The advantage of this specific configuration is to maximise the actuator force, elongation and linearity, chosen especially for audio applications.

To analyse the switching behaviour, which indicates the strength of the imprint of the piezoelectric material, current measurements were performed using a source meter (2450 SourceMeter, Keithley Instruments, Cleveland, OH, USA). The samples were initially depolarised by a decaying bipolar signal to assure a defined state [22]. For this purpose, a 10 V peak AC signal with a frequency of 1 kHz was applied, and the signal reduced in 1 V steps every second. After the depolarisation, the samples were subjected to a DC voltage that was ramped down from 30 V to −30 V and back up to 30 V again, during which the electric current was measured. The voltage steps were fixed to 250 mV with a delay time of 10 ms between the voltage step and the current acquisition. It should be noted that the switching current amplitudes depend on the measurement delay and the chosen voltage steps. Since the total generated surface charges caused by the dipole switching is constant for a given device and the electrical current is defined as charges per time, the measuring speed changes the measured current amplitude. In addition, if wider voltage steps are chosen, more displacement happens during one step which leads to more charges and results in a higher switching current. For these reasons, the voltage steps and the delay time were kept constant to compare the measurements. The measured electric current is the sum of the displacement current of the dipole switching, the leakage current and the charging current of the capacitor. The leakage current could be neglected due to the high electrical resistance of piezoceramics and the lower applied electric field in comparison to the breakdown field [23,24]. The decaying charging current was visible at the first few voltage steps but became negligible for the voltage region of interest where the dipole switching occurred. To focus on the switching current of the dipoles the charging current was cut off (Figure S2 in Supplementary Materials) [25]. In addition, P/E hysteresis loops were measured with a Sawyer−Tower circuit and an oscilloscope (DSOX1204G, Keysight, Santa Rosa, CA, USA), applying a 30 V peak AC signal at 1 kHz to the setup using a function generator (PSV 500, Polytec, Waldbronn, Germany) in combination with an amplifier (2105 gradient amplifier, AE Techron, Elkhart, IN, USA), in order to determine the imprint [26]. The working principle of the Sawyer−Tower circuit is explained in Supplementary Materials (Section 4). The amplitude of the imprint $E_{imprint}$ was calculated by the arithmetic mean of the negative coercive field E_{c-} and the positive coercive field E_{c+} [27]:

$$E_{imprint} = \frac{E_{c+} - |E_{c-}|}{2} . \tag{1}$$

In order to assess the temperature stability of the devices, measurements of the hysteresis loops and switching currents were performed at different temperatures. In addition, the piston deflection was used as a measure of performance of the fabricated cantilever structures in this research. The setup used to measure deflection and hysteresis loops under temperature is shown in Figure 1. The MEMS devices were fixed to a hotplate (Polyimide heating film, Thermo Tech, Rohrbach, Germany), with a thin metal plate between the hotplate and the device to ensure homogeneous heating. Micropositioner needles (XYZ 300 TR, Quarter Research & Development, Bend, OR, USA) used to electrically contact the sample provided sufficient pressure to generate good contact between the MEMS and the plate beneath. The hotplate was powered using a DC power supply (PWS4323, Tektronix, Berkshire, UK), while the temperature was measured by a self-sticking thermocouple type k. To evaluate the impact of the temperature on the MEMS performance, piston deflection measurements were performed using a laser doppler vibrometer (PSV 500, Polytec, Waldbronn, Germany), applying a 1 kHz 10 V peak AC signal with a 10 V DC bias for a duration of 10 min at room temperature, at 100 °C and again at room temperature, respectively. Measurements were made in 1 min intervals. To ensure a temperature equilibrium, a 30 min stabilization time was chosen before measurements at each different temperature. The 10 V DC bias was added to prevent any switching from occurring during the measurements, considering the results shown in this work. Switching current measurements were performed using the source meter as described in the previous paragraph under the same three temperature conditions, to analyse the influence of the temperature on the switching behaviour.

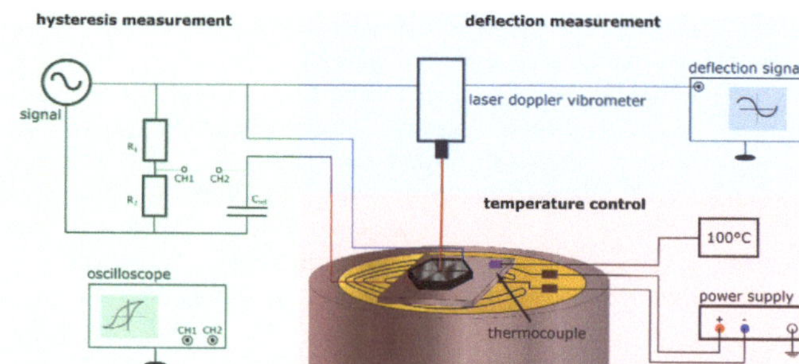

Figure 1. Setup for measuring the polarisation versus the electric field hysteresis loop of the piezoelectric device with a Sawyer−Tower circuit (green). As a measure of performance, the deflection of the MEMS was measured by a laser doppler vibrometer (blue). To obtain the temperature stability of the used devices, a hotplate was heated up by a DC power supply, and the temperature monitored by a thermocouple (red). Micropositioners are not shown for visibility.

To reveal the advantages of a strong imprint regarding possible operating signals, the piston deflection of the MEMS devices was measured at room temperature using a laser doppler vibrometer, applying different DC bias voltages with a constant 10 V peak AC driving voltage at 1 kHz. The applied DC voltage was reduced from 10 V DC to -9 V DC in 1 V intervals. These operation signals revealed the MEMS behaviour at unipolar, semi-bipolar and bipolar actuation. As a quantitative measure of nonlinearity, the total harmonic distortion (THD), which represents the ratio between the sum of displacement of the first five higher harmonics and the displacement of the fundamental actuation frequency, was calculated as:

$$THD_{\%} = \frac{\sqrt{\sum_{n=1}^{5} A_n^{2}}}{A_0} \cdot 100 \, ,$$

(2)

where A_n are the amplitudes of the higher harmonics of the fundamental actuation frequency and A_0 is the amplitude of the fundamental actuation frequency [28,29].

3. Results and Discussion

The bell-shaped switching current and the polarisation hysteresis loop from a device measured at room temperature are shown in Figure 2. Measurements of other devices showed the same characteristics, the results thereof are listed in Supplementary Materials (Sections 3 and 4). By using Equation (1) for the shifted hysteresis loop shown in Figure 2, a built-in field of -1.83 MV/m results. The imprint of the measured switching current can be calculated analogously, where the coercive fields are given by the position of the two current peaks. This gives a value of -2.25 MV/m for the built-in electric field. The calculated imprints are in the range of other reported values [13,14,27]. The mismatch of the coercive fields could be explained by the different characteristics of the presented measurement methods. The hysteresis was measured with a true AC signal with a frequency of 1 kHz, while the switching current was measured quasi-statically by a DC voltage ramp and a delay time between the steps for high current resolution. An investigation in the frequency dependence of the methods showed that the frequency using the Sawyer−Tower circuit had no appreciable influence on the coercive fields between 10 Hz and 1 kHz (Section 6 in Supplementary Materials). This extends the stable range of 15 to 200 Hz reported by Liu et al. [30], based on $BaTiO_3$/BTO samples. However, the imprint estimated from the peak positions of the switching current was found to vary by roughly 0.5 MV/m when varying the duration of an entire IV loop between 5 and 1000 s (Section 6 in Supplementary Materials). The reason for this dependency on the measurement speed could lie in the switching dynamics of the measured devices; however, a detailed investigation escapes the scope of this work. Furthermore, the individual switching current peaks display a degree of asymmetry, which affects how accurate the estimation of the coercive field is using the peak position.

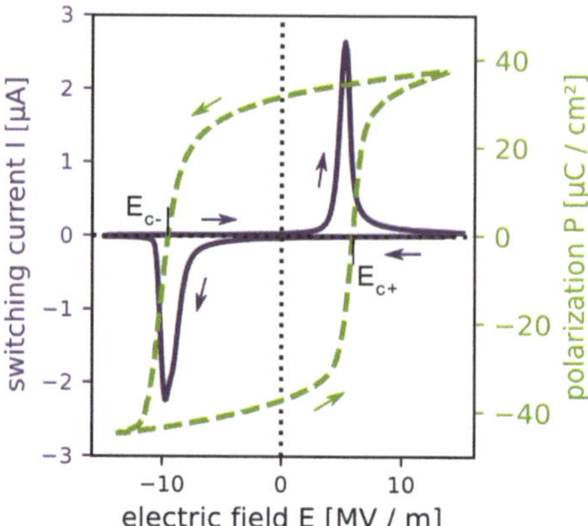

Figure 2. Shifted switching current (solid violet) and polarisation P (dashed green) as a function of the applied electrical field E to the PZT structures revealing a built-in electrical field, also called an imprint. The measurement direction is indicated by arrows. The strength of the imprint can be calculated by the negative (E_{c-}) and positive (E_{c+}) coercive fields.

The impact the temperature has on the switching behaviour can be seen in Figure 3a, showing the switching currents of a measured device at room temperature, at 100 °C and again at room temperature after exposure to 100 °C. Lower voltages are needed to switch the dipoles at higher temperature compared to room temperature. The imprint reduces from −2.25 MV/m at room temperature to a value of −1.62 MV/m at 100 °C. These results fit to the work of Pintilie et al. [31], where they recorded a decrease of the built-in electric field at higher temperatures. At higher temperatures, the Schottky barrier height decreases which leads to a decrease of the built-in electric field [31,32]. Akkopru-Akgun et al. [27] studied the mechanism and origin of the imprint with Nb-doped and Mn-doped PZT films. They hypothesized that charges from the electrodes are injected and trapped in the interface region due to Schottky emission, which could be the origin of the imprint. This coupling between the Schottky emission and the built-in electric field could be an explanation for the reduction of the imprint at higher temperature in this work. Furthermore, the I/V curve at 100 °C shows that the peak heights of the switching current are asymmetrical. This effect was also reported in the work of Chirila et al. [33], where PZT/SRO/STO/(Si) structures showed an increasing asymmetrical behaviour with increasing temperature. They observed that the dielectric constant has an asymmetric voltage dependency regarding the measuring voltage. This suggests that more free charge carriers are present on one PZT−electrode interface than on the other. These free charge carriers compensate the polarisation charges, which is reflected by a lower switching current [33]. Due to epitaxial growth, the defect density of the bottom interface is lower than the top one, which could cause the asymmetical current peaks [31]. The bell-shaped peak at positive applied field diverges, but the area under the peaks is of the same order of magnitude. Nevertheless, the switching characteristics recover after the high temperature to the same as initially measured, which shows the stability of the imprint. The deflection measurement results shown in Figure 3b agree with the data from the switching behaviour. The peak deflection after the high temperature equals the initial deflection. At 100 °C the peak deflection is higher than at room temperature, which is explained by the softening of the polymeric insulation layer on the cantilevers and the connectors between the cantilever tips and the piston with temperature [34]. The shift of the resonance frequency at different temperatures, which implies a softening of the polymer, is in accordance with the higher deflection with temperature (Section 7 in Supplementary Materials).

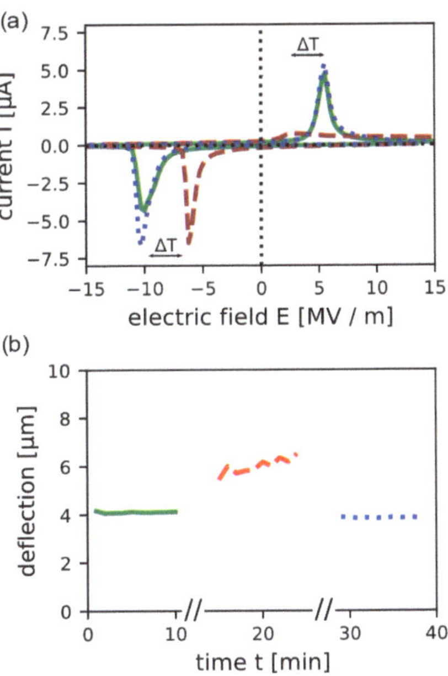

Figure 3. Temperature-dependent switching behaviour of the piezoelectric PZT layer and the deflection of the MEMS device. Measurements at room temperature (solid green line) as the initial, at 100 °C (dashed red line) and at room temperature after heating the sample (dotted blue line) were performed. (**a**) The switching currents as a function of the electric field where a decrease of the imprint at higher temperature is observable; (**b**) the peak deflection before, at and after heating up the sample as a function of time.

The piston deflection signals measured at different DC bias voltages are shown in Figure 4a. The nonlinearities start to rise when the applied voltage is in the range of the switching voltage. Outside of the switching voltage range, the applied bias voltage does not have an appreciable influence on the MEMS behaviour. In Figure 4b, the peak-peak deflection amplitude and the THD of the deflection signal, calculated using Equation (2), are shown as a function of the DC bias. At the operation signal of 10 V peak AC and -8 V DC, the negative electric field amplitude $-E_{max}$ has a value of -8 MV/m, which is in the range of the negative coercive field $-E_c$. Consequently, the dipoles begin to switch polarisation. When lowering the DC bias past the switching behaviour, the applied voltage signal and the deflection signal are in phase. Figure 4c,d shows the measured MEMS device in the laser doppler setup, where the six trapezoidal cantilevers are connected with a central piston.

In summary, the imprint of fabricated PZT ferroelectric thin films and its temperature stability were studied. Due to the epitaxial growth of the thin films, a built-in electric field of -1.83 MV/m occurred in the ferroelectric layer. The imprint could originate from internal strain-gradients which induce an electric field by the flexoelectric effect, charge defects at the PZT interfaces or from the asymmetric electrodes. Further investigations must be made to distinguish the emergence of the built-in field. Nevertheless, the imprint of the fabricated MEMS devices investigated in this work was found to have a temperature dependence and to recover to its initial value after cooling down. The present study showcases beneficial operation possibilities for MEMS with a strong built-in electric field. With an imprint bipolar, unipolar or semi-bipolar driving signals can be used for actuators without unwanted switching behaviour. By lowering the DC bias voltage, the leakage currents and the power consumption are reduced. In addition, if the amplitude of the driving voltage is lower than the battery voltage of the device, there is no need for additional DC−DC converters to generate the DC bias, which reduces cost and space.

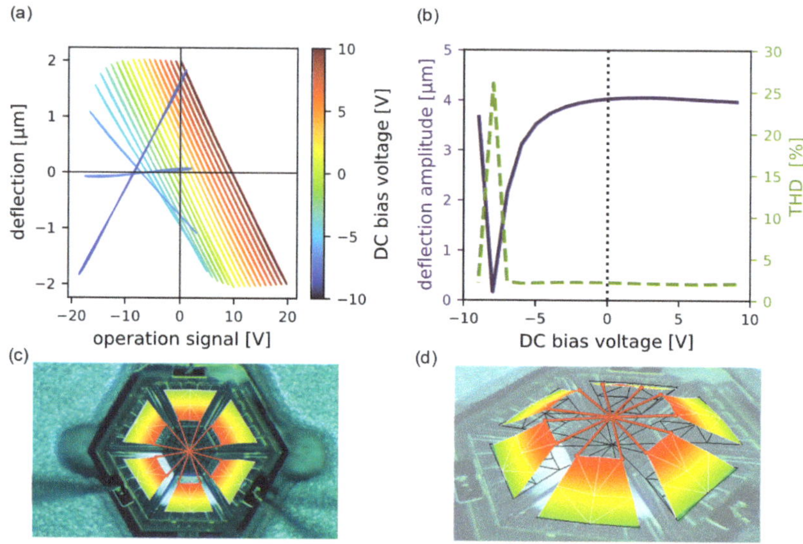

Figure 4. The performance of the MEMS device under different DC bias voltages and a constant applied AC driving voltage: (**a**) the deflection signals at different DC bias voltages; (**b**) the THD (dashed red line) and the peak deflection (solid blue line) as a function of the bias voltages; pictures from the laser doppler vibrometer of the measured MEMS device (**c**) and during operation (**d**).

Supplementary Materials: The following supporting information can be downloaded at: https://www.mdpi.com/article/10.3390/mi13101705/s1. See Supplementary Materials for further information about the fabricated MEMS structure, raw data of the measured switching curves and P/E hysteresis loops of five measured devices, the background of the Sawyer−Tower circuit for measuring the P/E hysteresis loops, further information on the laser doppler vibrometer measurements, the dependence of switching current and Sawyer−Tower measurements on the measurement speed and the temperature dependency of the MEMS resonance frequency. Refs. [22,23,26,34] have been cited in supplementary materials.

Author Contributions: Conceptualization, A.R.C., C.N. and M.T.; methodology, M.T. and P.H.; software, M.T., P.H. and S.H.; validation, A.R.C. and C.N.; formal analysis, A.R.C., C.N., P.H., S.H. and M.T.; investigation, M.T. and P.H.; resources, A.R.C.; data curation, M.T. and P.H.; writing—original draft preparation, M.T.; writing—review and editing, A.R.C., C.N., P.H., S.H. and M.T.; visualization, M.T. and P.H.; supervision, A.R.C. and C.N.; project administration, A.R.C.; All authors have read and agreed to the published version of the manuscript.

Funding: This research received no external funding.

Data Availability Statement: If any data is needed the corresponding author can be contacted.

Conflicts of Interest: The authors declare no conflict of interest.

References

1. Choi, W.J.; Jeon, Y.; Jeong, J.-H.; Sood, R.; Kim, S.G. Energy harvesting MEMS device based on thin film piezoectric cantilevers. *Electroceramics* **2006**, *17*, 543. [CrossRef]
2. Tian, X.; Wang, H.-g.; Wang, H.; Wang, Z.; Sun, Y.; Zhu, J.; Zhao, J.; Zhang, S.; Yang, Z. Design and test of a piezoelectric micropump based on hydraulic amplification. *AIP Adv.* **2021**, *11*, 065230. [CrossRef]
3. Li, M.; Ling, J.; He, Y.; Javid, U.A.; Xue, S.; Lin, Q. Lithium niobate photonic-crystal electro-optic modulator. *Nat. Commun.* **2020**, *11*, 4123. [CrossRef]

4. Shi, M.; Holmes, A.S.; Yeatman, E.M. Piezoelectric wind velocity sensor based on the variation of galloping frequency with drag force. *Appl. Phys. Lett.* **2020**, *116*, 264101. [CrossRef]

5. Hu, Y.; Lin, S.; Ma, J.; Zhang, Y.; Li, J.; Wen, J. Piezoelectric inertial rotary actuator operating in two-step motion mode for eliminating backward motion. *Appl. Phys. Lett.* **2020**, *117*, 031902. [CrossRef]

6. Shilpa, G.D.; Sreelakshmi, K.; Ananthaprasad, M.G. PZT thin film deposition techniques, properties and its application in ultrasonic MEMS sensors: A review. *IOP Conf. Ser. Mater. Sci. Eng.* **2016**, *149*, 012190. [CrossRef]

7. Wang, X.; Wang, F.; Guo, L.Q.R.; Li, B.; Chen, D.; Zou, H. Orientation transition, dielectric, and ferroelectric behaviors of sol-gel derived PZT thin films deposited on Ti–Pt alloy layers: A Ti content-dependent study. *Ceram* **2020**, *46*, 10256. [CrossRef]

8. Ma, Y.; Song, J.; Wang, X.; Liu, Y.; Zhou, J. Synthesis, Microstructure and Properties of Magnetron Sputtered Lead Zirconate Titanate (PZT) Thin Film Coatings. *Coatings* **2021**, *11*, 944. [CrossRef]

9. Nguyen, M.D. Ferroelectric and Piezoelectric properties of epitaxial PZT films and devices on silicon. Ph.D. Thesis, University of Twente, Twente, The Netherlands, 2010.

10. Nguyen, M.D.; Nazeer, H.; Dekkers, M.; Blank, D.H.A.; Rijnders, G. Optimized electrode coverage of membrane actuators based on epitaxial PZT thin films. *Smart Mater. Struct.* **2013**, *22*, 085013. [CrossRef]

11. Pandey, S.K.; James, A.R.; Prakash, C.; Goel, T.C.; Zimik, K. Electrical properties of PZT thin films grown by sol-gel and PLD using a seed layer. *Mater. Sci. Eng. B* **2004**, *112*, 96. [CrossRef]

12. Araujo, E.B.; Lima, E.C.; Bdikin, I.K.; Kholkin, A.L. Imprint effect in PZT thin films at compositions around the morphotropic phase boundary. *Ferroelectrics* **2016**, *498*, 18. [CrossRef]

13. Lee, J.; Choi, C.H.; Park, B.H.; Noh, T.W.; Lee, J.K. Built-in voltages and asymmetric polarization switching in Pb(Zr,Ti)O3 thin film capacitors. *Appl. Phys. Lett.* **1998**, *72*, 3380. [CrossRef]

14. Catalan, G.; Lubk, A.; Vlooswijk, A.H.G.; Snoeck, E.; Magen, C.; Janssens, A.; Rispens, G.; Rijnders, G.; Blank, D.H.A.; Noheda, B. Flexoelectric rotation of polarization in ferroelectric thin films. *Nat. Mater.* **2011**, *10*, 963. [CrossRef] [PubMed]

15. Gruverman, A.; Rodriguez, B.J.; Kingon, A.I.; Nemanich, R.J.; Tagantsev, A.K.; Cross, J.S.; Tsukada, M. Mechanical stress effect on imprint behavior of integrated ferroelectric capacitors. *Appl. Phys. Lett.* **2003**, *83*, 728. [CrossRef]

16. Afanasjev, V.P.; Petrov, A.A.; Pronin, I.P.; Tarakanov, E.A.; Kaptelov, E.J.; Graul, J. Polarization and self-polarization in thin PbZr$_{1-x}$Ti$_x$O$_3$ (PZT) films. *J. Phys. Condens. Mat.* **2001**, *13*, 8755. [CrossRef]

17. Misirlioglu, I.B.; Okatan, M.B.; Alpay, S.P. Asymmetric hysteresis loops and smearing of the dielectric anomaly at the transition temperature due to space charges in ferroelectric thin films. *J. Appl. Phys.* **2010**, *108*, 034105. [CrossRef]

18. Qu, T.L.; Zhao, Y.G.; Xie, D.; Shi, J.P.; Chen, Q.P.; Ren, T.L. Resistance switching and white-light photovoltaic effects in BiFeO3/Nb–SrTiO3 heterojunctions. *Appl. Phys. Lett.* **2011**, *98*, 173507. [CrossRef]

19. Kholkin, A.L.; Brooks, K.G.; Taylor, D.V.; Hiboux, S.; Setter, N. Self-polarization effect in Pb(Zr,Ti)O3 thin films. *Integr. Ferroelectr.* **1998**, *22*, 525. [CrossRef]

20. Lucke, P.; Bayraktar, M.; Schukkink, N.; Yakshin, A.E.; Rijnders, G.; Bijkerk, F.; Houwman, E.P. Influence of DC Bias on the Hysteresis, Loss, and Nonlinearity of Epitaxial PbZr$_{0.55}$Ti$_{0.45}$O$_3$ Films. *Adv. Electron. Mater.* **2021**, *7*, 2100115. [CrossRef]

21. Priyadarsini, S.; Das, J.K.; Dastidar, A. Analysis of MEMS cantilever geometry for designing of an array sensor. *Int. Conf. Signal Processing Commun. Power Embed. Syst. (SCOPES)* **2016**, *1*, 625.

22. Delimova, L.A.; Yuferev, V.S. Transient carrier transport and rearrangement of space charge layers under the bias applied to ferroelectric M/PZT/M structures. *J. Phys. Conf. Ser.* **2019**, *1400*, 055003. [CrossRef]

23. Nie, C.; Chen, X.F.; Feng, N.B.; Wang, G.S.; Dong, X.L.; Gu, Y.; He, H.L.; Liu, Y.S. Effect of external fields on the switching current in PZT ferroelectric ceramics. *Solid State Commun.* **2010**, *150*, 101. [CrossRef]

24. Balke, N.; Granzow, T.; Rödel, J. Degradation of lead-zirconate-titanate ceramics under different dc loads. *J. Appl. Phys.* **2009**, *105*, 104105. [CrossRef]

25. Hu, L.; Dalgleish, S.; Matsushita, M.M.; Yoshikawa, H.; Awaga, K. Storage of an electric field for photocurrent generation in ferroelectric-functionalized organic devices. *Nat. Commun.* **2014**, *5*, 3279. [CrossRef]

26. Hafner, J.; Benaglia, S.; Richheimer, F.; Teuschel, M.; Maier, F.J.; Werner, A.; Wood, S.; Platz, D.; Schneider, M.; Hradil, K.; et al. Multi-scale characterisation of a ferroelectric polymer reveals the emergence of a morphological phase transition driven by temperature. *Nat. Commun.* **2021**, *12*, 152. [CrossRef]

27. Akkopru-Akgun, B.; Zhu, W.; Lanagan, M.T.; Trolier-McKinstry, S. The effect of imprint on remanent piezoelectric properties and ferroelectric aging of PbZr$_{0.52}$Ti$_{0.48}$O$_3$ thin films. *J. Am. Ceram. Soc.* **2019**, *102*, 5328. [CrossRef]

28. Morris, D.J.; Youngsman, J.M.; Anderson, M.J.; Bahr, D.F. A resonant frequency tunable, extensional mode piezoelectric vibration harvesting mechanism. *Smart Mater. Struct.* **2008**, *17*, 065021. [CrossRef]

29. Uzun, Y.; Kurt, E.; Kurt, H.H. Explorations of displacement and velocity nonlinearities and their effects to power of a magnetically-excited piezoelectric pendulum. *Sens. Actuator A Phys.* **2015**, *224*, 119. [CrossRef]

30. Liu, F.; Fina, I.; Bertacco, R.; Fontcuberta, J. Unravelling and controlling hidden imprint fields in ferroelectric capacitors. *Sci. Rep.* **2016**, *6*, 25028. [CrossRef]

31. Pintilie, L.; Vrejoiu, I.; Hesse, D.; LeRhun, G.; Alexe, M. Ferroelectric polarization-leakage current relation in high quality epitaxial Pb(Zr,Ti)O$_3$ films. *Phys. Rev. B* **2007**, *75*, 104103. [CrossRef]

32. Xia, F.; Zhang, Q.M. Schottky emission at the metal polymer interface and its effecton the polarization switching of ferroelectric poly(vinylidenefluoride-trifluoroethylene) copolymer thin films. *Appl. Phys. Lett.* **2004**, *85*, 1719. [CrossRef]

33. Chirila, C.; Boni, A.G.; Pasuk, I.; Negrea, R.; Trupina, L.; Rhun, G.L.; Yin, S.; Vilquin, B.; Pintilie, I.; Pintilie, L. Comparison between the ferroelectric/electric properties of the $PbZr_{0.52}Ti_{0.48}O_3$ films grown on Si (100) and on STO (100) substrates. *J. Mater. Sci.* **2015**, *50*, 3883. [CrossRef]
34. Tsai, F.Y.; Blanton, T.N.; Harding, D.R.; Chen, S.H. Temperature dependence of the properties of vapor-deposited polyimide. *J. Appl. Phys.* **2003**, *93*, 3760. [CrossRef]

Article

Total Harmonic Distortion of a Piezoelectric MEMS Loudspeaker in an IEC 60318-4 Coupler Estimation Using Static Measurements and a Nonlinear State Space Model

Romain Liechti [1,2,*], Stéphane Durand [2], Thierry Hilt [1], Fabrice Casset [1], Christophe Poulain [1], Gwenaël Le Rhun [1], Franklin Pavageau [1], Hugo Kuentz [1] and Mikaël Colin [1]

[1]   University Grenoble Alpes, CEA, Leti, F-38000 Grenoble, France; thierry.hilt@cea.fr (T.H.); fabrice.casset@cea.fr (F.C.); christophe.poulain@cea.fr (C.P.); gwenael.le-rhun@cea.fr (G.L.R.); franklin.pavageau@cea.fr (F.P.); hugo.kuentz@cea.fr (H.K.); mikael.colin@cea.fr (M.C.)

[2]   Laboratoire d'Acoustique de l'Université du Mans, LAUM-UMR 6613 CNRS, Le Mans Université, F-72085 Le Mans, France; stephane.durand@univ-lemans.fr

*   Correspondence: Romain.Liechti@cea.fr

Abstract: We propose a method to evaluate the Total Harmonic Distortion generated by a cantilever-based PZT loudspeaker inside an IEC 60318-4 coupler. The model is validated using experimental data of a commercial loudspeaker. Using the time domain equations of the equivalent electrical circuit of the loudspeaker inside the coupler and a state space formulation, the acoustic pressure response is calculated and compared to the measurement of the manufacturer. Next, the stiffness, transduction and capacitance nonlinear functions are evaluated with a Double-Beam Laser Interferometer (DBLI) and a nanoindenter on test devices and on the commercial loudspeaker. By introducing the nonlinear functions into the model as amplitude-dependent parameters, the THD generated by the loudspeaker is calculated and compared to the value provided by the manufacturer. The good agreement between the measurement and the simulation could allow for a rather quick simulation of the performance of similarly designed loudspeakers at the early stage of the design, by only estimating the static linearity of the main nonlinearity sources.

Keywords: MEMS; piezoelectric; loudspeaker; total harmonic distortion; state space; IEC 711; thin film; PZT

Citation: Liechti, R.; Durand, S.; Hilt, T.; Casset F.; Poulain C.; Le Rhun G.; Pavageau F.; Kuentz H.; Colin M. Total Harmonic Distortion of a Piezoelectric MEMS Loudspeaker in an IEC 60318-4 Coupler Estimation Using Static Measurements and a Nonlinear State Space Model. *Micromachines* **2021**, *12*, 1437. https://doi.org/10.3390/mi12121437

Academic Editor: Libor Rufer

Received: 27 October 2021
Accepted: 20 November 2021
Published: 24 November 2021

Publisher's Note: MDPI stays neutral with regard to jurisdictional claims in published maps and institutional affiliations.

1. Introduction

Loudspeakers are used to converting an electrical signal into sound waves, as accurately as possible, with a sufficiently high sound pressure level. This is usually achieved with a piston-like membrane, behaving as a first-order oscillating system, actuated by an electromechanical transducer. Driven by the forever growing industry of portable and connected devices, research was carried out in order to make the loudspeaker compatible with microfabrication processes, in order to reduce the manufacturing tolerances, the cost and the thickness of the devices. An interesting performance was demonstrated with Micro Electromechanical Systems (MEMS) electrodynamic loudspeakers [1,2] and pieozoelectric loudspeakers [3–7]. Nevertheless, the achieved performance is not competitive compared to non-MEMS loudspeaker. However, as in-ear sound generation requires less mechanical displacement, MEMS loudspeakers have shown interesting results in terms of frequency response and sound pressure level when evaluated in couplers [8–10]. A few MEMS loudspeakers for in-ear applications have demonstrated to fulfill the industry required performances, with electrostatic [8] and piezoelectric transduction [9,10].

The second most important characteristic of a loudspeaker after its frequency response is its linearity. This characteristic is often evaluated using the Total Harmonic Distortion (THD). The THD is the ratio of the power of all harmonics over the total power. Studies have shown that depending on the order of the harmonics, the human ear can be sensitive to a THD as low as 0.1% [11–14]. Electrostatic transduction is inherently nonlinear, and the reachable displacement is not sufficient to advantageously replace non-MEMS electrodynamic loudspeakers. The most promising transduction is the piezoelectric transduction and more specifically the one using PZT (lead zirconate titanate) actuators, providing a high transduction factor for low actuation voltages in the case of thin films. Unfortunately, thin-film PZT shows a ferroelectric and electrostrictive behavior, creating a wide variety of nonlinearities [15–18].

To model the nonlinearities of classical electrodynamic loudspeakers, state space models, port-Hamiltonian systems, Hammerstein models and power series have been extensively used [19–25]. State space models are the most commonly used models to simulate nonlinear effects in electrodynamic loudspeaker and have shown accurate results. However, for piezoelectric transduction, such a model has not been reported yet. Recently, a model for the simulation of nonlinear effects in coupler for electrostatic transduction was reported [26]. Accurately, modeling the nonlinear behavior of loudspeakers from static nonlinear measurements of the transduction transfer function could be an efficient way for estimating the total harmonic distortion, without using expensive prototypes. In addition, an accurate model of the nonlinear behavior of the loudspeaker could help developing methods to actively reduce the nonlinearities, as it has been extensively performed with electrodynamic non-MEMS loudspeakers [27]. In addition, accurately simulating the nonlinearities generated by a loudspeaker could be used to simulate the subjective audio quality of the loudspeaker in the early phases of the design, using models of the perceptual components of the loudspeaker [28–31]. The aim of this study is to provide such a model to evaluate the distortion generated by a cantilever-based piezoelectric MEMS loudspeaker using static measurements and a lumped equivalent circuit in order to shorten the linearity optimization step of this type of loudspeakers in the design phase. Parameters are adapted to compare the results with the measurements of the Usound UT-P 2018, in order to validate the model.

2. Nonlinear State Space Model

As most loudspeakers behave as a simple mechanical oscillator for a wide range of frequencies, they can be well approximated as a first-order mechanical oscillator, two transformers and an electrical circuit. In the case of piezoelectric loudspeakers, the usual gyrator used for the electromechanical transduction is replaced by a transfomer, and the electrical circuit is a simple capacitor. The equivalent electrical circuit of the loudspeaker detailed in [32] is presented in Figure 1, where R_g is the output resistance of the amplifier and the resistance of the connection lines, C_p is the piezoelectric capacitance, γ is the transduction ratio of the piezoelectric layer, R_{ms} represents the viscous losses of the mechanical oscillator, M_{ms} is the moving mass of the loudspeaker, C_{ms} is the compliance of the loudspeaker and S_d is the effective radiating area of the membrane.

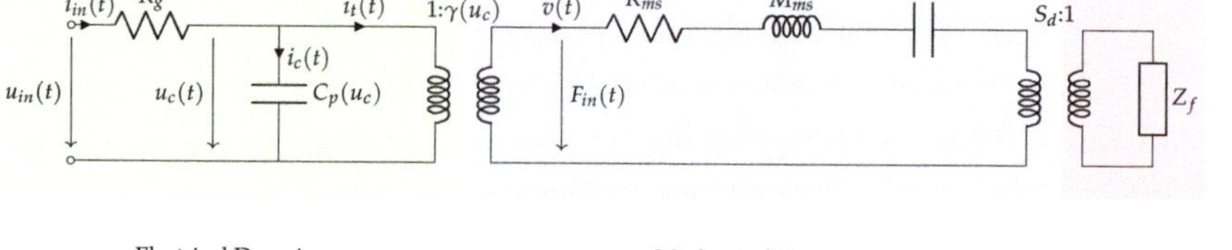

Figure 1. Equivalent electrical circuit of the loudspeaker.

In the case of cantilever-based piezoelectric loudspeakers, the transduction factor γ can be estimated using the unimorph piezoelectric cantilever model described in [33]. A schematic representation of a unimorph cantilever is depicted in Figure 2 with the relevant dimensions and axes. Considering the cantilever clamped on one end and guided on the other end and using a null displacement at the tip of the cantilever, the blocked force F_{bl} generated by the actuators as a function of the input voltage can be written:

$$F_{bl} = \frac{6Wt_p}{4s_{11}^p L} \frac{AB(B+1)}{AB+1} \left(1 - \frac{k_p}{k_p + k_a}\right) e_{31}(s_{11}^p + s_{12}^p)u_c \tag{1}$$

where W is the width of the actuator, t_p is the thickness of the piezoelectric layer, s_{11}^p and s_{12}^p are the elastic compliance of the piezoelectric layer at constant electric field in the 1 and 2 direction, L is the length of the cantilever, t_m is the thickness of the elastic layer, $A = s_{11}^p/s_{11}^m$, $B = t_m/t_p$, and k_a and k_p are the apparent stiffnesses of the active and passive parts of length L and L_p of the actuators. The transduction factor γ is the the ratio of the blocked force F_{bl} over the voltage u_c.

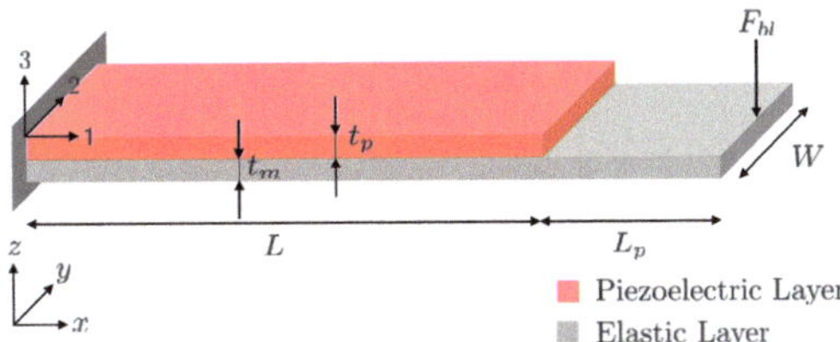

Figure 2. Schematic representation of a unimorph piezoelectric cantilever.

In our case, the load impedance Z_f is the acoustical impedance seen by the front side of the membrane and is equal to the acoustical impedance of the coupler. Due to the size of diaphragm of MEMS loudspeakers, which are mostly smaller than 1 cm^2, the back acoustic radiation impedance of the loudspeaker is not considered, due to its insignificant effect on the frequency response. A coupler is a mechanical device used to simulate the acoustical impedance of the average human ear, following the definition given in the IEC standard IEC 60318-4 [34]. The known electrical equivalent circuit is depicted in Figure 3.

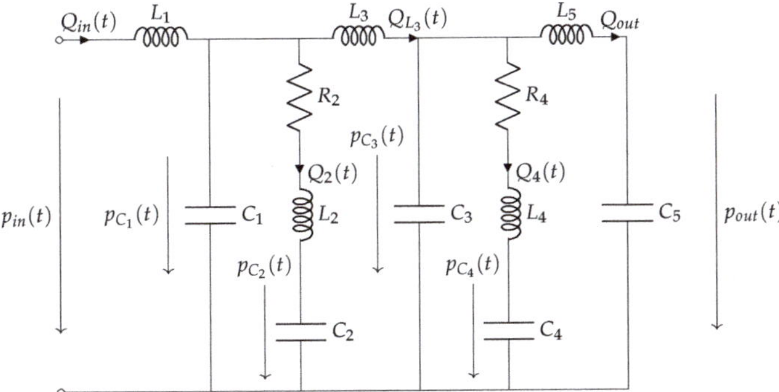

Figure 3. Lumped equivalent network of the coupler.

Using the circuit resulting of the combination of Figures 1 and 3, time domain differential equations can be derived. For the system to be fully defined, one state variable is needed for each capacitive or inductive element. The displacement x is also chosen, because it is used to evaluate the apparent stiffness of the actuators. The thirteen equations of the state variables are given in Appendix A. Equations (A1)–(A13) can be used in a matrix form and solved by using a differential approximation scheme [35]. The state space equation then writes:

$$\frac{\partial}{\partial t}\mathbf{x} = A\mathbf{x} + Bu_{in}(t) \tag{2}$$

with

$$\mathbf{x} = \begin{bmatrix} u_c(t) \\ v(t) \\ x(t) \\ F_{C_{ms}}(t) \\ p_{C_1}(t) \\ Q_2(t) \\ p_{C_2}(t) \\ Q_{L_3}(t) \\ p_{C_3}(t) \\ Q_4(t) \\ p_{C_4}(t) \\ Q_{out}(t) \\ p_{out}(t) \end{bmatrix}, \tag{3}$$

$$A = \begin{bmatrix}
-\frac{1}{R_g C_p(u_c)} & -\frac{\gamma(u_c)}{C_p(u_c)} & 0 & 0 & 0 & 0 & 0 & 0 & 0 & 0 & 0 & 0 & 0 \\
\frac{\gamma(u_c)}{M_{ms}+S_d^2 L_1} & -\frac{R_{ms}}{M_{ms}+S_d^2 L_1} & 0 & -\frac{1}{M_{ms}+S_d^2 L_1} & -\frac{S_d}{M_{ms}+S_d^2 L_1} & 0 & 0 & 0 & 0 & 0 & 0 & 0 & 0 \\
0 & 1 & 0 & 0 & 0 & 0 & 0 & 0 & 0 & 0 & 0 & 0 & 0 \\
0 & \frac{1}{C_{ms}(x)} & 0 & 0 & 0 & 0 & 0 & 0 & 0 & 0 & 0 & 0 & 0 \\
0 & \frac{S_d}{C_1} & 0 & 0 & 0 & -\frac{1}{C_1} & 0 & -\frac{1}{C_1} & 0 & 0 & 0 & 0 & 0 \\
0 & 0 & 0 & 0 & \frac{1}{L_2} & -\frac{R_2}{L_2} & -\frac{1}{L_2} & 0 & 0 & 0 & 0 & 0 & 0 \\
0 & 0 & 0 & 0 & 0 & \frac{1}{C_2} & 0 & 0 & 0 & 0 & 0 & 0 & 0 \\
0 & 0 & 0 & 0 & \frac{1}{L_3} & 0 & 0 & 0 & -\frac{1}{L_3} & 0 & 0 & 0 & 0 \\
0 & 0 & 0 & 0 & 0 & 0 & 0 & \frac{1}{C_3} & 0 & -\frac{1}{C_3} & 0 & -\frac{1}{C_3} & 0 \\
0 & 0 & 0 & 0 & 0 & 0 & 0 & 0 & \frac{1}{L_4} & -\frac{R_4}{L_4} & -\frac{1}{L_4} & 0 & 0 \\
0 & 0 & 0 & 0 & 0 & 0 & 0 & 0 & 0 & 0 & \frac{1}{C_4} & 0 & 0 \\
0 & 0 & 0 & 0 & 0 & 0 & 0 & 0 & \frac{1}{L_5} & 0 & 0 & 0 & -\frac{1}{L_5} \\
0 & 0 & 0 & 0 & 0 & 0 & 0 & 0 & 0 & 0 & 0 & \frac{1}{C_5} & 0
\end{bmatrix} \tag{4}$$

and

$$B = \begin{bmatrix} \frac{1}{R_g C_p(u_c)} \\ 0 \\ 0 \\ 0 \\ 0 \\ 0 \\ 0 \\ 0 \\ 0 \\ 0 \\ 0 \\ 0 \\ 0 \end{bmatrix}. \tag{5}$$

Equation (2) is of the form:

$$\dot{\mathbf{x}} = \mathbf{A}\mathbf{x} + \mathbf{B}\mathbf{u} \tag{6}$$

and thus, matrices can be discretized using a bilinear transform:

$$\mathbf{A}_d = \left(\mathbf{I} - \frac{1}{2}\mathbf{A}T_s\right)^{-1}\left(\mathbf{I} + \frac{1}{2}\mathbf{A}T_s\right) \tag{7}$$

and

$$\mathbf{B}_d = \left(\mathbf{I} - \frac{1}{2}\mathbf{A}T_s\right)^{-1}\mathbf{B}T_s \tag{8}$$

with $T_s = 1/f_s$ the sampling period. Using matrices A_d and B_d and an input signal, the output can be computed:

$$\mathbf{x}(k+1) = \mathbf{A}_d\mathbf{x}(k) + \mathbf{B}_d\mathbf{u}(k) \tag{9}$$

From the computed output pressure in the state vector $\mathbf{x}$, the THD, which is the ratio between the sum of effective values of the harmonics and the effective value of the fundamental harmonic and other harmonics, is calculated using:

$$\text{THD} = 100 \cdot \frac{\sqrt{\sum_{h=2}^{H} v_h^2}}{\sqrt{\sum_{h=1}^{H} v_h^2}} \tag{10}$$

3. Static Measurements

As other loudspeakers state space models, this model uses nonlinear functions as input parameters. The nonlinear functions can be measured dynamically using adaptive state space models, calculated analytically, computed with finite element models or determined from static measurement. In this paper, we propose to use static measurements of the nonlinear transfer functions, measured on a commercial loudspeaker [36] and on PZT test devices. Since internal dimensions of the loudspeaker are unknown and difficult to measure, the lumped element parameters of the equivalent circuit C_{ms}, S_d, M_{ms} and C_p are taken directly from the data sheet.

3.1. Nonlinear Electromechanical Function

The $e_{31,f}$ piezoelectric coefficient of thin-film PZT is extracted from the $d_{33,f}$ measurement, with the procedure described in [37], as a function of the applied electric field, using the DBLI, on a standard PZT test device. As the piezoelectric coefficient e_{31} is calculated for a piezoelectric material on wafer with this specific machine, it would be difficult to measure it on the loudspeaker directly. The parameter e_{31} is known to be highly dependent on the applied electric field, due to intrinsic and extrinsic various effects, such as electrostriction, ferroelectric behavior and the domain movements. With the complete physical interactions being not completely understood, these nonlinearities are often simulated using phenomenological models, whose parameters are deduced from measurements [38,39]. The parameter $e_{31,f}$ and the fitted logarithmic function are depicted in Figure 4.

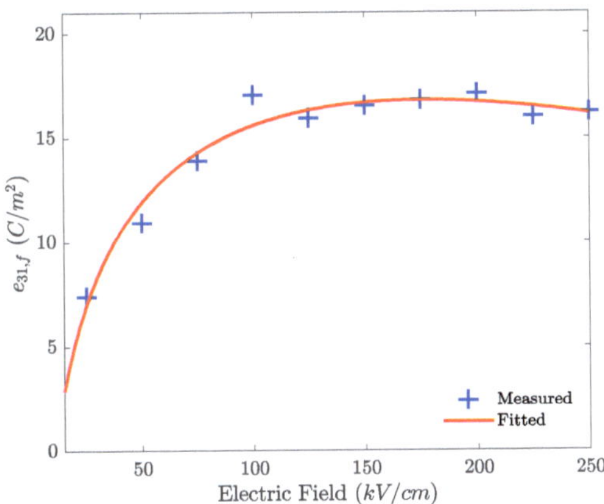

Figure 4. Measured and fitted function for the nonlinear transverse piezoelectric coefficient.

As it has been reported in multiple papers, the transversal piezoelectric coefficient of piezoelectric thin film increases for low electric fields, until it converges to the maximum value, due to the alignment of the domains of the material [40–44]. After the saturation of the rotation of the domains, the piezoelectric coefficient decreases slowly. The fitted function of $e_{31,f}$ is then used in Equation (1) to calculate the nonlinear transduction factor $\gamma(u_c)$. Depending on the actuators geometry, the fitted $e_{31,f}$ function can be used in any other electromechanical transduction function.

3.2. Nonlinear Stiffness

Using the setup depicted in Figure 5, the nonlinear relation between the force and the displacement of the commercial loudspeaker, not subjected to any voltage, is measured.

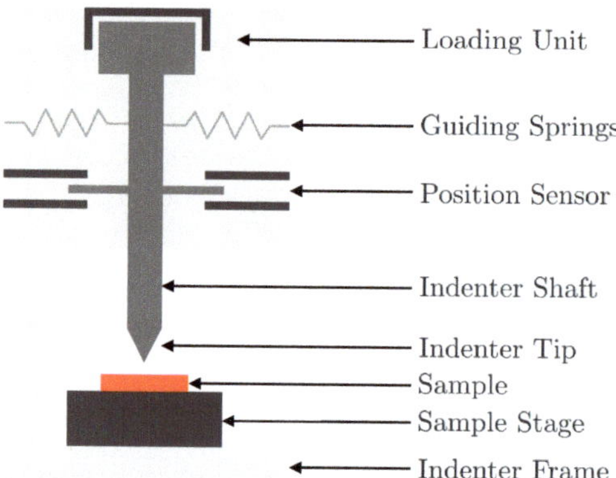

Figure 5. Schematic representation of a nanoindenter.

The commercial loudspeaker is placed on the sample loader. The nanoindenter, equipped with a spherical tip of radius 50 µm, applies a precise and well-known force on the moving part of the sample, until reaching the previously defined maximum displacement of 30 µm. After a processing of the raw data by the embedded software, the curve of the force applied on the sample and of the displacement of the shaft is retrieved and presented in Figure 6.

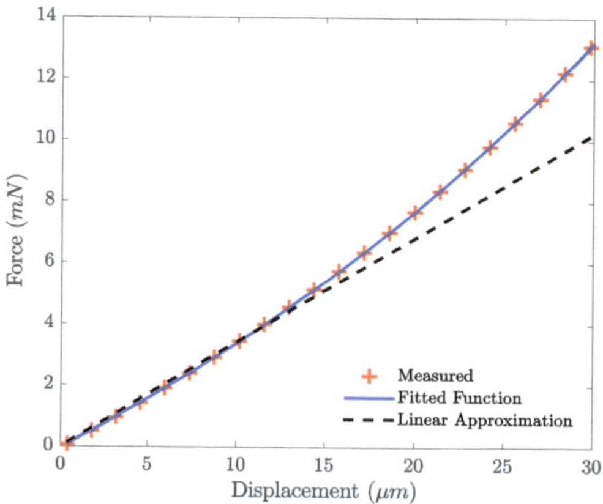

Figure 6. Nonlinear stiffness of the loudspeaker.

A third-order polynomial function, known to accurately represent the nonlinear behavior of clamped-guided structures, is used as the nonlinear fitting function. The linear approximation is calculated from the V_{AS}, which is the equivalent compliance volume value given in the loudspeaker data sheet, using:

$$C_{ms} = \frac{V_{AS}}{\rho_0 c_0^2 S_d^2} \tag{11}$$

with ρ_0 the density of air at room temperature, c_0 the speed of sound in air at room temperature and S_d the effective radiating surface of the loudspeaker [45]. From the fitted function, the nonlinear compliance $C_{ms}(x)$ can be calculated. This nonlinear function can also be calculated knowing the geometry of the loudspeaker and a finite element model or analytically.

3.3. Capacitance

As the PZT thin film is ferroelectric, the capacitance created by the piezoelectric layer between the electrodes varies with the applied electric field. The nonlinear relation was measured on a layer of PZT using the AixACCT TF Analyzer 2000. The measured variation of the capacitance and the fitted Lorentz function are presented in Figure 7.

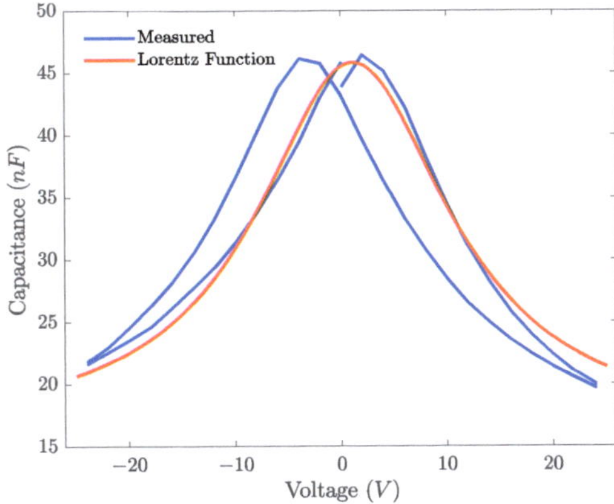

Figure 7. Non linear capacitance of the loudspeaker.

From the fitted Lorentzian function, the nonlinear capacitance $C_p(u_c)$ can be calculated. The curves can be adjusted in order to match the value of the data sheet of the loudspeaker.

4. Results

Using Equation (9) and a scaled dirac whose amplitude at the first sample is $f_s u_p/2$ as an input signal, the linear frequency response of the loudspeaker inside the coupler is computed. The computed frequency response is compared to the frequency response of the commercial loudspeaker, in Figure 8. The parameters for the circuits presented on Figures 1 and 3 are derived or directly taken from the data sheet of the loudspeaker [36] and taken from the literature for the coupler [46]. Although the influence of R_g is negligible, it is set to 1 mΩ to have a properly scaled $\mathbf{A}_d$ matrix. The value of M_{ms} is computed from the previously determined value of C_{ms} and the value of the resonance frequency given in the data sheet using:

$$M_{ms} = \frac{1}{4\pi^2 f_s^2 C_{ms}} \tag{12}$$

The value of R_{ms} is determined using the quality factor of the data sheet using:

$$R_{ms} = \frac{1}{Q}\sqrt{\frac{M_{ms}}{C_{ms}}} \tag{13}$$

All parameters are summarized in Table 1.

Table 1. Table of lumped elements circuits parameters.

Parameter	Value	Unit
R_g	1	mΩ
C_p	39	nF
R_{ms}	5×10^{-3}	Ns/m
M_{ms}	1.16	mg
C_{ms}	3×10^{-3}	m/N
S_d	12	mm^2
L_1	82.9	Pas^2/m^3
L_2	9400	Pas^2/m^3
L_3	130.3	Pas^2/m^3
L_4	983.8	Pas^2/m^3
L_5	133.4	Pas^2/m^3
C_1	1×10^{-12}	Pa/m^3
C_2	1.9×10^{-12}	Pa/m^3
C_3	1.5×10^{-12}	Pa/m^3
C_4	2.1×10^{-12}	Pa/m^3
C_5	1.517×10^{-12}	Pa/m^3
R_1	50.6×10^6	Pas/m^3
R_2	31.1×10^6	Pas/m^3

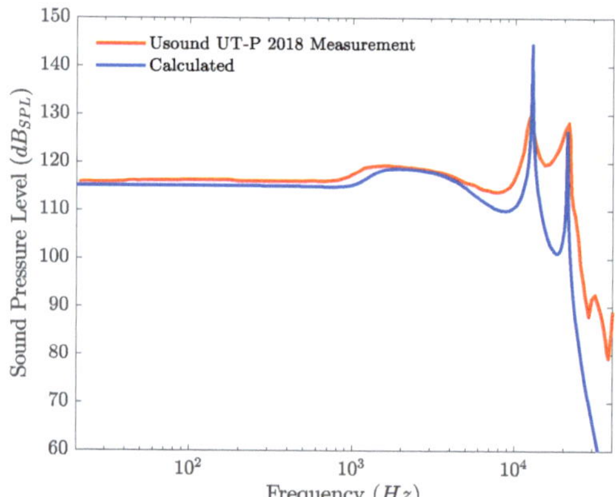

Figure 8. Simulated frequency response of the commercial loudspeaker and measured frequency response of the loudspeaker adapted from [36].

Despite some high-frequency mismatches, due to the modal behavior of the membrane above 20 kHz, and a slight difference between the quality factors of the coupler resonances at, respectively, 12,710 and 20,970 Hz, the simulated frequency response gives a good approximation of the manufacturer measurement.

Using the identified nonlinear functions for parameters γ, K_{ms} and C_p and a pure sine of 1 kHz of amplitude 1.41 V as an input signal, the Fourier transform of the output signal can be calculated and is presented in Figure 9, for different input signals, as well as the linear frequency response.

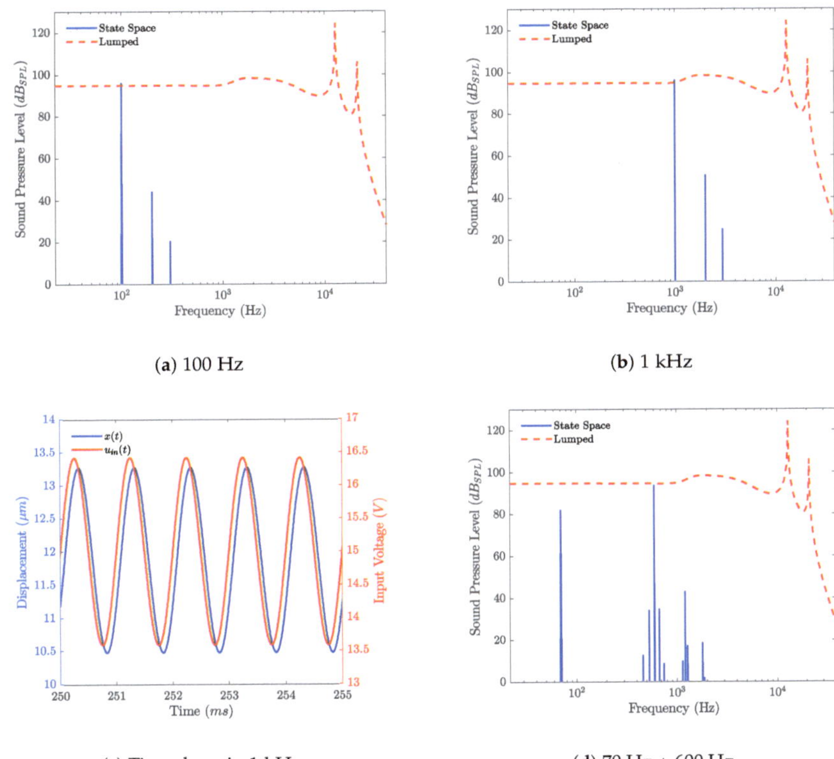

(a) 100 Hz

(b) 1 kHz

(c) Time domain 1 kHz

(d) 70 Hz + 600 Hz

Figure 9. Nonlinear loudspeaker model responses to different input signals with the linear frequency response as a reference in frequency and time domain.

In Figures 9a,b, the response of the loudspeaker to a pure sine of, respectively, 100 Hz and 1 kHz are given, and one can see the second and third harmonic clearly appear. In Figure 9c, the displacement and input voltage in time domain are given. In Figure 9d, the response of the loudspeaker to a sum of two pure sines with $f_1 = 70$ Hz and $f_2 = 600$ Hz, is depicted, and one can see that the intermodulation appears with the components of frequencies $f_2 \pm f_1$, $f_2 \pm 2f_1$, $2f_2 \pm f_1$ and $3f_2 \pm f_1$.

The THD is computed point by point using a pure sine as the input signal for 200 points. The simulated THD is compared to the THD measured by the manufacturer in Figure 10. The pressure signal is low pass filtered at 16 kHz before the THD estimation, as it is described in the data sheet of the loudspeaker. There is a good agreement between the absolute level of the THD between 40 and 200 Hz, where the difference between the two curves is below 20%. Above 500 Hz, the difference between the simulated and measured frequency responses is the cause of the difference between the THD curves and around the distortion peak at about 6.4 kHz, the difference is maximum and above 100%. Qualitatively, the shape of the curve around the resonance of the loudspeaker has a similar shape, capturing the effect of the resonance of the loudspeaker and the one of the coupler on the THD.

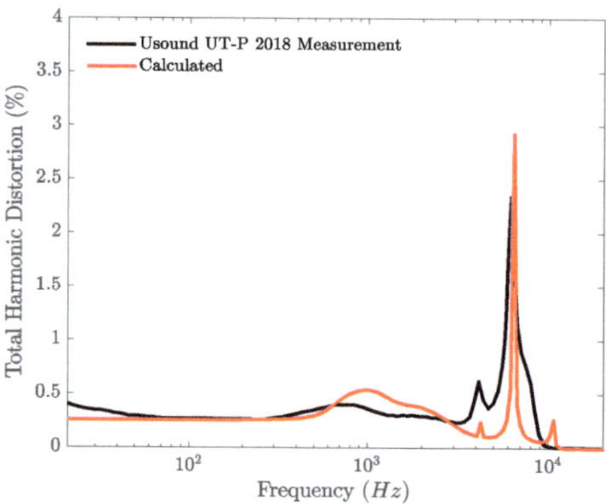

Figure 10. Simulated THD of the commercial loudspeaker and measured THD of the loudspeaker adapted from [36].

In Figure 11a, the respective contributions of $\gamma(u_c)$, $C_{ms}(x)$ and $C_p(u_c)$ to the total harmonic distortion are shown. As the displacement, the harmonic distortion generated by $C_{ms}(x)$ is maximum below the resonance frequency of the loudspeaker and negligible above the resonance frequency. The harmonic distortion generated by the transduction factor $\gamma(u_c)$ is modulated by the frequency response of the loudspeaker but remains high in the whole audio bandwidth. The contribution of $C_p(u_c)$ is three orders of magnitude below the other ones and then negligible. In Figure 11b, the THD for three different input levels is presented.

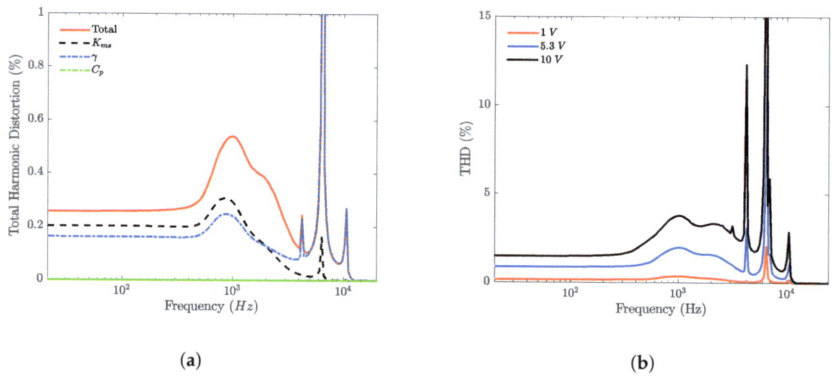

(a)   (b)

Figure 11. THD generated by individual parameters (**a**) and THD as a function of input level (**b**).

5. Conclusions

The model presented in this paper allows one to evaluate the THD generated by a cantilever-based MEMS loudspeaker, by only measuring the nonlinear relations between the stiffness and the displacement, the capacitance and the electric field, and the force and the electric field. It also could allow one to use predistortion techniques to reduce the said total harmonic distortion in MEMS loudspeakers. Although open-loop correction of loudspeaker nonlinearities is tedious for non-MEMS loudspeakers, due to large manufacturing discrepancies, this technique could be more suited for MEMS design, due to the tightest tolerances of the manufacturing processes. This model is also very efficient in terms of computational power compared to a finite element modeling solution and could be used in the early stages of piezoelectric MEMS loudspeakers development, to evaluate the linear performance of the loudspeaker. The discrepancies between the model and the measurement can be attributed to the assumptions made in the model of the loudspeaker. For example, the nonlinear behavior of the PZT is known to widely vary with frequency and to create hysteresis. In addition, other different nonlinear parameters were not taken into account, such as the nonlinearities due to high velocity in the coupler slits, but can be added easily to the model.

Author Contributions: Conceptualization, R.L., T.H., F.C. and S.D.; methodology, R.L.; validation, R.L.; formal analysis, R.L.; investigation, C.P., F.P., H.K., G.L.R. and R.L.; data curation, R.L.; writing—original draft preparation, R.L.; writing—review and editing, R.L., T.H., F.C., S.D., G.L.R. and C.P.; supervision, T.H., F.C., S.D. and M.C.; project administration, M.C.; funding acquisition, T.H. and M.C. All authors have read and agreed to the published version of the manuscript.

Funding: This research received no external funding.

Institutional Review Board Statement: Not applicable

Informed Consent Statement: Not applicable

Conflicts of Interest: The authors declare no conflict of interest.

Abbreviations

The following abbreviations are used in this manuscript:

THD	Total Harmonic Distortion
MEMS	Micro Electromechanical Systems
IEC	International Electrotechnical Commission
DUT	Device Under Test
DBLI	Double-Beam Laser Interferometer
PZT	Lead Zirconate Titanate

Appendix A. State Space Equations

$$\frac{\partial u_c(t)}{\partial t} = \frac{1}{R_g C_p(u_c)} u_{in}(t) - \frac{1}{R_g C_p(u_c)} u_c(t) - \frac{\gamma(u_c)}{C_p(u_c)} v(t) \tag{A1}$$

$$\frac{\partial v(t)}{\partial t} = \frac{\gamma(u_c)}{M_{ms} + S_d^2 L_1} u_c(t) - \frac{R_{ms}}{M_{ms} + S_d^2 L_1} v(t) \\ - \frac{1}{M_{ms} + S_d^2 L_1} F_{C_{ms}}(t) - \frac{S_d}{M_{ms} + S_d^2 L_1} p_{C_1}(t) \tag{A2}$$

$$\frac{\partial F_{C_{ms}}}{\partial t} = \frac{1}{C_{ms}(x)} v(t), \tag{A3}$$

$$\frac{\partial x(t)}{\partial t} = v(t), \tag{A4}$$

$$\frac{\partial p_{C_1}(t)}{\partial t} = \frac{S_d}{C_1} v(t) - \frac{1}{C_1} Q_2(t) - \frac{1}{C_1} Q_{L_3}(t) \tag{A5}$$

$$\frac{\partial Q_2(t)}{\partial t} = \frac{1}{L_2} p_{C_1}(t) - \frac{1}{L_2} p_{C_2}(t) - \frac{R_2}{L_2} Q_2(t) \tag{A6}$$

$$\frac{\partial p_{C_2}(t)}{\partial t} = \frac{1}{C_2} Q_2(t) \tag{A7}$$

$$\frac{\partial Q_{L_3}(t)}{\partial t} = \frac{1}{L_3} p_{C_1}(t) - \frac{1}{L_3} p_{C_3}(t) \tag{A8}$$

$$\frac{\partial p_{C_3}(t)}{\partial t} = \frac{1}{C_3} Q_{L_3}(t) - \frac{1}{C_3} Q_4(t) - \frac{1}{C_3} Q_{out}(t) \tag{A9}$$

$$\frac{\partial Q_4(t)}{\partial t} = \frac{1}{L_4} p_{C_3}(t) - \frac{1}{L_4} Q_4(t) - \frac{1}{L_4} p_{C_4}(t) \tag{A10}$$

$$\frac{\partial p_{C_4}(t)}{\partial t} = \frac{1}{C_4} Q_4(t) \tag{A11}$$

$$\frac{\partial Q_{out}(t)}{\partial t} = \frac{1}{L_5} p_{C_3}(t) - \frac{1}{L_5} p_{out}(t) \tag{A12}$$

$$\frac{\partial p_{out}(t)}{\partial t} = \frac{1}{C_5} Q_{out}(t) \tag{A13}$$

References

1. Shahosseini, I.; Lefeuvre, E.; Moulin, J.; Martincic, E.; Woytasik, M.; Lemarquand, G. Optimization and Microfabrication of High Performance Silicon-Based MEMS Microspeaker. *IEEE Sens. J.* **2013**, *13*, 273–284. [CrossRef]
2. Shahosseini, I. Vers des Micro-Haut-Parleurs à Hautes Performances Électroacoustiques en Technologie Silicium. Ph.D. Thesis, Université Paris Sud, Paris, France, 2012; p. 206.
3. Wang, H.; Feng, P.X.L.; Xie, H. A Dual-Electrode MEMS Speaker Based on Ceramic PZT with Improved Sound Pressure Level by Phase Tuning. In Proceedings of the 2021 IEEE 34th International Conference on Micro Electro Mechanical Systems (MEMS), Gainesville, FL, USA, 25–29 January 2021; pp. 701–704. [CrossRef]
4. Wang, Q.; Yi, Z.; Ruan, T.; Xu, Q.; Yang, B.; Liu, J. Obtaining High SPL Piezoelectric MEMS Speaker via a Rigid-Flexible Vibration Coupling Mechanism. *J. Microelectromech. Syst.* **2021**, *30*, 725–732. [CrossRef]
5. Ko, S.C.; Kim, Y.C.; Lee, S.S.; Choi, S.H.; Kim, S.R. Micromachined piezoelectric membrane acoustic device. *Sens. Actuators A Phys.* **2003**, *103*, 130–134. [CrossRef]
6. Wang, H.; Chen, Z.; Xie, H. A high-SPL piezoelectric MEMS loud speaker based on thin ceramic PZT. *Sens. Actuators A Phys.* **2020**, *309*, 112018. [CrossRef]
7. Wang, H.; Li, M.; Yu, Y.; Chen, Z.; Ding, Y.; Jiang, H.; Xie, H. A Piezoelectric MEMS Loud Speaker Based on Ceramic PZT. In Proceedings of the 2019 20th International Conference on Solid-State Sensors, Actuators and Microsystems & Eurosensors XXXIII (Transducers & Eurosensors XXXIII), Berlin, Germany, 23–27 June 2019; pp. 857–860. [CrossRef]
8. Kaiser, B.; Langa, S.; Ehrig, L.; Stolz, M.; Schenk, H.; Conrad, H.; Schenk, H.; Schimmanz, K.; Schuffenhauer, D. Concept and proof for an all-silicon MEMS micro speaker utilizing air chambers. *Microsyst. Nanoeng.* **2019**, *5*, 43. [CrossRef]
9. Stoppel, F.; Eisermann, C.; Gu-Stoppel, S.; Kaden, D.; Giese, T.; Wagner, B. Novel membrane-less two-way MEMS loudspeaker based on piezoelectric dual-concentric actuators. In Proceedings of the 2017 19th International Conference on Solid-State Sensors, Actuators and Microsystems (Transducers), Kaohsiung, Taiwan, 18–22 June 2017; pp. 2047–2050. [CrossRef]
10. Stoppel, F.; Mannchen, A.; Niekiel, F.; Beer, D.; Giese, T.; Wagner, B. New integrated full-range MEMS speaker for in-ear applications. In Proceedings of the 2018 IEEE Micro Electro Mechanical Systems (MEMS), Belfast, UK, 21–25 January 2018; pp. 1068–1071. [CrossRef]
11. Fielder, L.; Benjamin, E. Subwoofer Performance for Accurate Reproduction of Music. *J. Audio Eng. Soc.* **1988**, *36*, 443–456.
12. Vanderkooy, J.; Krauel K.B. Another View of Distortion Perception. In *Audio Engineering Society Convention 133*; Audio Engineering Society: New York, NY, USA, 2012; p. 4.
13. Larson, J.; DellaSala, G. A Tutorial on the Audibility of Loudspeaker Distortion at Bass Frequencies. In *Audio Engineering Society Convention 143*; Audio Engineering Society: New York, NY, USA, 2017; p. 10.
14. Temme, S.; Brunet, P.; Qarabaqi, P. Measurement of Harmonic Distortion Audibility Using a Simplified Psychoacoustic Model-Updated. In *Audio Engineering Society Conference: 51st International Conference: Loudspeakers and Headphones*; Audio Engineering Society: New York, NY, USA, 2013; p. 9.
15. Damjanovic, D. Hysteresis in Piezoelectric and Ferroelectric Materials. In *The Science of Hysteresis*; Elsevier: Amsterdam, The Netherlands, 2006; pp. 337–465. [CrossRef]
16. Jayendiran, R.; Arockiarajan, A. Non-linear electromechanical response of 1–3 type piezocomposites. *Int. J. Solids Struct.* **2013**, *50*, 2259–2270. [CrossRef]
17. Royston, T.J.; Houston, B.H. Modeling and measurement of nonlinear dynamic behavior in piezoelectric ceramics with application to 1–3 composites. *J. Acoust. Soc. Am.* **1998**, *104*, 2814–2827. [CrossRef]

18. Lucke, P.; Bayraktar, M.; Birkhölzer, Y.A.; Nematollahi, M.; Yakshin, A.; Rijnders, G.; Bijkerk, F.; Houwman, E.P. Hysteresis, Loss and Nonlinearity in Epitaxial $PbZr_{0.55}Ti_{0.45}O_3$ Films: A Polarization Rotation Model. *Adv. Funct. Mater.* **2020**, *30*, 2005397. [CrossRef]
19. Rébillat, M.; Hennequin, R.; Corteel, E.; Katz, B.F. Identification of cascade of Hammerstein models for the description of nonlinearities in vibrating devices. *J. Sound Vib.* **2011**, *330*, 1018–1038. [CrossRef]
20. Falaize, A.; Hélie, T. Passive modelling of the electrodynamic loudspeaker: From the Thiele–Small model to nonlinear port-Hamiltonian systems. *Acta Acust.* **2020**, *4*, 1. [CrossRef]
21. Falaize, A.; Papazoglou, N.; Hélie, T.; Lopes, N. Compensation of loudspeaker's nonlinearities based on flatness and port-Hamiltonian approach. In *22ème Congrès Français de Mécanique*; Association Française de Mécanique: Lyon, France, 2015; p. 11.
22. Novak, A.; Simon, L.; Lotton, P.; Merit, B.; Gilbert, J. Nonlinear Analysis and Modeling of Electrodynamic Loudspeakers. In *10ème Congrès Français d'Acoustique*; Société Française d'Acoustique: Paris, France, 2010; p. 6.
23. Klippel, W. Large Signal Performance of Tweeters, Micro Speakers and Horn Drivers. In *Audio Engineering Society Convention 118*; Audio Engineering Society: New York, NY, USA, 2005; p. 19.
24. Klippel, W. Loudspeaker Nonlinearities–Causes, Parameters, Symptoms. In *Audio Engineering Society Convention 119*; Audio Engineering Society: New York, NY, USA, 2005; p. 36.
25. King, A.; Agerkvist, F. Fractional Derivative Loudspeaker Models for Nonlinear Suspensions and Voice Coils. *J. Audio Eng. Soc.* **2018**, *66*, 525–536. [CrossRef]
26. Monsalve, J.M.; Melnikov, A.; Kaiser, B.; Schuffenhauer, D.; Stolz, M.; Ehrig, L.; Schenk, H.A.G.; Conrad, H.; Schenk, H. Large-Signal Equivalent-Circuit Model of Asymmetric Electrostatic Transducers. *IEEE/ASME Trans. Mechatron.* **2021**, 1–11. [CrossRef]
27. Klippel, W. Optimal Design of Loudspeakers with Non-Linear Control. In *Audio Engineering Society Conference: 32nd International Conference: DSP For Loudspeakers*; Audio Engineering Society: New York, NY, USA, 2007; p. 11.
28. Moore, B.C.J.; Tan, C.T. Perceived naturalness of spectrally distorted speech and music. *J. Acoust. Soc. Am.* **2003**, *114*, 408–419. [CrossRef]
29. Tan, C.T.; Moore, B.C.J. The Effect of Nonlinear Distortion on the Perceived Quality of Music and Speech Signals. *J. Audio Eng. Soc.* **2003**, *51*, 20.
30. Moore, B.C.J. Computational models for predicting sound quality. *Acoust. Sci. Technol.* **2020**, *41*, 75–82. [CrossRef]
31. Olsen, S.L.; Agerkvist, F.; MacDonald, E.; Stegenborg-Andersen, T.; Volk, C.P. Modelling the Perceptual Components of Loudspeaker Distortion. In *Audio Engineering Society Convention 140*; Audio Engineering Society: New York, NY, USA, 2016; p. 9.
32. Liechti, R.; Durand, S.; Hilt, T.; Casset, F.; Dieppedale, C.; Verdot, T.; Colin, M. A Piezoelectric MEMS Loudspeaker Lumped and FEM models. In Proceedings of the 2021 22nd International Conference on Thermal, Mechanical and Multi-Physics Simulation and Experiments in Microelectronics and Microsystems (EuroSimE), Online, 19–21 April 2021; pp. 1–8. [CrossRef]
33. Wang, Q.M.; Du, X.H.; Xu, B.; Cross, L.E. Electromechanical coupling and output efficiency of piezoelectric bending actuators. *IEEE Trans. Ultrason. Ferroelectr. Freq. Control* **1999**, *46*, 638–646. [CrossRef]
34. International Electrotechnical Commission. *Electroacoustics: Simulators of Human Head and Ear. Part 4, Partie 4*; OCLC: 954221482; International Electrotechnical Commission: Geneva, Switzerland, 2010.
35. Fadali, M.S.; Visioli, A. *Digital Control Engineering: Analysis and Design*, 2nd ed.; Academic Press, Elsevier: Amsterdam, The Netherlands, 2013.
36. Usound. Achelous UT-P 2018 Datasheet. 2020. Available online: https://www.usound.com/wp-content/uploads/2020/01/20 01_Achelous-UT-P-2018-Datasheet.pdf (accessed on 23 November 2021).
37. Sivaramakrishnan, S.; Mardilovich, P.; Schmitz-Kempen, T.; Tiedke, S. Concurrent wafer-level measurement of longitudinal and transverse effective piezoelectric coefficients ($d_{33,f}$ and $e_{31,f}$) by double beam laser interferometry. *J. Appl. Phys.* **2018**, *123*, 014103. [CrossRef]
38. Marton, P.; Rychetsky, I.; Hlinka, J. Domain walls of ferroelectric $BaTiO_3$ within the Ginzburg-Landau-Devonshire phenomenological model. *Phys. Rev. B* **2010**, *81*, 144125. [CrossRef]
39. Mehling, V.; Tsakmakis, C.; Gross, D. Phenomenological model for the macroscopical material behavior of ferroelectric ceramics. *J. Mech. Phys. Solids* **2007**, *55*, 2106–2141. [CrossRef]
40. Tsujiura, Y.; Kawabe, S.; Kurokawa, F.; Hida, H.; Kanno, I. Comparison of effective transverse piezoelectric coefficients $e_{31,f}$ of $Pb(Zr,Ti)O_3$ thin films between direct and converse piezoelectric effects. *Jpn. J. Appl. Phys.* **2015**, *54*, 10NA04. [CrossRef]
41. Völker, B.; Kamlah, M.; Wang, J. Phase-field Modeling of Ferroelectric Materials. In Proceedings of the COMSOL Conference 2009, Milanop, Italy, 14–16 October 2009; p. 28.
42. Tan, G.; Maruyama, K.; Kanamitsu, Y.; Nishioka, S.; Ozaki, T.; Umegaki, T.; Hida, H.; Kanno, I. Crystallographic contributions to piezoelectric properties in PZT thin films. *Sci. Rep.* **2019**, *9*, 7309. [CrossRef] [PubMed]
43. Wang, Y.; Cheng, H.; Yan, J.; Chen, N.; Yan, P.; Ouyang, J. Nonlinear electric field dependence of the transverse piezoelectric response in a (001) ferroelectric film. *Scr. Mater.* **2020**, *189*, 84–88. [CrossRef]
44. Le Rhun, G.; Dieppedale, C.; Wague, B.; Querne, C.; Enyedi, G.; Perreau, P.; Montmeat, P.; Licitra, C.; Fanget, S. Transparent PZT MIM Capacitors on Glass for Piezoelectric Transducer Applications. In Proceedings of the 2019 20th International Conference on Solid-State Sensors, Actuators and Microsystems & Eurosensors XXXIII (Transducers & Eurosensors XXXIII), Berlin, Germany, 23–27 June 2019; pp. 1800–1802. [CrossRef]

45. Kleiner, M. *Electroacoustics*; CRC Press: Boca Raton, FL, USA, 2013; p. 620.
46. Jønsson, S.; Liu, B.; Nielsen, L.B.; Schuhmacher, A. Simulation of Couplers. In *Audio Engineering Society Workshop 7*; Audio Engineering Society: New York, NY, USA, 2003; p. 29.

micromachines

Article

Novel Fabrication Technology for Clamped Micron-Thick Titanium Diaphragms Used for the Packaging of an Implantable MEMS Acoustic Transducer

Lukas Prochazka [1,*], Alexander Huber [1], Michael Schneider [2], Naureen Ghafoor [3], Jens Birch [3] and Flurin Pfiffner [1]

1   Department of Otorhinolaryngology, Head & Neck Surgery, University Hospital Zurich, University of Zurich, 8091 Zurich, Switzerland; alex.huber@usz.ch (A.H.); flurin.pfiffner@usz.ch (F.P.)
2   SwissNeutronics AG, 5313 Klingnau, Switzerland; michael.schneider@swissneutronics.ch
3   Department of Physics, Chemistry and Biology (IFM), Linköping University, 581 83 Linköping, Sweden; naureen.ghafoor@liu.se (N.G.); jens.birch@liu.se (J.B.)
*   Correspondence: lukas.prochazka@usz.ch; Tel.: +41-(0)44-255-5823

Abstract: Micro-Electro-Mechanical Systems (MEMS) acoustic transducers are highly sophisticated devices with high sensing performance, small size, and low power consumption. To be applied in an implantable medical device, they require a customized packaging solution with a protecting shell, usually made from titanium (Ti), to fulfill biocompatibility and hermeticity requirements. To allow acoustic sound to be transferred between the surroundings and the hermetically sealed MEMS transducer, a compliant diaphragm element needs to be integrated into the protecting enclosure. In this paper, we present a novel fabrication technology for clamped micron-thick Ti diaphragms that can be applied on arbitrary 3D substrate geometry and hence directly integrated into the packaging structure. Stiffness measurements on various diaphragm samples illustrate that the technology enables a significant reduction of residual stress in the diaphragm developed during its deposition on a polymer sacrificial material.

Keywords: implantable acoustic transducer; packaging; titanium/platinum multi-layer diaphragm; polymer sacrificial material; DC magnetron sputtering; intrinsic stress; stress measurement; stress relief

Citation: Prochazka, L.; Huber, A.; Schneider, M.; Ghafoor, N.; Birch, J.; Pfiffner, F. Novel Fabrication Technology for Clamped Micron-Thick Titanium Diaphragms Used for the Packaging of an Implantable MEMS Acoustic Transducer. *Micromachines* **2022**, *13*, 74. https://doi.org/10.3390/mi13010074

Academic Editor: Libor Rufer

Received: 30 November 2021
Accepted: 28 December 2021
Published: 31 December 2021

Publisher's Note: MDPI stays neutral with regard to jurisdictional claims in published maps and institutional affiliations.

1. Introduction

The fast growth of the telecommunication market over the last two decades has enabled high investments in the research and development of MEMS transducers such as gyros, accelerometers, and acoustic transducers. The result are off-the-shelf MEMS transducers with a miniature size, low cost, and low power consumption, but at the same time with a sensing performance that meets the requirements of most electronic devices for daily use [1]. In the medical field, implantable sensor solutions play an increasingly important role, but for most applications, the market is too small to justify the required high investment in the development of application-specific, high-performance MEMS transducers that are designed to meet the stringent requirements for the biocompatibility and reliability of implantable devices [2,3]. If so, an eligible solution could be to use off-the-shelf MEMS transducer chips and the application-specific integrated circuit (ASIC) and package them in a customized, biocompatible, and hermetic enclosure that protects the delicate MEMS parts and electronics against the harsh environment in a living subject and vice versa. However, such an approach involves several design difficulties if applied to an acoustic MEMS transducer.

Most acoustic MEMS transducers use a thin flexing element (usually a clamped diaphragm) to capture the smallest pressure fluctuations in the surrounding fluid (receiver) or to radiate sound into the surroundings (transmitter) [4,5]. As soon as the transducer is installed in a hermetic enclosure, the direct access to the surroundings is interrupted. To overcome this problem, a flexible diaphragm-like element must be integrated into the protecting shell that can transmit acoustic sound energy into the transducer's interior or the surroundings (cf. Figure 1a). We call this design element the protective diaphragm (PD). The PD adds to the mechanical impedance of the MEMS transducer and hence reduces the receiving (or transmitting) sensitivity. Therefore, it must be designed to exhibit maximum compliance and must be arranged close to the transducer's diaphragm to reduce the volume that couples the two flexible elements. Another difficulty of a hermetic encapsulation of the acoustic transducer is that static pressure equalization between the surroundings and the interior of the packaging structure cannot occur. Excessive loading of the PD by ambient pressure changes can lead to a varying or not tolerable loss in sensing performance. A deeper discussion of that problem and a proposal for a system that can maintain pressure equalization in a hermetically sealed microphone system are given in [6].

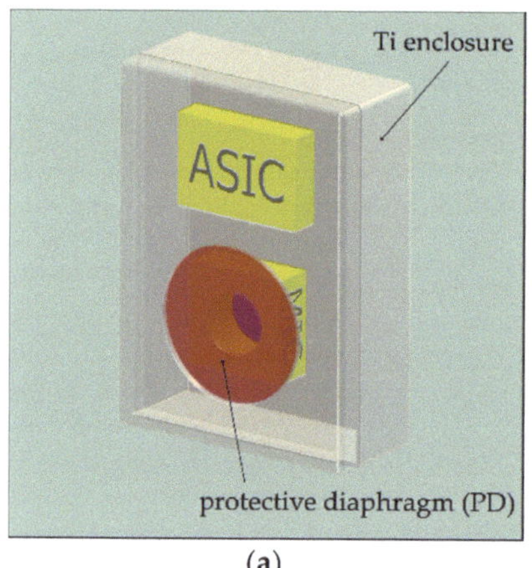

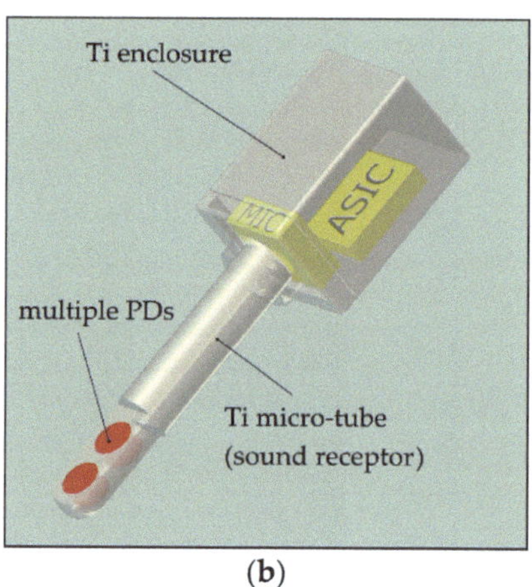

Figure 1. Schematic drawings of a MEMS microphone and the corresponding ASIC packaged in a biocompatible and hermetic Ti enclosure with a PD that can transfer the sound energy to the MEMS transducer. (**a**) A generic enclosure geometry with one larger PD located directly in front of the MEMS microphone diaphragm; (**b**) a more specific enclosure geometry that was developed for an intracochlear acoustic receiver [6]. Four PDs are integrated into a tube-like part of the enclosure to allow the recording of the liquid-borne sound pressure in the human inner ear with sufficiently high receiving sensitivity.

To maximize the PD's mechanical compliance, several design parameters characterizing the PD geometry (size and thickness) and material (Young's modulus) can be adjusted under the consideration of different design constraints. The PD area is often restricted by the miniature size of the transducer. The requirement for a long-term hermetic encapsulation of a typical implantable transducer restricts the material choice to metals and ceramics and makes it difficult to reduce the PD thickness below 1 µm [7]. Fabrication-related properties of the PD, such as residual stress or the number of pinholes or voids within the PD structure, have either a direct impact on the PD stiffness or augment, for instance, the lower PD thickness limit.

A practical example of a packaging concept that illustrates these design guidelines is the intracochlear acoustic receiver (ICAR), which is currently under development by our group [6,8]. The receiver is designed to be used as an implantable microphone solution for a fully implantable cochlea implant. It is built upon a commercially available MEMS condenser microphone with a customized Ti enclosure that exhibits four circular PDs with a diameter of 0.6 mm and 1 µm thickness. Multiple PDs free of intrinsic stress are required to minimize the loss in sensing performance caused by the encapsulation of the

MEMS transducer. Using a theoretical model, a loss in sensitivity of approximately 10 dB was predicted below the resonance operation of the sensor at 3.5 kHz [6]. The inertia of the PDs that are in contact with the perilymph (similar to water) in the inner ear defines the resonance and limits the bandwidth of the ICAR to approximately 6 kHz. To enable sound pressure recording in the tiny human inner ear (cochlea), the PDs are arranged in a pair configuration and on opposite sides within the tip region of a tube-like part of the enclosure (cf. Figure 1b). Titanium is chosen as the enclosure material due to its excellent biocompatibility, hermeticity properties, and inherently higher fracture toughness than ceramics or typical semiconductor materials [9,10]. In addition, it allows the combination of micro- and macro-machining processes such as thin film deposition and etching combined with milling, turning, and welding processes.

The fabrication of micron-thick and sub-millimeter large Ti diaphragms suspended in a Ti structure with a complex geometry is a challenging task. Cold-rolled 1 µm thick Ti foils are commercially available (American Elements, Los Angeles, CA, USA), but the fixation of such a thin foil on the carrier by micro-laser welding or mechanical clamping or bonding is not feasible due to size, mechanical stability, and hermeticity constraints. Mechanical structures with such small dimensions are usually fabricated based on MEMS fabrication technology, which is mostly applied on semiconductor materials such as single-crystal silicon (SCS). MEMS condenser microphones, for instance, which exhibit a sub-micron thick sensing diaphragm sitting on an SCS support and separated from the rigid backplate by few microns, are fabricated based on bulk and surface micromachining technology [11,12].

Various research papers report bulk and surface micromachining processes for Ti as substrate material [13–17]. The most common bulk micromachining process for Ti that is currently widely commercially available is a wet etching technology, called photochemical machining (PCM) [15]. The process is isotropic and thus limits the minimum feature size to approximately the metal sheet thickness if the sheet needs to be etched through [18]. Another bulk micromachining technology for Ti was developed by Aimi et al. [13]. It is a highly anisotropic dry etching technique called the metal anisotropic reactive ion etching with oxidation (MARIO) process. The process allows users to etch straight sidewalls with a high-aspect-ratio and features on the micrometer scale using common dry etching equipment. Bulk micromachining should be better suited than surface micromachining to fabricate a highly compliant PD structure from a Ti substrate because surface micromachined mechanical structures often exhibit high residual stress and hence show high stiffness, high mechanical failure, and high batch to batch variation [19]. However, to obtain a 1 µm thick PD by bulk micromachining, the etch process needs to be stopped very precisely as soon as the targeted etch depth is reached. A time etch stop is not accurate enough to obtain uniform and repeatable structures with micron thickness. Etch stop techniques, as known from the wet etching of SCS (e.g., doping or electrochemical control [20]), which might provide the required etch stop accuracy, do not exist for the PCM of Ti to the best of our knowledge.

To fabricate highly accurate Ti structures at micrometer-scale, bulk and surface machining need to be combined, similar to the case of MEMS fabrication based on semiconductor materials. Such an approach applied to PD fabrication would start with an electrochemically polished Ti substrate with a reactively sputtered TiO_2 layer covered also with a sputtered structural Ti layer that forms the PD element. The PDs are released from the substrate's rear side using the MARIO process with the oxide layer as a precise etch stop. The remaining oxide layer is stripped to end up with a pure Ti PD. The process looks simple at first glance, but technologies such as the MARIO process and also the electrochemical polishing of Ti are typically not well established in commercial or research fabrication labs (clean rooms). Creating a more complex enclosure geometry with multiple PDs, as required for the ICAR, creates another difficulty associated with this fabrication approach that is limited to 2D substrates. Thermal compression bonding might be a solution to combine multiple already processed Ti layers, but the compatibility of that quite rough bonding process with the delicate PDs is questionable. To overcome the listed difficulties would require a high process development effort in well-equipped clean-room fabrication facilities, which is associated with high development costs. Therefore, we have developed an alternative fabrication approach for thin diaphragms on a Ti supporting structure that comprises commercially available fabrication processes and processes that can be performed outside of clean-room facilities. In contrast to dry etching techniques, the new fabrication technology can be applied on unpolished substrates with a complex 3D geometry and, therefore, provides more freedom in the design of packaging structures for acoustic MEMS transducers.

The approach is based on the deposition of the structural Ti layer forming the PD on a low-temperature decomposable polymer sacrificial material (SM) as a temporary support. So far, we could confirm the feasibility of producing robust 1 μm thick Ti diaphragms on unpolished substrates with 2D and 3D geometry using this new fabrication approach [6]. We could also show that the method incorporates a stress relief mechanism that originates from a change in the PD's surface geometry upon releasing from the SM.

The present paper introduces the novel diaphragm fabrication method applied on 2D Ti substrates that belong to a testbed developed for process development and the investigation and testing of the fabricated PDs. The other focus is related to the stress-relief mechanism, which is demonstrated and analyzed on numerous PDs. The paper concludes with a discussion of the results and ongoing studies that are being conducted to obtain robust and hermetic PDs with vanishing residual stress.

2. Materials and Methods

2.1. Diaphragm Fabrication

2.1.1. Process Overview

The process is applied on a substrate with pre-fabricated cavities or through-holes that define the outer size and geometry of the diaphragms. In the first step, the cavities are filled with a fully decomposable polymer sacrificial material (SM) such as polypropylene carbonate (QPAC, Empower Materials, New Castle, DE, USA) dissolved in a solvent (cf. Figure 2a). The solution is added to the cavities using standard adhesive dispensing equipment. For filling, the substrate is mounted on a heated carrier that properly seals the bottom side of the through-holes. Because most of the volume of the added SM (dissolved state) is lost during the curing process, the filling is conducted in multiple steps. The elevated processing temperature of 85 °C enhances the curing process and hence reduces the processing time. After the cavities are filled with SM, a thin metallic film, which forms the diaphragm, is deposited on the front surface of the substrate using physical vapor deposition (PVD) technology such as magnetron sputtering (cf. Figure 2b). Finally, the polymer SM is fully decomposed during thermal treatment in a vacuum or an inert gas atmosphere, releasing the diaphragms from the SM (cf. Figure 2c). The used SM (QPAC) requires a temperature of 350 °C for a full decomposition in various atmospheres, even in a vacuum, forming carbon dioxide and water.

2.1.2. Substrate

The development of the process for diaphragm fabrication was carried out by experimental means, requiring a larger amount of test samples. Accordingly, the substrate was designed to keep the complexity of substrate and diaphragm fabrication small and to enable the simple testing and inspection of the diaphragms. The substrate is a 30 mm × 30 mm large Ti plate with a thickness of 0.3 mm (cf. Figure 3a). It carries 57 through-holes with a circular geometry and a diameter of 0.6 mm. The through-holes are arranged in equally distributed groups of seven and five holes on the substrate. Each group is surrounded by a release trench which forms circular samples with diameters of 3, 4, and 5 mm. Only a narrow connecting bar holds the sample in place during processing on the substrate level and prevents excessive mechanical stress being caused by the handling of the thin substrate, which might potentially damage the delicate diaphragms. Different sample diameters were considered to account for sample integration on different testing and inspection equipment. The diaphragm size of 0.6 mm was chosen to match the corresponding dimensions of the implantable microphone (ICAR). The substrate was fabricated by photo etching (TiME process, Advanced Chemical Etching, Ltd., Telford, UK) of a 0.3 mm thick unalloyed commercially pure Ti sheet from grade 2. The substrate thickness of 0.3 mm results from the trade-off between the minimum feature size that can be etched for a specific substrate thickness and the need for a sufficiently high mechanical stiffness of the substrate. Figure 3b,c are optical microscope images showing the etch profiles of the through-holes with a uniform, slightly rounded edge geometry and sided walls with only a small etch cusp at half the profile depth.

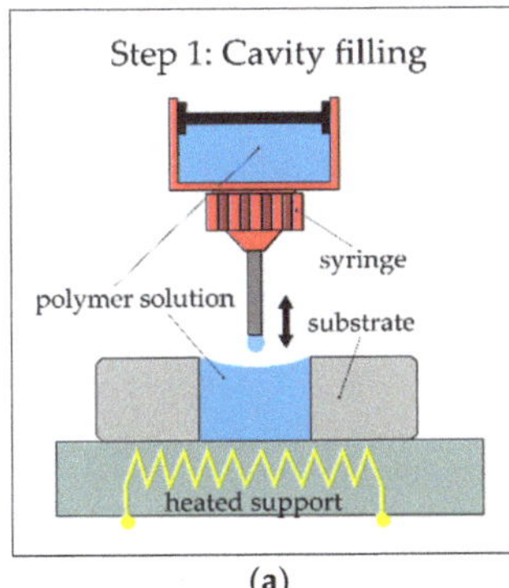

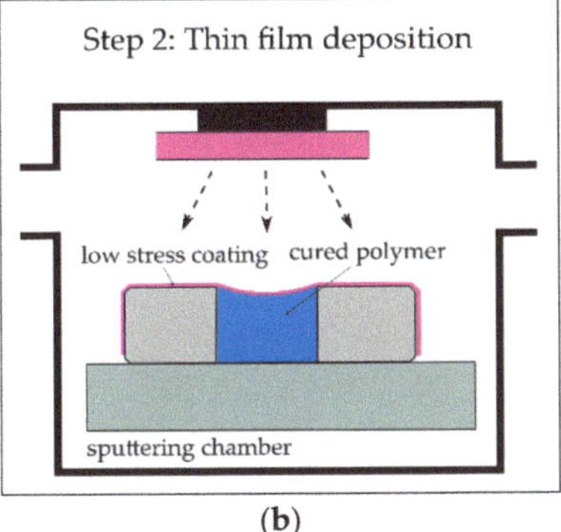

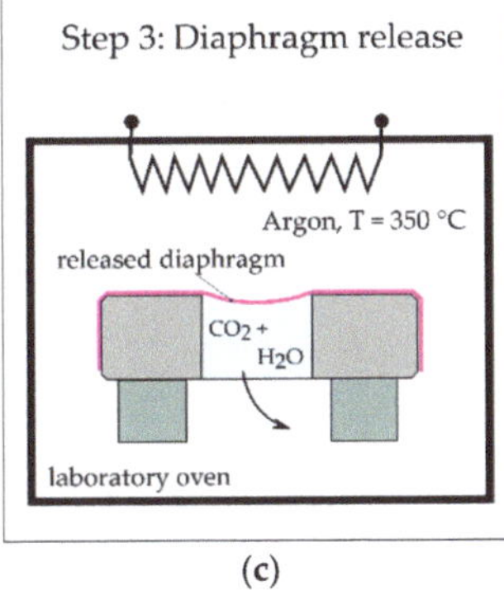

(**a**)   (**b**)   (**c**)

Figure 2. General overview of the diaphragm fabrication process. (**a**) Step 1, filling process: sacrificial material (SM) is added to the substrate; (**b**) Step 2, thin film deposition: metallic coating that represents the diaphragm is deposited on the substrate using chemical vapor deposition; (**c**) Step 3, diaphragm release: the diaphragm is released from the SM during a thermal treatment process.

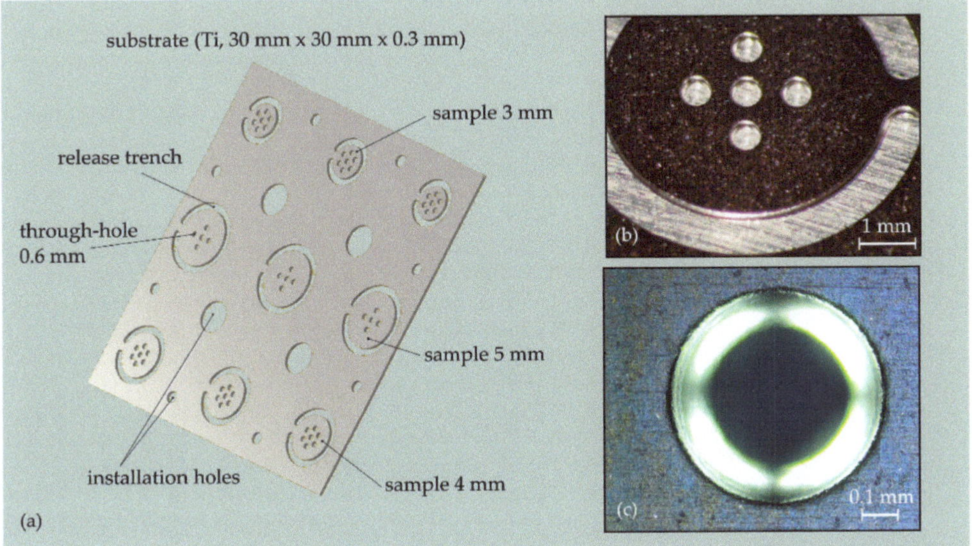

Figure 3. (**a**) CAD drawing of the substrate specifically designed for the development of the diaphragm fabrication process; (**b**) optical microscope image of one of the nine samples on a substrate containing five through-holes; (**c**) microscope image showing the etch profile of a 0.6 mm large through-hole.

2.1.3. Filling Process

The filling process was performed in a dust-free environment under a laminar airflow hood. QPAC 40 (Empower Materials, New Castle, DE, USA), a low-temperature decomposable polymer, was chosen as the SM for diaphragm fabrication. The granulate (20 wt.%) was dissolved in γ-Butyrolactone (TCI GmbH, Eschborn, Germany). The solution was filtered using a Whatman GD/X syringe filter (GE Healthcare, Chicago, IL, USA) with a regenerated cellulose membrane and a pore size of 0.45 µm. A pneumatic dispenser (PDS 7350, Poly Dispensing Systems, Orgeval, France) and a tiny dispensing needle (size 32 G) were used to control the amount of added QPAC solution. During the dispensing of the SM, the tip of the needle was partially inserted into the cavity or positioned slightly above the cavity opening. The needle position was set by visual observation through a surgical microscope (Wild M650, Leica AG, Heerbrugg, Switzerland) at the highest magnification ($40\times$). The up and down movement of the syringe was adjusted using a manually controlled motorized micro-positioning unit (M-443, Newport, Irvine, CA, USA + T-NA08A50, Zaber Technologies Inc., Vancouver, BC, Canada). Going from one cavity to the other, the substrate mounted on the carrier was manually moved in the lateral direction on a stable support plate.

Small doses of SM were successively added to each cavity to prevent the solution from flowing over the cavity boundary. The filling process was conducted at an elevated temperature of 80 °C to enhance the evaporation of the solvent, causing a high volume loss of the added material. The filling was therefore repeated several times before the actual curing was performed. During curing at 130 °C for 1 h in a vacuum (>-0.8 bar) atmosphere, the remaining solvent and entrained air were removed from the polymer. At 130 °C, QPAC40 becomes a melted mass of low viscosity, forming by capillary forces a free surface at the cavity rim, which is smooth, uniform, and has a concave shape. The surface curvature of the polymer filling is a critical parameter for stress relief in the diaphragm after it is released from the SM (cf. Section 3). During a repetitive filling and curing process (up to 10 cycles), an optimum surface curvature is adjusted by adding the right amount of SM to the cavity. The substrate was heated using two 100 Watt cartridge heaters installed in the aluminum substrate carrier at a maximum heating rate of approximately 10 °C/min (cf. Figure 4). The cooling rate was mainly controlled by the laminar airflow. A digital PID temperature controller (KS20, West Control Solutions, Kassel, Germany) and a PT1000 temperature probe were used to control a predefined substrate temperature.

Proper sealing of the bottom side of the through-holes is crucial to maintain high filling repeatability and short processing times. The sealing was achieved with a circular PTFE nub (height: 0.1 mm, diameter bottom area: 0.6 mm) pressed against the bottom rim of the through-hole (cf. Figure 4). PTFE was chosen to prevent adhesion between SM and the nub. In addition, PTFE shows high chemical and temperature resistance and, due to the low hardness, it can adapt to any irregularities of the rim geometry. The nub structure was created by adding a 0.1 mm thick PTFE foil as a separator between the substrate and a supporting plate. The latter plate has the same geometry as the actual substrate, but the through-holes are filled with high-temperature epoxy adhesive. The amount of the epoxy was precisely adjusted using the aforementioned dispensing system to create a surface topography on the supporting plate, which was used to stamp the corresponding nubs into the PTFE foil. A laser-cut stainless-steel plate (thickness 0.5 mm) uniformly clamped the whole stack of the individual plates down to the carrier using 16 M2.5 screws. The clamping force was induced at the border of each circular sample plate of the substrate using spring elements (lock washer) attached to the clamping plate. The support plate was reusable, whereas the PTFE foil was replaced when a new substrate was installed on the carrier.

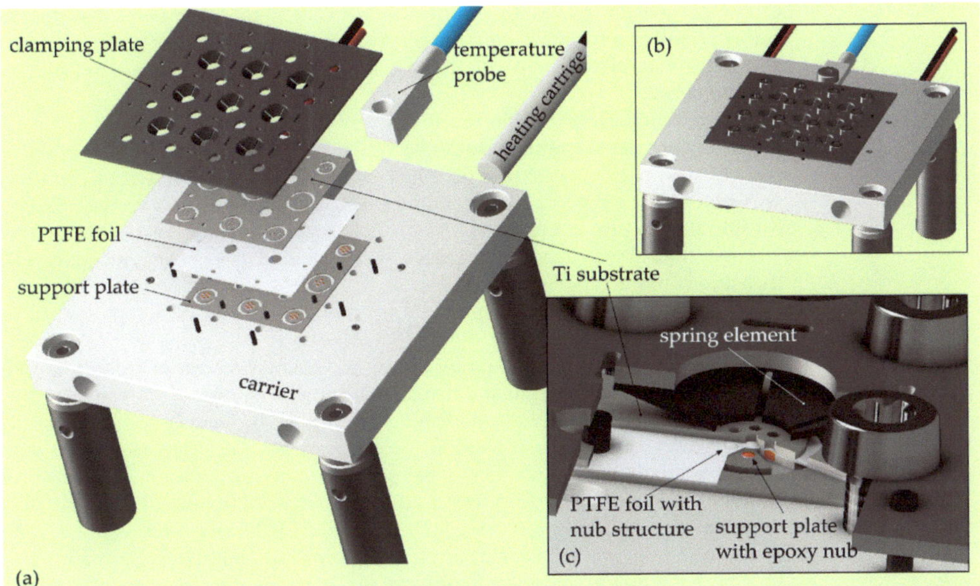

Figure 4. CAD drawings of the instrumented substrate carrier used for the filling process. (**a**) Exploded view showing the different plates for sealing and clamping of the substrate and the heating system of the carrier; (**b**) carrier with the substrate and heating system installed; (**c**) zoomed view on one sample of the substrate illustrates the sealing and clamping mechanism.

2.1.4. Thin Film Deposition

The diaphragms considered in the present study were made from a 1 μm thick Ti monolayer and titanium/platinum multilayer coatings (Ti/Pt ML). They were deposited on the pre-processed substrate by physical vapor deposition (PVD) in an industrial inline DC magnetron sputter facility Z600 (Leybold Heraeus, Steinhausen, Switzerland) at the company SwissNeutronics AG in Switzerland (Klingnau, Switzerland). To provide good thermalization of the substrate during the deposition, the substrate was screwed down on the carrier plate of the sputtering device. High-purity Ti (PK500 6 mm) and Pt (PK75 2 mm) targets were used for the deposition process. A distance of 61 mm between the substrate and cathode was maintained for all deposition runs. The argon pressure or flow rate were kept constant at 1.2×10^{-3} mbar or 55 sccm, respectively. These sputter gas settings were used in literature to produce 1 μm thick Ti layers with low tensile stress [21]. A deposition rate for Ti of approximately 1.2 nm/s and 1.5 nm/s for Pt were achieved with a sputtering power of 1000 W (Ti) and 200 W (Pt), respectively. To prevent excessive load acting on the growing film as a result of the thermal expansion of the SM, it was crucial to keep the substrate temperature during the initial phase of the deposition process well below 40 °C, which is the glass transition temperature of QPAC 40. Using a system for time-resolved substrate temperature monitoring on the Z600 sputtering facility enabled the evaluation of the required low-temperature deposition process. Reverse sputtering, which is described by Tsuchiya et al. 2005 [21] as a mechanism to reduce residual stress in thick (>0.5 μm) Ti films, is not compatible with a polymer SM due to the excessive heating of the substrate during this process step. The active cooling of the substrate might be a solution to mitigate that problem. However, Ti/Pt ML deposition is regarded as an alternative approach for lowering residual stress that is compatible with a low-temperature deposition process and fulfills biocompatibility requirements. In an ideal multilayer system, the two materials exhibit residual stress with opposite signs, a stress level, and layer composition that in combination produce a stress-free multi-layer coating [22]. The fact that zero stress prevails only for a specific design temperature is less critical for an implant that operates in the human body at a constant temperature. We want to emphasize that the investigations of residual stress formation in multilayer coatings are still ongoing and not the focus of the present paper.

For a comparison of the residual stress estimated from the measured diaphragm stiffness data, stress measurements based on the deflection method applied on an SCS test beam were conducted as well. The coated silicon test beam with a certain curvature was placed on a precision edge, and the resulting maximum gap was measured with a calibrated digital microscope. From the distance and the beam dimensions (length: 150 mm, thickness: 0.3 mm), the stress was determined using Stoney's equation [23] with a measurement uncertainty of 0.06 GPa.

2.1.5. Thermal Decomposition of the SM

The thermal decomposition of QPAC 40 was carried out in argon to prevent oxide formation on the diaphragm and hence, an additional source of stress generation. During the decomposition in an inert gas atmosphere, QPAC 40 went over into a cyclic carbonate vapor leaving minimal ash residues (<10 ppm) [24]. The decomposition process was completed at 350 °C. A stainless-steel pressure container with a cylindrical processing chamber ($\varnothing$50 mm × 80 mm) was used for the thermal treatment process. The chamber temperature was monitored using an integrated PT1000 temperature probe. A circulating air oven (SNOL 58/350 LSN11, Boldt Wärmetechnik, Biebertal, Germany) was used to heat the container to the desired temperature. Before heating, the processing chamber was purged with argon with repeated evacuation steps (10×) in between (60RVD, Sirio Dental SRL, Meldola FC, Italy). The substrate was heated up to 350 °C with a maximum heating rate of 2.8 °C/min followed by a holding phase of 2 h before the oven was switched off and the substrate was cooled down to ambient temperature with a maximum cooling rate of 1 °C/min. Before the cooling phase started, the purge cycle was repeated to remove all decomposition products from the processing chamber. Before the purge gas entered the chamber, it was heated up to the substrate temperature using a coil heat exchanger in the gas supply line that was situated together with the pressure container inside of the oven.

2.2. Diaphragm Inspection and Testing

2.2.1. Diaphragm Surface Shape Characterization

The surface geometry of the diaphragm was characterized by optical inspection and by measuring its curvature depth (CD) before and after the diaphragm was released from the SM. Both procedures were conducted using a Leica DMR light microscope (Leica, Heerbrugg, Switzerland) equipped with different polarization filters. The CD is defined as the vertical distance between the rim and the center of the diaphragm (cf. Figure 5). A frequently used measure for surface curvature is the radius of curvature (CR), which can be calculated from the CD using the formula in Figure 5, where r depicts the radius of the diaphragm.

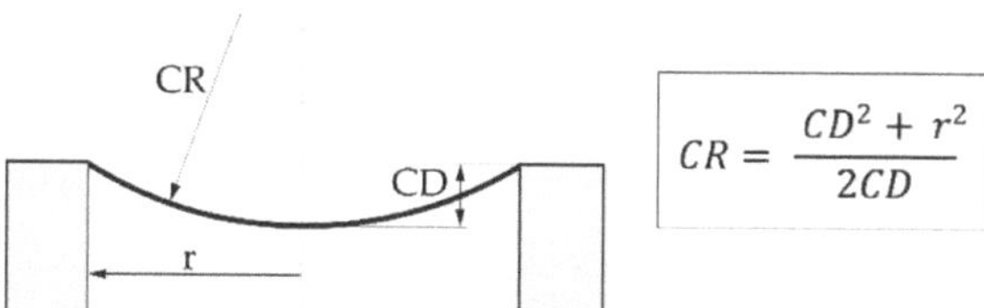

$$CR = \frac{CD^2 + r^2}{2CD}$$

Figure 5. Relationship between the curvature depth (CD) and the curvature radius (CR) of a clamped circular diaphragm with a radius r.

The CD of the diaphragm was measured by the focus variation (FV) technique using the light microscope. The FV technique uses the small depth of field (DOF) of high-magnification microscope optical systems to obtain the depth information of the object's surface topography. The DOF for an objective with a magnification of 20 and a numerical aperture of 0.5 is approximately 2 μm [25]. During CD measurement, the diaphragm's edge and the center were successively brought into focus by the visual observation of the corresponding roughness topography and by moving the vertical micro-positioning stage of the microscope. The corresponding vertical position was read with an optical incremental encoder (AEDB-9140, Broadcom, San José, CA, USA) and a LabView interface (Version 2017, National Instruments, Austin, TX, USA). Despite the very smooth diaphragm surface, an uncertainty of CD measurement below 5 μm was achieved using the FV technique.

2.2.2. Diaphragm Stiffness Measurement

The diaphragm stiffness K_m was determined by measuring the diaphragm response to acoustic stimulation using a single-point laser Doppler velocimeter (LDV, CLV-2534, Polytec GmbH, Waldbronn, Germany). The stiffness represents the ratio between the force acting on the diaphragm and the corresponding diaphragm dynamic center displacement w_c obtained from the LDV measurement (Equation (1)). The force is the product of the stimulation sound pressure p and the diaphragm's surface area A.

$$K_m = \frac{pA}{w_c} \tag{1}$$

The interrogation laser beam was directed normal to the front surface of the diaphragm and manually adjusted in the center under a surgical microscope. The acoustic sound impinged the diaphragm from the rear side of the sample (cf. Figure 6). To interface the individual samples with the sound source (ER-2 loudspeaker, Etymotic Research Inc., Elk Grove Village, IL, USA), the substrate was installed on a sample carrier with a pressure port underneath each sample. An airtight coupling between the sample and the carrier was achieved using a rubber X-ring seal underneath each sample and the same clamping plate as employed for sample filling. To minimize deformations of the sample from clamping and hence clamping-induced variations of the diaphragm stress, only a small clamping force was applied on the sample. A Luer-Lock pneumatic fitting was used to interface the sound delivery hose of the loudspeaker system with the pressure port of the sample under investigation. The sound pressure driving the diaphragm was monitored using a microphone (ER-7C, Etymotic Research Inc., Elk Grove Village, IL, USA) optimized for measurements in a human ear canal and thus well suited for measurements inside the closed cavity underneath the sample. The tiny probe tube of the microphone was fed through the pneumatic connector, and the tube's free end was positioned directly in front of the sample's rear side (cf. Figure 6). A multi-channel audio analyzer (APx585, Audio Precision Inc., Beaverton, OR, USA) was used to acquire the microphone and the LDV output signals and for outputting the signal that drove the loudspeaker via a power amplifier (RMX 850, QSC Audio Products LLC, Costa Mesa, CA, USA). The data acquisition and signal generation were controlled from a custom-built LabVIEW interface (Version 2017, National Instruments, Austin, TX, USA). To increase the robustness of the stiffness measurement, five frequency points between 1 and 5 kHz were considered to finally obtain an average stiffness value. The LDV signal representing a velocity was filtered using a digital narrow bandpass filter (third-order Butterworth, one-third octave band) before the RMS value was calculated and the velocity divided by the frequency to obtain the vibration amplitude. An excitation sound pressure between 115 and 125 dB SPL was applied to drive the diaphragms in the considered frequency range. With a proper installation of the sample (low clamping force) the setup allowed diaphragm stiffness measurement with an uncertainty smaller than 5% of the measured stiffness value.

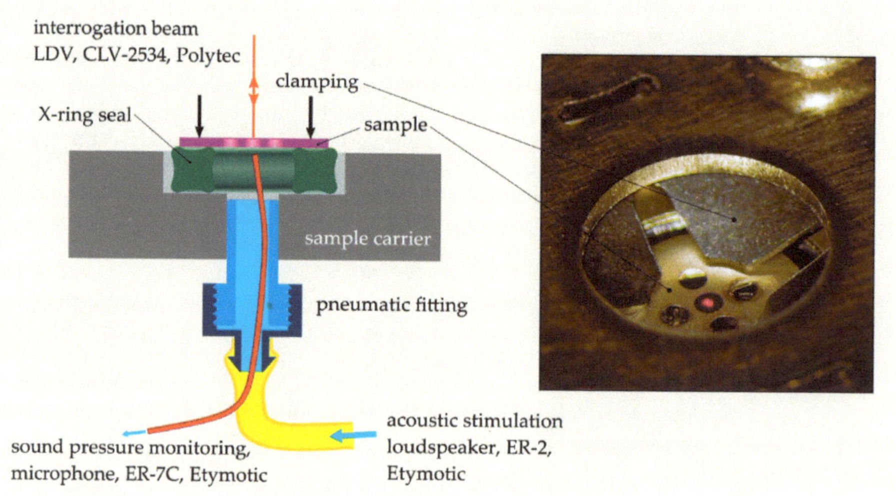

Figure 6. Schematic drawing of the setup for stiffness measurement on the fabricated diaphragms. The method relies on the measurement of the diaphragm response to acoustic stimulation using a single-point LDV system. The microscope image shows a sample with diaphragms clamped down on a rubber seal ring. The interrogation laser beam is positioned in the center of the diaphragm. Three of the five diaphragms are broken and thus sealed with epoxy adhesive.

2.2.3. Intrinsic Stress Estimation

The intrinsic stress in the diaphragm with a thickness t and radius r was estimated from the measured diaphragm stiffness and using a semi-empirical formula that represents the relationship between the pressure load p acting on a flat, circular diaphragm with intrinsic stress σ and the resulting diaphragm center deflection w_c (Equation (2) [26]).

$$\frac{pr^4}{Et^4} = \left(\frac{16}{3(1-v^2)} + \frac{4\sigma r^2}{Et^2} \right) \left(\frac{w_c}{t} \right) + \frac{2.83}{(1-v^2)} \left(\frac{w_c^3}{t^3} \right) \tag{2}$$

The Young's modulus and Poisson ratio of the diaphragm are denoted with E and v, respectively. The second term on the right side of Equation (2) accounts for a large deflection behavior ($w_c \gg t$), which does not prevail for the present application. Hence, the term is neglected. To obtain a formula (Equation (5)) that directly relates the stiffness and the intrinsic stress, the equation was rewritten using the following definitions (Equations (3) and (4)):

$$E' = \frac{E}{1-v^2} \tag{3}$$

$$D = \frac{E't^3}{12} = E'I \tag{4}$$

Equation (4) describes the flexural rigidity D of a plate, where E' denotes the effective Young's modulus (Equation (3)) and I the area moment of inertia per unit length.

$$\sigma = \frac{1}{4\pi t} \left(K_m - \frac{64\pi D}{r^2} \right) \tag{5}$$

Most of the fabricated diaphragms represent multi-layer coatings built from Ti ($E_{Ti} = 102$ GPa, $v_{Ti} = 0.33$) and Pt ($E_{Pt} = 172$ GPa, $v_{Pt} = 0.38$) with different layer designs and hence a different flexural rigidity $\overline{E'I}$. Based on the approach described by

Guo et al. [27], the parameter D was calculated for all layer designs considered in the present study (cf. Table 1). The chosen material properties are typical values for bulk Ti (grade 2) and Pt which might deviate from the corresponding properties of sputtered Ti and Pt films. Therefore, the present approach must be regarded as an estimate of the intrinsic stress in a clamped, circular diaphragm with flat surface geometry.

Table 1. Coating designs, sputter process parameters, and material properties of the diaphragms considered for the present study. For all sputter runs, argon at 55 sccm (0.9 mTorr) was used as a sputtering gas.

Sample No.	Coating	Number of Layers [#]	Power Ti/Pt Target [W]	Intrinsic Stress on SCS [MPa]	Flexural Rigidity [Nm $\times$ 10^{-9}] [1]
H2	425 nm Ti/150 nm Pt/425 nm Ti	3	1000/200	91	9.49
H3	460 nm Ti/90 nm Pt/460 nm Ti	3	1000/200	334	9.76
H4	425 nm Ti/150 nm Pt/425 nm Ti	3	1500/200	110	9.49
I3	227 nm Ti [30 nm Pt/227 nm Ti] 3	7	1000/200	345	9.76
I4	82 nm Ti [10 nm Pt/82 nm Ti] 10	21	1000/200	221	10.14
J1	300 nm Ti [10 nm Pt/60 nm Ti] 10	21	1000/200	115	9.94
J2	1000 nm Ti	1	1500	340	9.47

[1] The flexural rigidity was calculated based on the approach described by Guo et al. and assuming the following material properties for Ti and Pt: $E_{Ti} = 102$ GPa, $\nu_{Ti} = 0.33$, $E_{Pt} = 172$ GPa, $\nu_{Pt} = 0.38$

3. Results

Figure 7a,b shows microscope images of a typical 1 µm thick and 0.6 mm large Ti/Pt diaphragm fabricated on a 0.3 mm thick Ti substrate using the new developed fabrication technology. The right image is taken at higher magnification and zooms to the edge region of the diaphragm. The diaphragm has a flat geometry and an estimated surface roughness in the nanometer range (specular appearance under visible light). In contrast, the estimated surface roughness of the substrate is in the order of the diaphragm thickness. Preliminary tests have shown that the diaphragms can withstand static pressure loads of up to 1 atm and fulfill requirements for hermeticity according to typical helium leak tests (He leak rate $\leq 1 \times 10^{-10}$ mbar L/s).

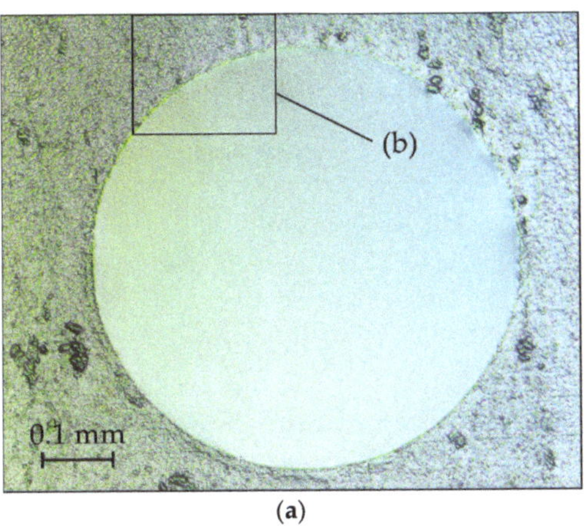

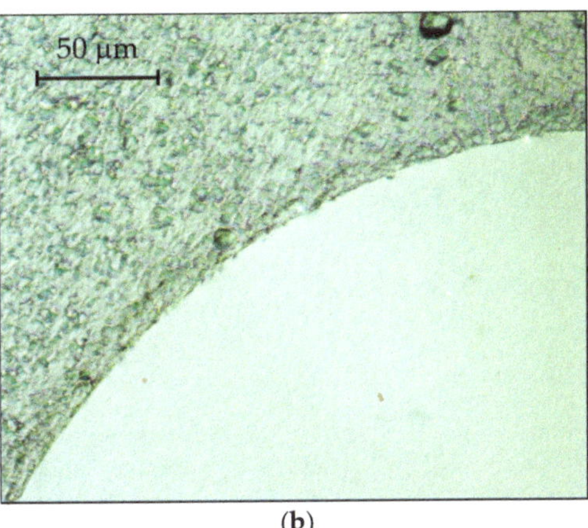

Figure 7. Microscope images of a typical Ti/Pt diaphragm supported by a commercial pure cold-rolled Ti plate: (**a**) image taken at a magnification of 100×; (**b**) zoomed view on the edge region of the diaphragm (magnification 200×).

The first eigenfrequency of a clamped circular Ti diaphragm with a 0.6 mm diameter and 1 µm thickness undergoing free vibration in air is located above 30 kHz [28]. Well below the resonance frequency, the diaphragm is purely stiffness driven; i.e., the vibration amplitude is directly related to the mechanical stiffness of the diaphragm. The cavity underneath the sample is sufficiently large to neglect any contribution to the diaphragm's stiffness. Figure 8a confirms the stiffness-controlled vibration behavior of the diaphragms by the flat frequency response to acoustic stimulation between 800 Hz and 5 kHz (maximum considered frequency). The data represent the diaphragm stiffness calculated according to Equation (1) and determined from LDV measurements on various diaphragms with different stiffness values, which were driven with sound pressure levels as depicted in Figure 8b. At low frequencies, the stiffness starts to deviate from a flat behavior. The falling stiffness is associated with a low signal-to-noise ratio that results from a decaying vibration velocity with decreasing frequency and stimulation strength (cf. Figure 8b). At frequencies below 1 kHz and low vibration velocities, the noise of the LDV increases with decreasing frequency, leading to the measurement of a higher diaphragm response and correspondingly lower stiffness values than expected [29]. Diaphragms with low stiffness show a flat response already above 200 Hz, whereas for high stiffness values, flat behavior arises first above 800 Hz. To ensure robust stiffness measurement even for very stiff diaphragms, only data between 1 kHz and 5 kHz were considered for averaging over the frequency band (cf. green area in Figure 8a,b). The upper frequency bound was set to maintain sound pressure levels above 110 dB SPL and to prevent resonance phenomena in the acoustic supply line of the loudspeaker.

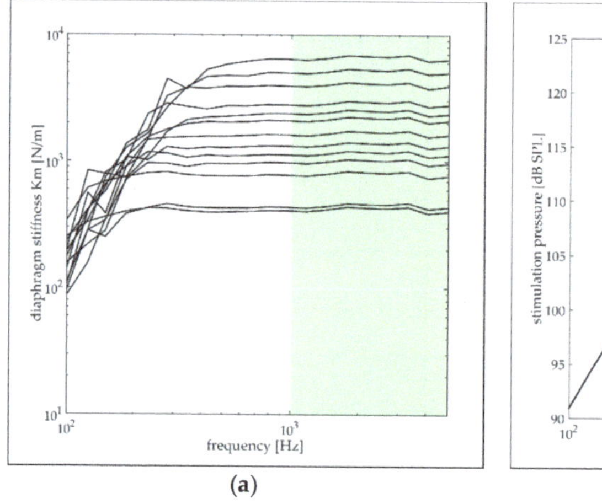
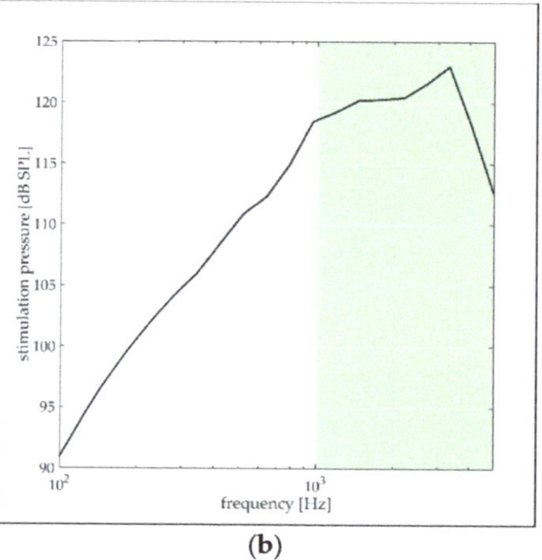

(a) (b)

Figure 8. (a) Diaphragm stiffness as a function of the stimulation frequency between 100 Hz and 5 kHz. The stiffness data were determined from vibration amplitude measurements (LDV) on the diaphragms driven with acoustic sound; (b) sound pressure levels driving the diaphragms between 100 Hz and 5 kHz. The green area in both figures indicates the frequency range considered for averaging the stiffness data over the frequency.

The diaphragm fabrication trials revealed that, depending on the surface shape of the SM that defines the shape of the deposited coating, diaphragms with a flat or curved geometry are formed. Moreover, it was observed that the noted surface geometry of the SM has a significant influence on the diaphragm's intrinsic stress. This dependence is illustrated in Figure 9 by data from diaphragms of two (H2 and H4) of the total seven substrates considered in the present study (cf. Table 1). The coating design of both substrates is identical. It represents a three-layer design composed of two Ti outer layers (425 nm) and one Pt middle layer (150 nm, cf. Table 1) with an overall thickness of 1 μm. The deposition process differs only in the sputter power for Ti, which was increased from 1 kW to 1.5 kW for substrate H4. The figure shows the averaged (over frequency) stiffness data K_m of the diaphragms plotted as a function of the curvature radius CRBR that characterizes the shape of the diaphragm before it is released from the SM. The curvature radius was determined from the measured CD data according to the equation depicted in Figure 5. The intrinsic stress calculated from the corresponding stiffness data using Equation (5) is shown on the right y-axis of Figure 9. The stress data are only valid for diaphragms with a flat surface geometry.

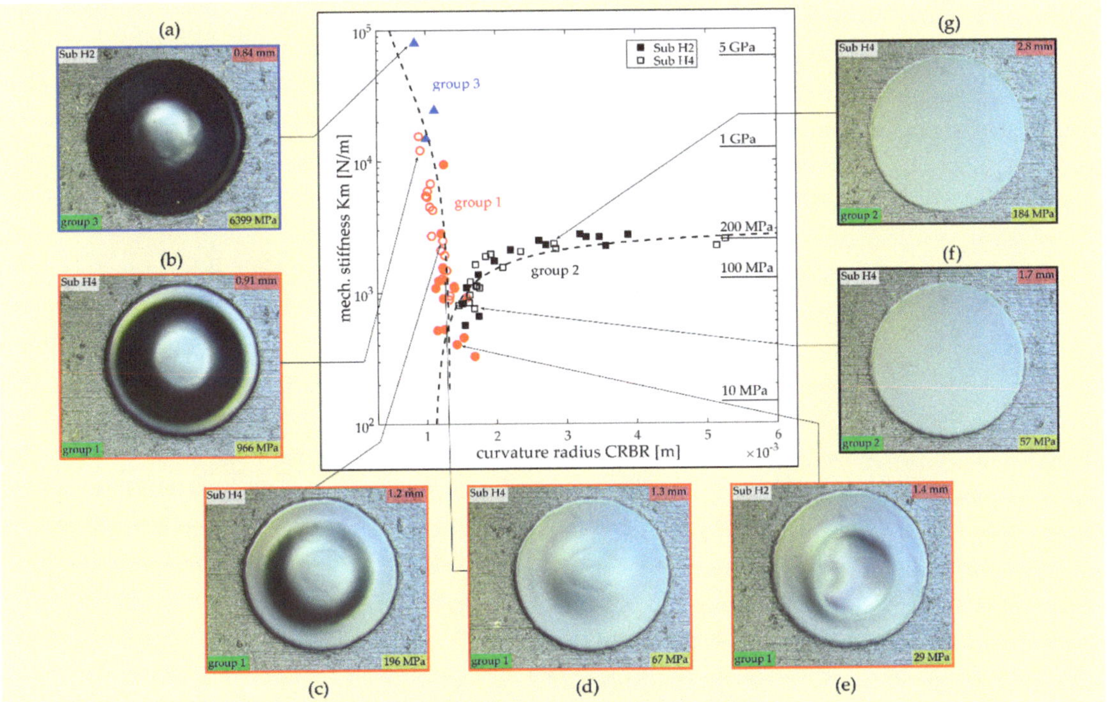

Figure 9. Diaphragm stiffness as a function of the diaphragm's curvature radius before it is released from the SM (CRBR) for diaphragms of the substrate H2 (filled markers) and H4 (empty markers). Group 1 diaphragms are indicated with red circles, group 2 with black squares, and group 3 with blue triangles. The stiffness-related intrinsic stress is indicated on the right y-axis (only valid for group 2 diaphragms). The dashed lines represent trendlines assuming linear behavior if the stiffness is considered in linear space. (**a–g**) Optical microscope images showing the diaphragm's surface geometry (after release) depending on the CRBR. The CRBR and the intrinsic stress values are indicated in the images with a red and yellow background color, respectively. The group name is highlighted green.

Diaphragms with a concave shape with high curvature before they are released from the SM (i.e., small CRBR) show very high stiffness. The stiffness rapidly drops with increasing CRBR and reaches minimum values for CRBRs of approximately 1.3 mm. With further increases in the CRBR value, the stiffness starts to rise again but with a smaller slope than for CRBRs below 1.3 mm. Above a CRBR of 3 mm, the stiffness takes a constant value of approximately 2500 N/m.

Considering the diaphragm's surface geometry after it is released from the SM allows the diaphragms to be classified into three groups. Group 2 represents the diaphragms that are flat after the release process (cf. Figure 9f,g). These diaphragms arise when the CRBR is higher than 1.3 mm. In group 3, the diaphragms remain curved with a concave shape even after they are released from the SM (cf. Figure 9a). They exhibit very high stiffness, which mainly originates from the reinforcement effect associated with a spherical thin-wall geometry. Finally, group 1 contains all diaphragms with a surface geometry that describes an intermediate state between a flat and a curved shape (cf. Figure 9b–e). Within that group, the diaphragm stiffness decreases with smaller CD (after release) and as the fraction of the flat diaphragm surface increases. The lowest stiffness values are reached at the transition between group 1 and group 2. It has to be emphasized that the estimated stress values are only valid for flat diaphragms (group 2). Considering the group 2 diaphragms (flat surface geometry), the varying stiffness with CRBR can be only attributed to varying intrinsic stress. The data of group 2 diaphragms indicate that the intrinsic stress in the diaphragm can be significantly reduced (from approximately 200 MPa to 30 MPa) if the surface geometry of the SM is adjusted for an optimum CRBR ($\approx$1.3 mm) during the filling process.

From Figure 9, it is obvious that the diaphragms of group 2 change the surface geometry while they are released from the SM. As depicted in Figure 10, this is also true for the diaphragms of the two other groups. The figure shows optical microscope images of representative diaphragms of each group before (first row) and after (second row) the release process. The corresponding CD is indicated by the red tag in each image. Before release, all diaphragms show a curved geometry with a CDBR between 25 μm and 54 μm, whereas after release the diaphragms adopt a geometry with smaller curvature. The curvature change (ΔCD) provoked by the release process varies in a broad range with the largest change for the group 2 diaphragm (from 25 μm to 0 μm) followed by group 1 (from 34 μm to 22 μm). For the diaphragm of group 3, only a small change in the CD was identified (from 54 μm to 50 μm).

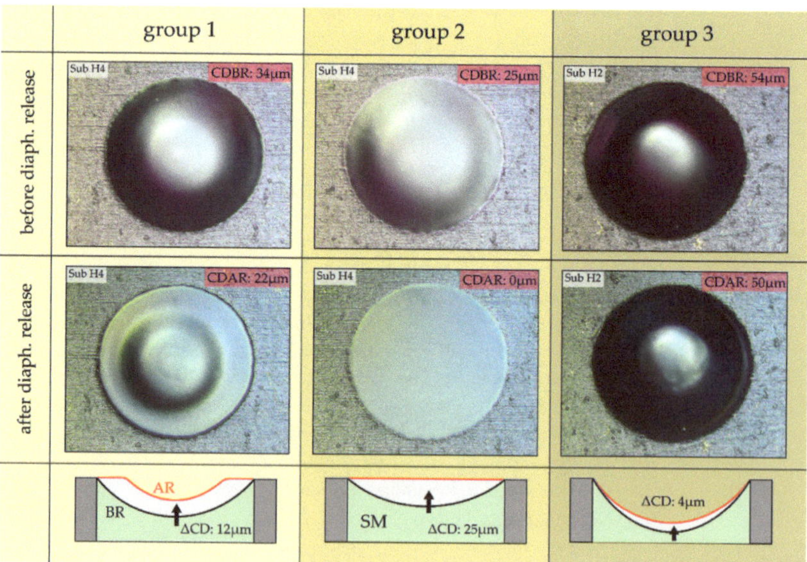

Figure 10. Optical microscope images of typical diaphragms of the three groups before (first row) and after (second row) they are released from the SM. The corresponding curvature depth is indicated with a red tag in the images. The bottom row shows schematic illustrations of the diaphragm profile geometry of each group before (BR) and after release (AR). Dimensions do not scale.

In Figure 11a, the ΔCD data of all diaphragms on substrate H2 and H4 are plotted against the CD before they are released from the SM. The data are again color-coded according to the three diaphragm groups. The effect of the ΔCD on the diaphragm stiffness is illustrated in Figure 11b. The ΔCD of group 2 diaphragms increases linearly with increasing CDBR up to the transition region (TR) where the group 2 diaphragms go into group 1. The TR is associated with the occurrence of the lowest intrinsic stress or stiffness, respectively. From there, the ΔCD starts to drop rapidly with further increases in the CDBR up to the diaphragms of group 3 where only a small ΔCD was identified. The high spreading of the group 1 data is explained with the uncertainty of CD measurement of 5 µm and the less defined surface geometry (curved and flat parts) after diaphragm release compared with a flat (group 2) or fully curved (group 3) diaphragm geometry. We assume that the intrinsic stress in the diaphragm formed during the deposition process is the driving force for the ΔCD causing full or partial stress relief while the diaphragm is released from the SM. Despite the expected full stress relief for the diaphragms of groups 1 and 3, they exhibit high stiffness as a result of the non-flat surface geometry. With an increasing CDBR and hence higher diaphragm stiffness, the stress relief mechanism loses its effectiveness as a driving force for ΔCD.

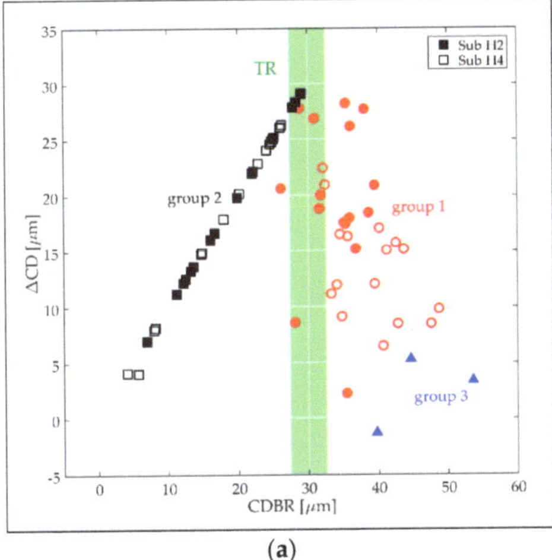

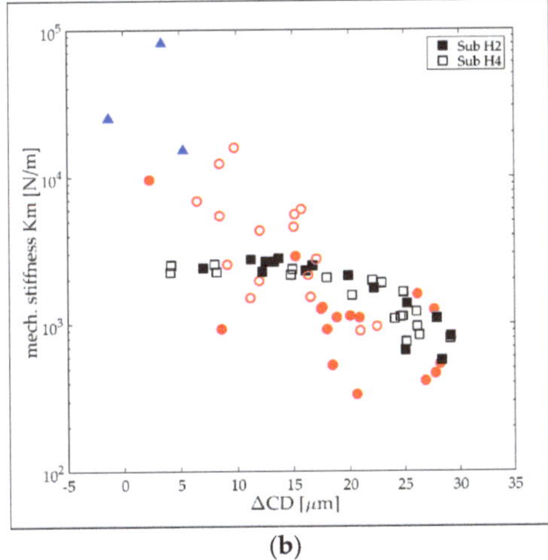

Figure 11. (**a**) Release induced change of the diaphragms' CD (ΔCD) as a function of the corresponding CD value before the release process (CDBR). Considered are again data of the diaphragms on substrate H2 (filled markers) and H4 (empty markers). The data are color-coded depending on the diaphragm group. The green shaded area represents the transition region between group 1 and 2 diaphragms; (**b**) mechanical diaphragm stiffness plotted versus ΔCD.

Different Ti/Pt multi-layer coatings were tested to reduce the intrinsic stress in the diaphragm by optimizing the coating design and the sputtering process. The goal was to verify if, in combination with the stress relief mechanism discussed above, nearly stress-free diaphragms can be fabricated. The coatings that were tested so far showed a tensile stress on SCS beams (reference method for stress characterization) between 90 and 345 MPa (cf. Table 1). The corresponding stress values obtained from measurements of the diaphragm stiffness are shown in Figure 12. The data are again plotted as a function of the CRBR to illustrate the effect of the different coating designs on the stiffness of diaphragms of the different groups. The stiffness of group 1 and 3 diaphragms (CRBR < 1.3 mm) is not affected by the coating design at all. This outcome appears to be reasonable, as the stiffness of these diaphragms is mainly determined by the curved geometry. Because of the stress relief mechanism, the contribution of intrinsic stress on the diaphragm's stiffness should be negligible. In contrast, the flat diaphragms (group 2) show clear variations in intrinsic stress between the different substrates. To compare the diaphragm stress with the measured stress on the SCS beams, diaphragms with CRBRs larger than 3 mm (stress relief mechanism should have a small effect) were considered. On substrate H3, similar stress levels were observed as on the corresponding SCS stripe, whereas the diaphragms of the substrates I3 and J2 showed considerably lower stress than the corresponding reference values (>340 MPa). In contrast, the low reference stress in the range of 100 MPa was not found on the diaphragms. The deviations are not surprising. In contrast to the SCS beams, the diaphragms were deposited on a polymer SM which could influence the crystal growth during deposition. It is also conceivable that, compared with SCS, a different adhesion between the coating and SM influences the evolution of intrinsic stress during the deposition process that differs from the adhesion on SCS and causes stress relief. At last, the diaphragms form a free-standing, clamped plate structure with boundary conditions that could influence stress formation. Considering the group 2 diaphragms closer to the TR, stress variations between the substrates could not be assigned to the specific sputter process but rather to the strong dependence of the CRBR on the diaphragm stress within the TR.

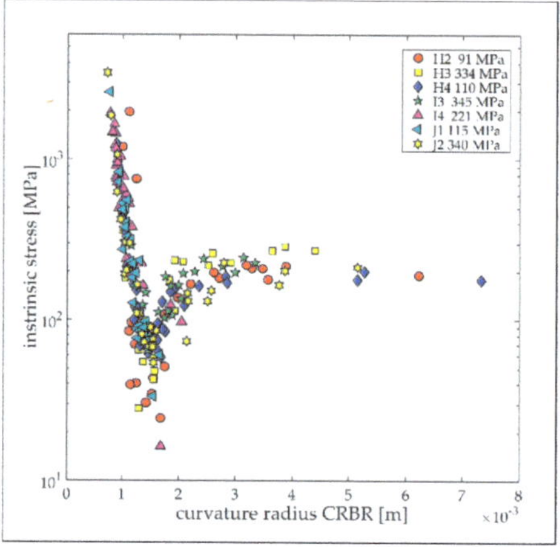

Figure 12. Intrinsic stress as a function of the CRBR for different mono- and multi-layer coatings. The stress values are only valid for diaphragms with a flat surface geometry (group 2, CRBR > 1.3 mm). The stress levels of the coatings deposited on SCS beams are indicated in the legend.

4. Discussion

The present study introduces a new fabrication process for micron-thick Ti and Ti/Pt multilayered diaphragms. In contrast to typical MEMS fabrication technology for diaphragm fabrication, the introduced fabrication process does not rely on anisotropic dry etching, which is not well established for the processing of Ti substrates. In addition, dry etching in combination with a sacrificial material layer as an etch stop requires highly polished substrates. Such substrates made from Ti are not off-the-shelf products such as silicon wafers; they have to be custom-made by chemical mechanical polishing (CMP), a process that is also not very well established for Ti in typical MEMS fabrication facilities.

We could show that the introduced fabrication process seems to be compatible with unpolished substrates with a surface roughness similar to the thickness of the diaphragm. The possibility of using untreated Ti substrates lowers the fabrication complexity and the associated costs but also provides more freedom during the design process of the diaphragm support structure (i.e., 3D geometries such as tubes, etc.). We assume that the higher roughness of the substrate might even improve the strength of the interface between the diaphragm and the support, providing a kind of interdigitation mechanism with an increased contact area at the interface. Currently, a study is ongoing to better understand and assess the interface between the diaphragm and the support using more dedicated inspection techniques, such as scanning and transmission electron microscopy in combination with static pressure loading and hermeticity tests.

Another advantage of the new diaphragm fabrication technology is the ease of defining a nonplanar surface shape of the diaphragm before it is released from the SM. Currently, the SM is applied as a solution with low viscosity where surface tension forces enforce the formation of a spherical surface shape with a curvature radius defined by the filling state. During the thermal decomposition of the SM, the diaphragm tends to adopt a flat surface geometry. It is assumed that the mechanism is driven by tensile intrinsic stress in the coating and results in full or partial stress relief while the diaphragm is released from the SM. A similar stress relief mechanism was reported by Wang et al. [30] on single deeply corrugated diaphragms (SDCD) fabricated from polysilicon on a silicon substrate. The design contained a flat membrane with suspending sidewalls forming a deep corrugation. They observed a zero-pressure deflection (cf. ΔCD) of the thin-wall structure when it was released from the sacrificial layer. Compared with a flat diaphragm design with equal intrinsic stress, the SDCD showed the considerably higher mechanical sensitivity of the transducer, which was increasing with higher corrugation depths. Wang et al. concluded that the suspending side walls have a relieving effect on the intrinsic stress in the sensing membrane. The group 2 diaphragms of the present study can be regarded as an SDCD before they are released from the SM. In contrast to the design of Wang et al., the group 2 diaphragms undergo a larger zero-pressure offset during their transition into a flat geometry state. Similar zero-pressure offset values as reported by Wang et al. were only observed on diaphragms of group 3 with CDBR values (cf. corrugation height) larger than 40 μm. Such diaphragms show very high stiffness as a result of the curved geometry, whereas the SDCD design benefits from the flat and stress-free section that is used for pressure sensing.

The CRBR transition region (TR) between group 1 and group 2 diaphragms (CRBR $\approx$ 1.3 mm) is narrow and is characterized by the occurrence of lowest stress values that increase rapidly on both sides of the TR (cf. Figure 13). The limited accuracy of CD measurement using the focusing technique and the number of available data do not currently allow a more precise characterization of the TR, such as the width and potential shifts in CR depending on the stress level of the coating. The coatings considered in the present study were mainly designed to obtain low stress. Hence, the largest stress differences between the coatings are not larger than 100 MPa (considering the stress of group 2 diaphragms at high CRBRs). It is expected that if the stress in the coating is further reduced, the TR will shift to higher CRBRs (cf. red curve in Figure 13). Higher stress levels should shift the TR in the opposite CRBR direction. Although diaphragms within the TR could benefit from a high stress relief, the steep gradients on both sides would make the control of the fabrication process difficult. The design point on the Km-CRBR characteristic curve must be situated on the right side of the TR (group 2 diaphragms) in a region where the slope of the curve complies with the allowable part-to-part variation. To shift the design point as close as possible to the TR and hence benefit from a high stress relief, the process for applying the SM on the substrate (i.e., cavities) needs to be optimized concerning a precise adjustment of the desired CRBR or CDBR, respectively.

However, to obtain diaphragms nearly free of intrinsic stress, the stress relief mechanism must be combined with other technologies for stress reduction in the coating, such as sputter process parameter optimization, multi-layer coatings, and thermal annealing. A potential multi-step strategy is depicted in Figure 13. In the first step, the intrinsic stress in the coating must be reduced by an optimization of the sputtering process and by the use of a multi-layer coating design. The minimum stress that can be achieved will result from a trade-off behavior between the stress and the mechanical strength of the diaphragm (pores, voids, and defects). Using the stress relief mechanism presented in the present study would be the second step to further reduce the stress in the diaphragm. Here, the potential of stress reduction depends on the aforementioned fabrication process considerations and the characteristics of the Km-CRBR curve at low-stress level values, which are currently only estimated. Thermal annealing is considered as the third step to end up with a stress-free diaphragm. Annealing could be an efficient way to reduce stress in a PD because the protective enclosure of the transducer in which the PD is integrated is free of electronic parts that would limit the process to low annealing temperatures. Currently, a study is ongoing to evaluate a Ti/Pt multi-layer coating with low stress, high mechanical strength, and long-term hermeticity. In addition, thermal annealing shall be applied on the diaphragm samples to verify if it is suitable to fully eliminate the remaining residual stress.

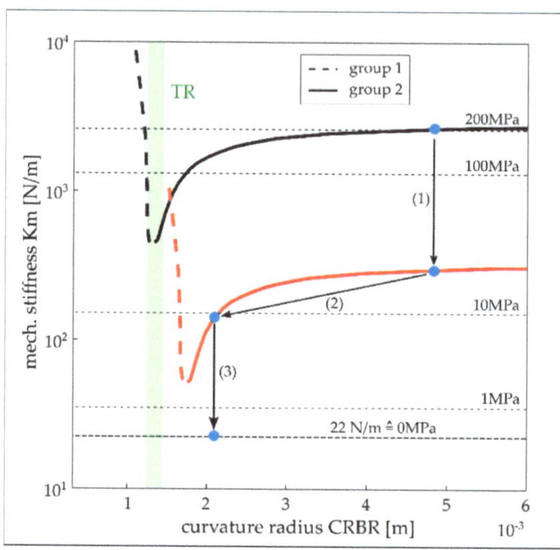

Figure 13. A possible strategy for the fabrication of stress-free diaphragms illustrated in a Km-CRBR plot: (1) Ti/Pt multi-layer coating optimized for low stress and high mechanical strength; (2) stress relief mechanism as a result of the adjustment of an optimum diaphragm surface shape before it is released from the SM; (3) thermal annealing. Remark: The stress values are valid for a diaphragm with 0.6 mm diameter and 1 mm thickness and made from an H2 coating. For zero intrinsic stress, the diaphragm exhibits a stiffness of 22 N/m.

As reported in Prochazka et al. [6], the diaphragm fabrication technology was already applied on the sound receptor of the ICAR, which represents a tube-like structure with a rectangular cross-section and multiple PDs arranged within the tip region on two opposite sides (cf. Figure 1b). The feasibility of critical process steps such as the more complex filling with SM and the release of gas products through a micro-channel during the thermal decomposition of the SM were verified by several tests. Further process optimizations are ongoing. In addition to the aforementioned evaluation of a process for PDs free of intrinsic stress, the filling process needs further optimization to make the process faster and more repeatable. Currently, a polymer solution with only 20 weight percent of QPAC is used as an SM for filling the cavities. The process must be conducted in multiple steps because most of the added QPAC volume is lost during the curing of the solution. Using a polymer melt instead of a solution might considerably reduce the required filling cycles, but is also associated with a more complex dispensing system required for a melt instead of a solution. Further improvement of the filling efficiency but also precision and repeatability might be achieved by an automated filling process using an appropriate robot system.

5. Conclusions

Overall, our results clearly show that using thin film deposition techniques on a low-temperature decomposable polymer as a SM seems to be a promising technology for the fabrication of thin-walled metallic encapsulations in packaging solutions. In particular, the technology is beneficial if the packaging solution incorporates substrate materials and a complex 3D geometry that are not in line with the typical requirements of classical MEMS fabrication processes such as the use of polished silicon 2D substrates. The present study demonstrates the capability of the technology for the fabrication of 1 μm thick and sub-millimeter large Ti (or Ti/Pt multi-layer) diaphragms on an unpolished commercially pure Ti substrate. Currently, the SM is applied on the substrate using adhesive dispensing technology, which suits application with a non-planar surface shape of the deposited diaphragm before it is released from the SM well. After the removal of the SM, the diaphragm takes a surface shape with a smaller curvature, causing a certain relief of intrinsic stress. We could determine a curvature range with the largest stress relief from stiffness measurements on numerous diaphragms with different surface curvatures before their release.

A next step would be to further optimize the fabrication process to obtain robust and hermetic diaphragms with near-zero intrinsic stress. These PDs are required for the encapsulation of implantable acoustic transducers, such as for a fully implantable cochlear implant system. Further research and development work is required to explore further advantages but also limitations of the present technology in comparison with the classical MEMS technologies for diaphragm fabrication based on bulk and surface micromachining. The focus will be laid on the mechanisms controlling the interface between the PD and the unpolished substrate surface and on alternative processes for the dispensing of the polymer SM to make the process more efficient, precise, and repeatable.

Author Contributions: Conceptualization, L.P., F.P. and A.H.; methodology, L.P., F.P., M.S., N.G. and J.B.; software, L.P.; validation, L.P., M.S., N.G., J.B. and F.P.; formal analysis, L.P., M.S., N.G., J.B. and F.P.; investigation, L.P., M.S., N.G. and F.P.; resources, A.H., F.P., M.S. and J.B.; data curation, L.P. and F.P.; writing—original draft preparation, L.P.; writing—review and editing, L.P., F.P., A.H., M.S., N.G. and J.B.; visualization, L.P.; supervision, A.H., F.P., M.S. and J.B.; project administration, L.P., F.P., M.S. and J.B.; funding acquisition, L.P., F.P., A.H., M.S. and J.B. All authors have read and agreed to the published version of the manuscript.

Funding: This research was funded by The European Union's Horizon 2020 program under the Eurostars (grant number E12584—TiSR); Cochlear AG, European Headquarters, Switzerland; Cochlear Technology Centre Belgium, Mechelen, Belgium.

Acknowledgments: The authors thank Francesca Harris and Stijn Eeckhoudt (Cochlear Technology Centre, Belgium), who provided insight and expertise that greatly assisted the research.

Conflicts of Interest: The authors declare no conflict of interest.

References

1. Bogue, R. Recent developments in MEMS sensors: A review of applications, markets and technologies. *Sens. Rev.* **2013**, *33*, 300–304. [CrossRef]
2. Sutanto, J.; Anand, S.; Sridharan, A.; Korb, R.; Zhou, L.; Baker, M.S.; Okandan, M.; Muthuswamy, J. Packaging and Non-Hermetic Encapsulation Technology for Flip Chip on Implantable MEMS Devices. *J. Microelectromech. Syst.* **2012**, *21*, 882–896. [CrossRef] [PubMed]
3. Meng, E.; Sheybani, R. Insight: Implantable medical devices. *Lab Chip* **2014**, *14*, 3233–3240. [CrossRef] [PubMed]
4. Shah, M.A.; Shah, I.A.; Lee, D.-G.; Hur, S. Design Approaches of MEMS Microphones for Enhanced Performance. *J. Sens.* **2019**, *2019*, 9294528. [CrossRef]
5. Wang, H.; Ma, Y.; Zheng, Q.; Cao, K.; Lu, Y.; Xie, H. Review of Recent Development of MEMS Speakers. *Micromachines* **2021**, *12*, 1257. [CrossRef] [PubMed]
6. Prochazka, L.; Huber, A.; Dobrev, I.; Harris, F.; Dalbert, A.; Röösli, C.; Obrist, D.; Pfiffner, F. Packaging Technology for an Implantable Inner Ear MEMS Microphone. *Sensors* **2019**, *19*, 4487. [CrossRef]
7. Jiang, G.; Zhou, D.D. Technology advances and challenges in hermetic packaging for implantable medical devices. In *Implantable Neural Prostheses 2*; Springer: Berlin/Heidelberg, Germany, 2009; pp. 27–61.
8. Pfiffner, F.; Prochazka, L.; Dobrev, I.; Klein, K.; Sulser, P.; Péus, D.; Sim, J.; Dalbert, A.; Röösli, C.; Obrist, D. Proof of Concept for an Intracochlear Acoustic Receiver for Use in Acute Large Animal Experiments. *Sensors* **2018**, *18*, 3565. [CrossRef]
9. Brunette, D.M.; Tengvall, P.; Textor, M.; Thomsen, P. *Titanium in Medicine: Material Science, Surface Science, Engineering, Biological Responses and Medical Applications*; Springer: Berlin/Heidelberg, Germany, 2001.
10. Donachie, M.J. *Titanium: A Technical Guide*; ASM international: Materials Park, OH, USA, 2000.
11. Brauer, M.; Dehé, A.; Bever, T.; Barzen, S.; Schmitt, S.; Füldner, M.; Aigner, R. Silicon microphone based on surface and bulk micromachining. *J. Micromechanics Microengineering* **2001**, *11*, 319–322. [CrossRef]

12. Scheeper, P.; Olthuis, W.; Bergveld, P. Fabrication of a subminiature silicon condenser microphone using the sacrificial layer technique. In Proceedings of the TRANSDUCERS'91: 1991 International Conference on Solid-State Sensors and Actuators, San Francisco, CA, USA, 24–27 June 1991; pp. 408–411.

13. Aimi, M.F.; Rao, M.P.; Macdonald, N.C.; Zuruzi, A.S.; Bothman, D.P. High-aspect-ratio bulk micromachining of titanium. *Nat. Mater.* **2004**, *3*, 103–105. [CrossRef]

14. Parker, E.R.; Thibeault, B.J.; Aimi, M.F.; Rao, M.P.; Macdonald, N.C. Inductively Coupled Plasma Etching of Bulk Titanium for MEMS Applications. *J. Electrochem. Soc.* **2005**, *152*, C675–C683. [CrossRef]

15. Allen, D.M. The state of the art of photochemical machining at the start of the twenty-first century. *Proc. Inst. Mech. Eng. Part B J. Eng. Manuf.* **2003**, *217*, 643–650. [CrossRef]

16. Zhao, G.; Shu, Q.; Tian, Y.; Zhang, Y.; Chen, J. Wafer level bulk titanium ICP etching using SU8 as an etching mask. *J. Micromech. Microeng.* **2009**, *19*, 095006. [CrossRef]

17. O'Mahony, C.; Hill, M.; Hughes, P.J.; Lane, W.A. Titanium as a micromechanical material. *J. Micromech. Microeng.* **2002**, *12*, 438–443. [CrossRef]

18. Allen, D.M. Photochemical Machining: From 'manufacturing's best kept secret' to a \$6 billion per annum, rapid manufacturing process. *CIRP Ann.* **2004**, *53*, 559–572. [CrossRef]

19. French, P.J.; Sarro, P.M. Surface versus bulk micromachining: The contest for suitable applications. *J. Micromech. Microeng.* **1998**, *8*, 45–53. [CrossRef]

20. Collins, S.D. Etch stop techniques for micromachining. *J. Electrochem. Soc.* **1997**, *144*, 2242. [CrossRef]

21. Tsuchiya, T.; Hirata, M.; Chiba, N. Young's modulus, fracture strain, and tensile strength of sputtered titanium thin films. *Thin Solid Films* **2005**, *484*, 245–250. [CrossRef]

22. Ennos, A.E. Stresses developed in optical film coatings. *Appl. Opt.* **1966**, *5*, 51–61. [CrossRef]

23. Stoney, G.G. The tension of metallic films deposited by electrolysis. In *Proceedings of the Royal Society of London. A. Mathematical and Physical Sciences*; The Royal Society: London, UK, 1909; pp. 172–175.

24. Empower Materials, I. QPAC 40 (PPC): Ash/Metal Impurity Levels after Decomposition. Available online: https://www.empowermaterials.com/low-impurity-levels-after-decomposition (accessed on 25 November 2021).

25. Sheppard, C. Depth of field in optical microscopy. *J. Microsc.* **1988**, *149*, 73–75. [CrossRef]

26. Jerman, J. The fabrication and use of micromachined corrugated silicon diaphragms. *Sens. Actuators Phys.* **1990**, *23*, 988–992. [CrossRef]

27. Guo, X.-G.; Zhou, Z.-F.; Sun, C.; Li, W.-H.; Huang, Q.-A. A simple extraction method of Young's modulus for multilayer films in MEMS applications. *Micromachines* **2017**, *8*, 201. [CrossRef] [PubMed]

28. Blevins, R.D. *Formulas for Natural Frequency and Mode Shape*; Krieger Publishing Company: Malabar, FL, USA, 1979.

29. O'Malley, P.F.; Vignola, J.F.; Judge, J.A. Amplitude and Frequency Dependence of the Signal-to-Noise Ratio in LDV Measurements. In Proceedings of the International Design Engineering Technical Conferences and Computers and Information in Engineering Conference, Washington, DC, USA, 26–29 August 2011; pp. 545–549.

30. Wang, W.; Lin, R.; Ren, Y. Design and fabrication of high sensitive microphone diaphragm using deep corrugation technique. *Microsyst. Technol.* **2004**, *10*, 142–146. [CrossRef]

Article

Acoustic Transmission Measurements for Extracting the Mechanical Properties of Complex 3D MEMS Transducers

Dennis Becker [1,*] , Moritz Littwin [1], Achim Bittner [1,*] and Alfons Dehé [1,2]

[1] Hahn-Schickard, Wilhelm-Schickard Straße 10, 78052 Villingen-Schwenningen, Germany;
     moritz.littwin@hahn-schickard.de (M.L.); alfons.dehe@hahn-schickard.de (A.D.)
[2] Georg H. Endress Chair of Smart Systems Integration, Department of Microsystems Engineering—IMEK,
     Albert-Ludwigs-Universität Freiburg, Georges-Köhler-Allee 103, 79110 Freiburg, Germany
*   Correspondence: dennis.becker@hahn-schickard.de (D.B.); achim.bittner@hahn-schickard.de (A.B.);
     Tel.: +49-7721-943-182 (D.B.)

Abstract: Recent publications on acoustic MEMS transducers present a new three-dimensional folded diaphragm that utilizes buried in-plane vibrating structures to increase the active area from a small chip volume. Characterization of the mechanical properties plays a key role in the development of new MEMS transducers, whereby established measurement methods are usually tailored to structures close to the sample surface. In order to access the lateral vibrations, extensive and destructive sample preparation is required. This work presents a new passive measurement technique that combines acoustic transmission measurements and lumped-element modelling. For diaphragms of different lengths, compliances between 0.08×10^{-15} and 1.04×10^{-15} m^3/Pa are determined without using destructive or complex preparations. In particular, for lengths above 1000 µm, the results differ from numerical simulations by only 4% or less.

Keywords: acoustic transmission; characterization; MEMS transducer; modeling; folded diaphragm

Citation: Becker, D.; Littwin, M.; Bittner, A.; Dehé, A. Acoustic Transmission Measurements for Extracting the Mechanical Properties of Complex 3D MEMS Transducers. *Micromachines* **2024**, *15*, 1070. https://doi.org/10.3390/mi15091070

Academic Editor: Libor Rufer

Received: 26 July 2024
Revised: 22 August 2024
Accepted: 23 August 2024
Published: 24 August 2024

1. Introduction

Micro-electro-mechanical systems (MEMS) enable the integration of miniaturized transducers for various sensing and actuation applications in different domains. Particularly in the acoustic field, a major drawback of miniaturized transducers is performance limitations due to smaller active areas and higher compliances. Innovative MEMS devices are being developed to overcome these limitations. While out-of-plane vibrating membranes are slit [1] or corrugated [2] to increase the displaced volume through higher deflections, three-dimensional approaches with in-plane vibrating structures enable larger active areas from small chip sizes [3,4].

The development of especially three-dimensional MEMS structures relies on efficient and precise characterization methods, e.g., to measure the mechanical or mechano-acoustical properties. In state-of-the-art measurement methods, bulge tests [5] or atomic force microscopy [6] statically deform near-surface structures in the vertical direction in order to determine mechanical stiffness parameters. For lateral motion transducers, these methods can be performed only with extensive sample preparation, where the structures of interest are extracted and mounted to vibrate out-of-plane. However, this changes the behavior of the sample to be characterized and therefore no longer represents the boundary conditions of the intended application. Alternatively, optical measurement techniques such as laser Doppler vibrometry (LDV) or digital holographic microscopy (DHM) are an effective way to characterize MEMS components with out-of-plane and in-plane vibrations [7,8]. The quality and precision of such techniques depend on the structures of interest and their motion being close to the surface and therefore optically accessible. In the case of the folded MEMS diaphragm from [4], the lateral motions of the vertical sidewalls are optically inaccessible because they are deep within the high-aspect-ratio trenches of the diaphragm.

This paper presents a new measurement technique that allows the determination of the mechanical properties of three-dimensional folded MEMS diaphragms. This technique utilizes acoustic transmission measurements (ATMs) to characterize the dynamic mechanical behavior of such membranes without extensive sample preparation. ATMs have been used in several publications to determine the acoustic absorption properties of macroscopic samples such as building materials [9] or structural skin elements [10]. Unlike there, this work passively actuates a MEMS diaphragm, which on the other hand radiates an acoustic signal into a receiving chamber. Combining this method with lumped-element modelling (LEM) allows the determination of mechanical properties on a miniaturized scale for MEMS transducer diaphragms with hidden or deep vertical elements. This will provide an efficient and application-oriented measurement method that will advance the latest developments in three-dimensional acoustic MEMS transducers.

2. Experimental Method

The presented measurement concept utilizes ATMs in order to characterize deep vertical structures of MEMS transducers. This allows a non-destructive characterization of mechanical properties that are essential for the development and optimization of MEMS transducers such as microphones or loudspeakers. The idea of this measurement method is to use two chambers, which are separated by the diaphragm to be characterized. Both chambers contain microphones (Figure 1). A reference loudspeaker is placed in the transmission chamber, where it is acoustically coupled to the MEMS diaphragm to excite it. This creates a damped pressure wave in the second receiving chamber. Parameter fitting with LEM determines the mechanical parameters of the diaphragm from the difference between the two sound pressures.

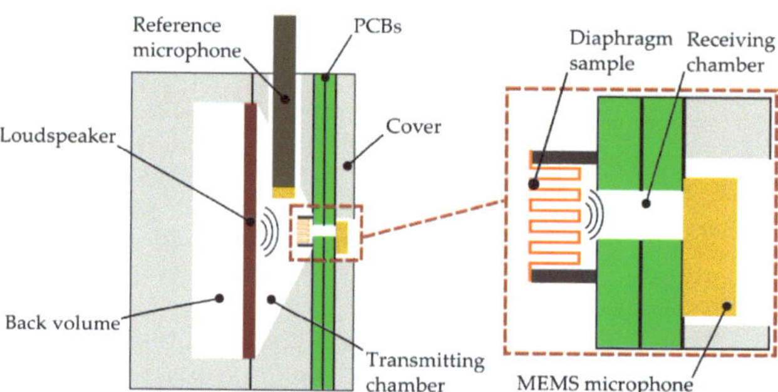

Figure 1. Schematic illustration of the experimental setup.

Figure 1 shows the realization of this measurement concept. The transmitting chamber is a conical volume inside an aluminum block. It contains an electrodynamic loudspeaker (GF0401M) from CUI Devices in Lake Oswego, OR, US and a $\frac{1}{4}$-inch pressure microphone set (46BP-1) from GRAS Sound & Vibration in Holte, Denmark with a flat frequency response up to 80 kHz [11]. A defined back volume is added to prevent external influences on the acoustic measurements and loudspeaker behavior. The diaphragm sample is glued to a carrier PCB with a hole for the sound port.

The volume of the receiving chamber is as small as possible in order to maximize the received SPL, allowing the characterization of low-compliance MEMS samples. This is achieved by using a MEMS microphone (IM69D130) from Infineon in Neubiberg, Germany. The ports of the PCBs thereby define the volume of the receiving chamber. A cover is utilized to mount the components together to ensure an acoustically sealed setup.

To reveal its influence on the acoustic signal caused by resonances or passively vibrating structures, the measurement setup is characterized in advance. Ensuring that only the diaphragm characteristics are measured, a chip frame without the diaphragm is mounted on the PCB as shown in Figure 2a. This configuration analyzes the total influences of all the elements expected to affect the diaphragm itself. Both microphones measure the pressure radiated from the reference loudspeaker.

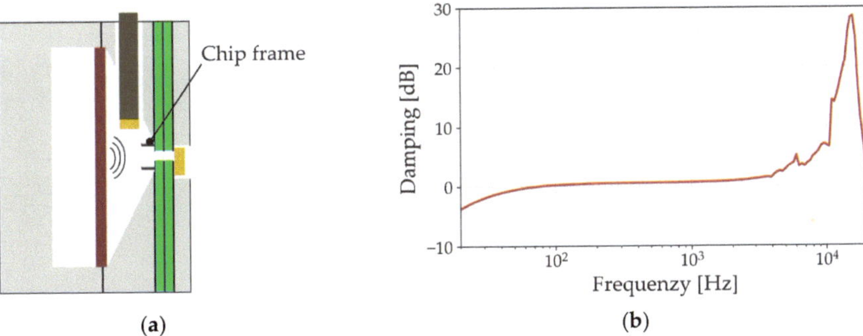

Figure 2. Schematic illustration of (**a**) the measurement setup without a folded diaphragm; (**b**) its measured acoustic damping behavior.

Subtracting the received signal from the transmitted signal reveals the damping caused by the setup, as shown in Figure 2b. This graph shows that between 100 Hz and 3 kHz, the setup does not affect the acoustic measurements. Below 100 Hz, the signal is damped towards lower frequencies due to the characteristic sensitivity of the MEMS microphone, which is integrated into the receiving chamber [12]. For higher frequencies, the setup shows a resonance at 16.6 kHz. These effects are considered later to avoid misinterpretation of the measurement data from the folded diaphragm samples.

3. Acoustic Measurements

These acoustic experiments are carried out with folded MEMS diaphragms, which are mounted on carrier PCBs. Deep reactive ion etching (DRIE) of a 100-silicon wafer defines the folded structure of the diaphragm. Chemical vapor deposition (CVD) ensures a conformal coverage of the deep structures with the thin film stack of 400 nm SiO_2, 1000 nm n-doped poly-crystalline silicon (Poly-Si), and 110 nm Si_3N_4 [4]. For these experiments, three samples each of 500 µm, 1000 µm, and 2000 µm long diaphragms with 210 µm deep structures are fabricated. The upper and lower bridges are 80 µm and 90 µm wide. Figure 3 shows a cross-section of such a diaphragm, imaged by scanning electron microscopy (SEM). The enlarged section proves that the diaphragm is completely released. This ensures that no residual silicon (Si) will affect the mechanical behavior and therefore the results of the subsequent measurements in this work.

An Audio Precision audio analyzer APx525 is used to perform the measurements. As an example, Figure 4a shows the measured SPLs in both chambers of the three 2000 µm long samples. The SPL in C_1 includes the results in the transmitting chamber for a loudspeaker drive voltage of 160 mVp. The measured values reflect the frequency response of the reference loudspeaker with its resonance at 1.1 kHz, as well as the flat response at lower frequencies and the SPL drop after the resonance. For the receiving chamber C_2, the results show the signal radiated by the folded diaphragm. It can be seen that the fundamental behavior of the reference speaker is transmitted through the sample. The measurement setup causes the drop towards low frequencies and the increase in SPL above 10 kHz (Figure 2).

Subtracting the SPL values of C_1 and C_2 determines the damping characteristics of the folded diaphragms (Figure 4b). This also eliminates the characteristic of the measurement setup to extract the effective transmission of the folded diaphragm only, revealing a flat response of the diaphragm up to 10 kHz. At higher frequencies, the folded diaphragm dampens the setup resonance, as shown by the negative peak in the effective transmission curve. Nevertheless, the mechanical behavior of the diaphragm is expected to be flat in this region. For this reason, this measurement setup should only be used for frequencies below 10 kHz to perform the parameter extraction of the mechanical diaphragm properties. No further resonances of the specimens are detected in this frequency range. Due to this flat behavior, the extraction in this work is performed representatively at a single frequency of 1 kHz.

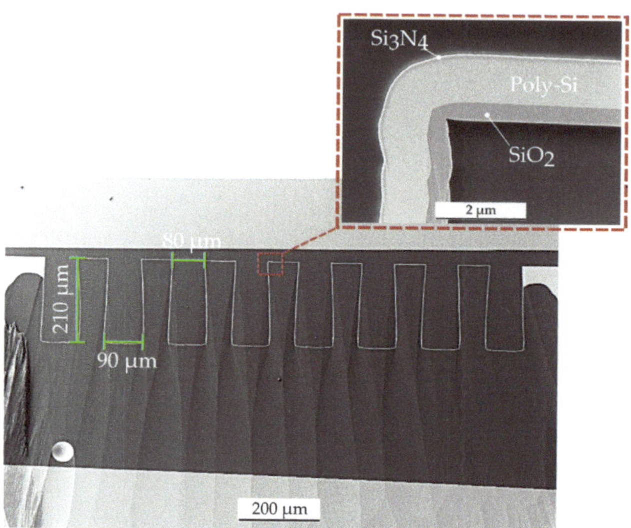

Figure 3. SEM micrograph of a cross-section of a polished, folded diaphragm.

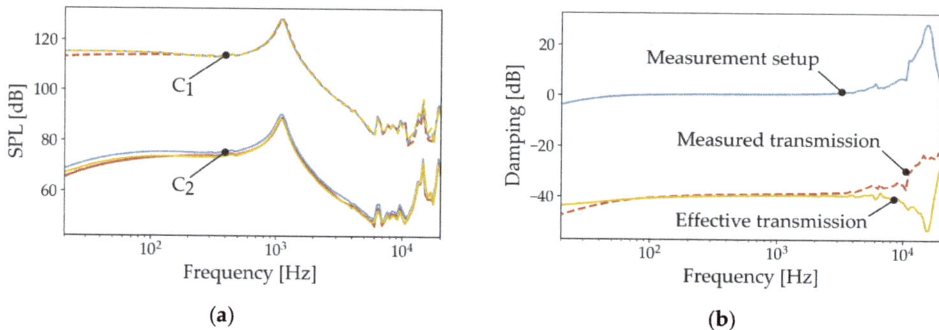

Figure 4. Results from the acoustic measurements showing (**a**) the measured SPL in both chambers for three 2000 µm long diaphragms; (**b**) the determination of their damping characteristics.

4. Lumped-Element Modelling

As mentioned above, LEM is used to extract mechanical parameters from the ATMs. A valid representation of the transducer requires its characteristic length scale being smaller than the wavelength of interest. In the case of the folded MEMS diaphragm, the 200 µm high sidewalls mainly define the mechanical behavior of the transducer. In the acoustic frequency range, wavelengths of interest are between 17.15 m and $17.15 \cdot 10^{-3}$ m. This proves the validity of such a model for frequencies from 20 Hz to 20 kHz. A circuit with equivalent elements models the measurement setup, including the diaphragm sample, in the impedance analogy. As the LEM method is primarily used for linear behaving systems, the behavior of the folded diaphragm and the measurement setup are determined in advance with regard to their linearity. Figure 5 therefore shows the radiated SPL at 1 kHz for all diaphragm lengths for different SPLs radiated by the loudspeaker.

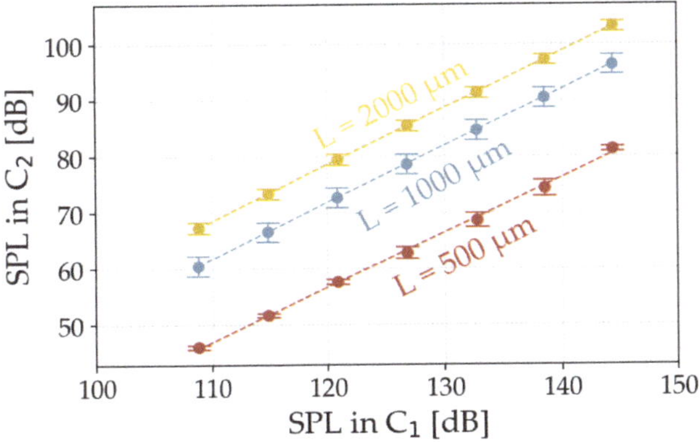

Figure 5. Radiated SPL at 1 kHz through the folded diaphragm for different input pressures.

This plot concludes that the analyzed diaphragms behave linearly over this range of applied input pressures. Mechanical damping and non-linear stiffening effects therefore do not affect the mechanical vibration of the diaphragm. As a result, LEM is well suited for modelling the setup and extracting the mechanical properties.

The equivalent circuit model of the setup including the folded diaphragm is shown in Figure 6a. The system is modelled in the acoustic domain. Both chambers are represented by the acoustic compliances $C_{a,1}$ and $C_{a,2}$. Equation (1) defines them with the chamber volume V, the density of the gas ρ_0, and the speed of sound c [13].

$$C_a = \frac{V}{\rho_0 c^2} \, , \tag{1}$$

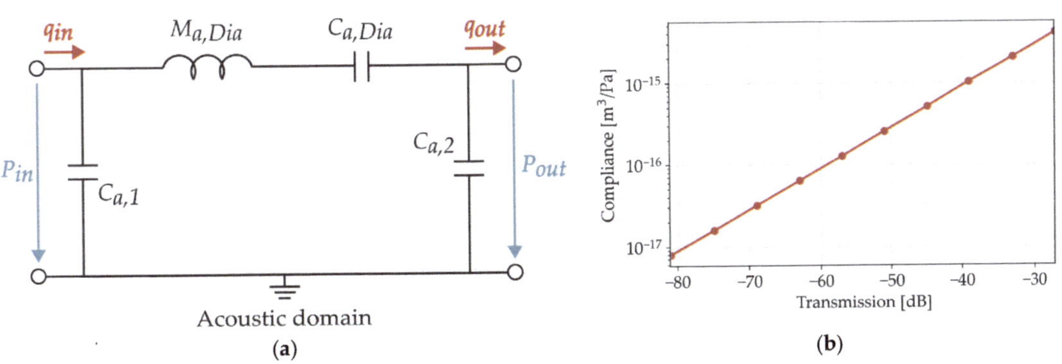

Figure 6. Illustration of (**a**) the lumped-element model of the measurement setup; (**b**) the resulting compliance extraction for 2000 μm long diaphragms.

A single sidewall of the folded MEMS diaphragm is modelled as a mass-spring-damper system, assuming a piston-like motion. Based on the results in Figure 5, this model neglects the mechanical damping in the range relevant to this work. The equivalent circuit of the folded diaphragm is composed as a simplified series circuit of its mass $M_{a,Dia}$ and its compliance $C_{a,Dia}$. Figure 4 shows that the resonances of the diaphragms are above the frequency range that can be measured with this setup. Since the influence of the mass is evident only in the resonance, it is calculated from the geometric parameters and material properties from the literature, which are summarized in Table 1. Within the measured frequency range, the compliance of the diaphragm is dominant and can be deduced from the measured transmission characteristics.

Table 1. Material properties of the diaphragm layer stack.

Thin-Film Material	Thickness [nm]	Density [kg/m^3]
SiO$_2$	400	2180 [14]
Poly-Si	1000	2300 [15]
Si$_3$N$_4$	110	3187 [16]

The loudspeaker in the transmitting chamber radiates the pressure P_{in} and the volume velocity q_{in}. Accordingly, the folded diaphragm radiates P_{out} and q_{out} into the receiving chamber. Fitting its value until the same simulated transmission characteristics are obtained between P_{in} and P_{out} determines $C_{a,Dia}$. Since the model represents only one sidewall of the transducer, it multiplies the radiated pressure by the total number of sidewalls, assuming a constant behavior along the diaphragm. Figure 6b shows a linear correlation between the fitted acoustic compliance and the resulting transmission characteristics. In the case of a 2000 µm long folded diaphragm, the transmission SPL_{C2}–SPL_{C1} is a damping of -41 ± 0.9 dB. According to the LE model, this results in an acoustic compliance of $1.04\,^{+1.09}_{-0.99}\cdot10^{-15}$ m^3/Pa.

5. Numerical Simulations

To validate the extracted values from the transmission measurements, finite element (FE) simulations are performed and compared with the LEM results. For this purpose, a three-dimensional model of the folded diaphragm is set up. In order to introduce the boundary conditions of the diaphragm clamping along the chip frame, the end faces are provided with a fixed support function in the y-direction. Only one lamella is meshed, whereas the others are mirrored via symmetry functions. The model applies a full surface pressure to one side of the diaphragm to simulate the input signal of the reference loudspeaker inside the transmission chamber. Figure 7 shows an example of the resulting lateral deflection of a 2000 µm long diaphragm along the x-axis of the diaphragm. For an acoustic input SPL of 94 dB, each vertical sidewall has a peak deflection of 0.33 nm for 500 µm and 0.36 nm for 1000 µm and 2000 µm long diaphragms. This causes a volume displacement in the receiving chamber.

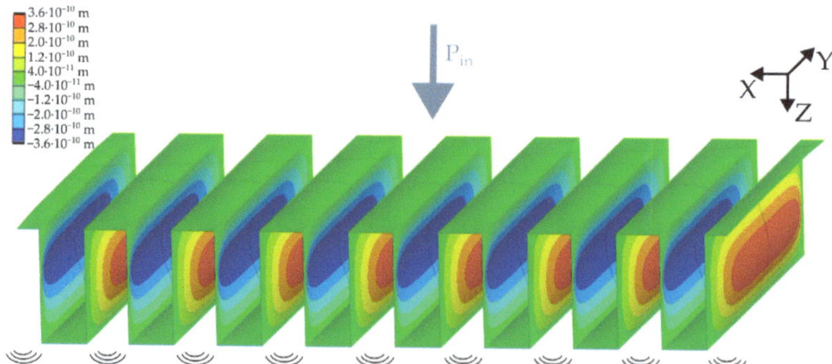

Figure 7. Numerical simulation of the deflected folded diaphragm with input pressure P_{in}.

To calculate the expected SPL L_{FEM} from the numerical model, Equation (2) determines the pressure difference Δp caused by the folded diaphragm with the adiabatic index k, the volume V_{C2}, and the ambient air pressure p_0. It is assumed that the wavelengths are longer than the largest dimension of C_2 and that there is no leakage [3].

$$\Delta p = \frac{k \cdot p_0 \cdot (N \cdot A_{Sw} \cdot w_{avg})}{V_{C2}}, \tag{2}$$

Here, the multiplication of the number of active sidewalls N, the sidewall area A_{Sw}, and the averaged displacement of a single sidewall w_{avg} equals the total displaced volume of the folded diaphragm. In Equation (3), the SPL is calculated with the ratio of Δp to the reference pressure p_{ref} of 20 µPa [13].

$$L_{FEM} = 20{\cdot}\log\left(\frac{\Delta p}{p_{ref}} {\cdot} \frac{1}{\sqrt{2}}\right) \text{dB} ,\tag{3}$$

To validate the results, the difference between the simulated input and output SPLs is determined and compared to the measured transmission of the folded diaphragms. For the folded diaphragms, simulated transmission values of -54.5 dB, -46.2 dB, and -39.8 dB are expected at an input SPL of 94 dB.

6. Results and Discussion

In order to extract their acoustic compliances, this work performs acoustic transmission measurements on the three-dimensional folded diaphragms shown in Figure 3. Three lengths of 500 µm, 1000 µm, and 2000 µm are fabricated with three samples each. For validation, each diaphragm geometry is numerically simulated using the model from Section 5. The resulting compliances for the three different diaphragm lengths are shown in Figure 8. Theoretically, the acoustic compliance of the diaphragm should increase linearly with its length and therefore its effective area. In comparison with the numerical model, this behavior is given for diaphragm lengths above 1500 µm. Below this, the compliance starts to decrease non-linearly. A similar behavior can be observed for the compliances extracted from the experimental setup. For the 1000 µm and 2000 µm long diaphragms, the measured results differ from the simulated values by only 3.7% and 4.0%. The shorter 500 µm diaphragm also shows that the simulated and measured acoustic compliance decreases non-linearly compared to the theoretical model. However, the extracted compliance from the measurements is almost half of that from the numerical simulations, meaning that the characterized samples transmit 6 dB less of the acoustic input signal than expected from the models.

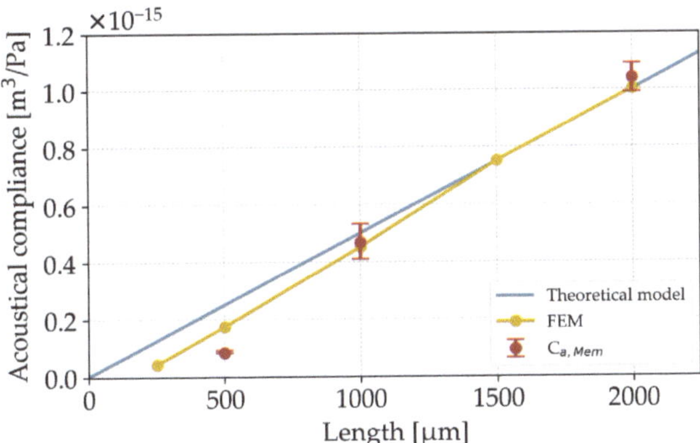

Figure 8. Comparison of extracted acoustic compliances from transmission measurements, numerical simulations, and theory.

The three-dimensional folded design, particularly its clamping within the chip frame, causes the non-linear behavior of the acoustic compliance for shorter diaphragm lengths. As described in previous work, the aspect ratios of the deep etched trenches are the key to achieving lateral sidewall deflections and hence volume displacement [4]. For longer diaphragms, the height of these structures is the main determinant of their compliance. As the sidewall length decreases, the influence of the stiff clamping at the front faces of the trenches begins to dominate the behavior of the diaphragm. This causes a lower transmitted signal than theoretically expected due to the smaller displacement of the sidewalls. In the case of the 500 μm long diaphragm, the numerically simulated signal is higher than the measured one. This could be caused by the simplified representation of the diaphragm suspension in the FE model. A detailed implementation of the chip frame could increase the accuracy of the numerical simulations.

The ATM approach provides a fast method to characterize MEMS transducers with hidden or optically difficult-to-access structures, such as the vertical sidewalls of the three-dimensional folded diaphragm. This makes it possible to characterize these folded MEMS diaphragms without the need for extensive and destructive sample preparations. In contrast to state-of-the-art measurement methods, such diaphragms can now be analyzed with mechanical boundary conditions close to the final application. The resulting determination of mechanical parameters is a significant input for the development and simulation of systems with these structures, allowing the early design towards desired applications. However, the measured acoustic signals are always the sum of all vibrating structures within the sample. Characterizing the setup in advance increases the accuracy by eliminating its influence on the measurements. The volumes of the chambers have no direct influence on the measurement results, as this method determines the differential pressure for the parameter extraction. However, it is important to determine the dimensions of the chambers so that the MEMS sample of interest can still generate an acoustically measurable signal. To reduce processing time, LEM or FE simulations are built assuming that all active elements of a folded transducer behave the same and are independent of the adjacent structures. As the vertical structures close to the chip frame are clamped differently to those in the center, it can be assumed that they also show a different behavior. This could lead to a less accurate determination of an effective acoustic compliance, especially for a diaphragm with a lower number of vertical structures. For this reason, future work on this topic should focus on quantifying the influence of the diaphragm suspension by characterizing samples with varying numbers of vertical structures per diaphragm. The results should then be implemented into the lumped-element model as a secondary compliance. In addition to the diaphragm clamping, following studies should analyze the general measurement accuracy of this method. The manufacturing tolerances and the tolerances of the microphones should be taken into account, and their influence on the results should be described. The compliance values obtained in this work are compared with state-of-the-art MEMS transducer diaphragms in Table 2. Characterization of the dynamic mechano-acoustical behavior of the folded diaphragm samples shows that increasing their length also leads to higher acoustic compliances. In the case of the 2000 μm long samples, the results are in the same order of magnitude as those obtained with state-of-the-art diaphragms.

Table 2. Comparison of determined acoustic compliances to state-of-the-art transducer diaphragms.

Transducer	Acoustic Compliance [m³/Pa]	Transducer Area [mm²]
Folded diaphragm (2000 μm)	400	2.4
Folded diaphragm (1000 μm)	1000	1.2
Folded diaphragm (500 μm)	110	0.6
[17]	0.026×10^{-15}	2.25
[18] [1]	14.4×10^{-15}	1.44
[19] [1]	34.75×10^{-15}	20.25
[20] [1]	1.4×10^{-15}	0.57

[1] Calculated from given mechanical compliance values.

The results of the three variants with different lengths allow a demonstration of the advantages of this new measurement method and to validate the experiments in this work. Figure 8 shows that three-dimensional transducers such as the folded diaphragm can be successfully characterized using this technique. This is also expected to be the case for other diaphragm geometries besides the demonstrated folded structure. Future work will include the testing of more diaphragm variants in order to consolidate the understanding of their behaviors and limitations. In addition to the number and height of lamellas, other lengths below 1000 µm and especially below 500 µm will be investigated in order to determine the exact influence of the clamping within the chip frame. This will result in more accurate results for shorter samples. Lengths above 1000 µm will also be studied to confirm the small deviations from the numerical simulations and to find critical constraints for longer diaphragms. Nevertheless, the current results of this work show that this measurement technique is a useful tool to support the research and development of MEMS diaphragms.

The principle of this measurement method makes it easy to use in practical applications. Differential pressure measurements allow a high degree of freedom in chamber design and sample integration. The chamber volumes are dimensioned in such a way that loudspeakers and microphones can be integrated without leakage paths and that the acoustic signals can be measured. Leakage in the measurement setup or in the MEMS membrane will falsify the results. In this case, the microphone in the receiving chamber would also measure the characteristics of the reference loudspeaker. It must also be possible to simulate the system with LEM to enable a valid parameter extraction. As described in Section 4, the dimensions of interest of MEMS diaphragms limit the valid frequency range for this measurement method.

7. Conclusions

In this work, a new measurement technique has been developed utilizing non-destructive ATMs and LEM to determine the mechanical properties of three-dimensional folded MEMS transducers. While state-of-the-art optical methods are not suitable for such structures, the acoustic approach of this work allows the extraction of their dynamic behavior without extensive sample preparation. The setup consists of two chambers, which are separated by the sample to be characterized. It is acoustically characterized in advance in order to quantify and later on subtract its influence from the measurement results. An equivalent circuit model is used to extract the mechanical properties by parameter fitting. It shows that the diaphragm behavior is mainly defined by its compliance, while the mass can be calculated from its geometry, and the damping can be neglected. FE simulations of the folded diaphragm validate the measured and extracted values. For the longer samples (1000 µm and 2000 µm), the results are in very good agreement with the numerical simulations. For shorter diaphragms, the deviation between measurement and simulation increases non-linearly. This is due to the additional stiffness of the chip frame suspension, which dominates the mechanical behavior of shorter lengths. Further investigation and adaptation of the model with a detailed reconstruction of the chip frame could improve the accuracy of the parameter extraction.

The three-dimensional folded design offers a wide range of possibilities to tailor its behavior for various applications. The new measurement method presented in this paper plays a key role in this development, advancing it through its simplicity and efficiency.

Author Contributions: Conceptualization, D.B. and A.D.; methodology, D.B and A.D.; software, D.B. and M.L.; validation, D.B., A.B. and A.D.; formal analysis, D.B.; investigation, D.B, M.L., A.B. and A.D.; data curation, D.B. and M.L.; writing—original draft preparation, D.B.; writing—review and editing, D.B., A.B. and A.D.; visualization, D.B. and M.L.; supervision, A.B. and A.D.; project administration, D.B. All authors have read and agreed to the published version of the manuscript.

Funding: This research received no external funding.

Data Availability Statement: The original contributions presented in the study are included in the article; further inquiries can be directed to the corresponding authors.

Conflicts of Interest: The authors declare no conflicts of interest.

References

1. Stoppel, F.; Männchen, A.; Niekiel, F.; Beer, D.; Giese, T.; Wagner, B. New integrated full-range MEMS speaker for in-ear applications. In Proceedings of the 2018 IEEE Micro Electro Mechanical Systems (MEMS), Belfast, UK, 21–25 January 2018; pp. 1068–1071. [CrossRef]

2. Fuldner, M.; Dehé, A.; Lerch, R. Analytical analysis and finite element simulation of advanced membranes for silicon microphones. *IEEE Sens. J.* **2005**, *5*, 857–863. [CrossRef]

3. Kaiser, B.; Langa, S.; Ehrig, L.; Stolz, M.; Schenk, H.; Conrad, H.; Schenk, H.; Schimmanz, K. Concept and proof for an all-silicon MEMS micro speaker utilizing air chambers. *Microsyst. Nanoeng.* **2019**, *5*, 43. [CrossRef] [PubMed]

4. Becker, D.; Bittner, A.; Dehé, D. Three-dimensional folded MEMS manufacturing for an efficient use of area. In Proceedings of the MikroSystemTechnik Kongress 2023, Dresden, Germany, 23–25 October 2023; pp. 307–310, ISBN 978-3-8007-6203-3.

5. Neggers, J.; Hoefnagels, J.P.M.; Geers, M.G.D. On the validity regime of the bulge equations. *J. Mater. Res.* **2012**, *5*, 1245–1250. [CrossRef]

6. Martins, P.; Delobelle, P.; Malhaire, C.; Brida, S.; Barbier, D. Bulge test and AFM point deflection method, two technics for the mechanical characterization of very low stiffness freestanding films. *Eur. Phys. J. Appl. Phys.* **2009**, *45*, 10501. [CrossRef]

7. Petitgrand, S.; Bosseboeuf, A. Simultaneous mapping of out-of-plane and in-plane vibrations of MEMS with (sub)nanometer resolution. *J. Micromech. Microeng.* **2004**, *14*, S97–S101. [CrossRef]

8. Ennen, M.; Lherbette, M.L. In-Plane and Out-Of-Plane Analyses of Encapsulated Mems Device by IR Laser Vibrometry. In Proceedings of the 2022 IEEE 35th International Conference on Micro Electro Mechanical Systems (MEMS), Tokyo, Japan, 9–13 January 2022; pp. 814–817. [CrossRef]

9. Del Rey, R.; Alba, J.; Rodríguez, C.J.; Bertó, L. Characterization of New Sustainable Acoustic Solutions in a Reduced Sized Transmission Chamber. *Buildings* **2019**, *9*, 60. [CrossRef]

10. Urbán, D.; Roozen, N.B.; Jandák, V.; Brothánek, M.; Jiříček, O. On the Determination of Acoustic Properties of Membrane Type Structural Skin Elements by Means of Surface Displacements. *Appl. Sci.* **2021**, *11*, 10357. [CrossRef]

11. GRAS: 46BP-1 1/4″ LEMO Pressure Standard Microphone Set. Available online: https://www.grasacoustics.com/products/measurement-microphone-sets/traditional-power-supply-lemo/product/688-46bp-1 (accessed on 10 June 2024).

12. Infineon: IM69D130 High Performance Digital XENSIVTM MEMS Microphone. Available online: https://www.infineon.com/cms/en/product/sensor/mems-microphones/mems-microphones-for-consumer/im69d130/ (accessed on 10 June 2024).

13. Beranek, L.; Mellow, T. *Acoustics: Sound Fields, Transducers and Vibration*, 2nd ed.; Elsevier Inc.: San Diego, CA, USA, 2019; p. 106, ISBN 978-0-12-815227-0.

14. El-Kareh, B. Thermal Oxidation and Nitridation. In *Fundamentals of Semiconductor Processing Technology*, 1st ed.; Springer: Boston MA, USA, 1995; pp. 38–95. [CrossRef]

15. MKS Instruments: Polycrystalline Silicon Thin Films. Available online: https://www.mks.com/n/polycrystalline-silicon-thin-films (accessed on 10 June 2024).

16. Pierson, H.O. The CVD of Ceramic Materials: Nitrides. In *Handbook of Chemical Vapour Deposition (CVD): Principles, Technology, and Applications*, 2nd ed.; Noyes Publications: New York, NY, USA, 1999; p. 281. [CrossRef]

17. Arya, D.S.; Prasad, M.; Tripathi, C.C. Design and modeling of a ZnO-based MEMS acoustic sensor for aeroacoustic and audio applications. In Proceedings of the 2015 2nd International Symposium on Physics and Technology of Sensors (ISPTS), Pune, India, 8–10 March 2015; pp. 278–282. [CrossRef]

18. Dehé, A.; Wurzer, M.; Füldner, M.; Krumbein, U. The Infineon Silicon Microphone. In Proceedings of the 16th Sensors and Measurement Technology International Conference, Nürnberg, Germany, 14–16 May 2013; pp. 95–99. [CrossRef]

19. Gazzola, C.; Zega, V.; Corigliano, A.; Lotton, P.; Melon, M. Lumped-Parameters Equivalent Circuit for Piezoelectric MEMS Speakers Modeling. In Proceedings of the 10th Convention of the European Acoustics Association (Forum Acusticum), Turin, Italy, 11–15 September 2023. Available online: https://univ-lemans.hal.science/hal-04266242 (accessed on 10 June 2024).

20. Shubham, S.; Seo, Y.; Naderyan, V.; Song, X.; Frank, A.J.; Johnson, J.T.M.G.; da Silva, M.; Pedersen, M. A Novel MEMS Capacitive Microphone with Semiconstrained Diaphragm Supported with Center and Peripheral Backplate Protrusions. *Micromachines* **2022**, *13*, 22. [CrossRef] [PubMed]

Article

Road to Acquisition: Preparing a MEMS Microphone Array for Measurement of Fuselage Surface Pressure Fluctuations

Thomas Ahlefeldt *, **Stefan Haxter**, **Carsten Spehr**, **Daniel Ernst** and **Tobias Kleindienst**

German Aerospace Center (DLR), Bunsenstr. 10, D-37073 Göttingen, Germany; stefan.haxter@dlr.de (S.H.); carsten.spehr@dlr.de (C.S.); daniel.ernst@dlr.de (D.E.); tobias.kleindienst@dlr.de (T.K.)
* Correspondence: thomas.ahlefeldt@dlr.de

Abstract: Preparing and pre-testing experimental setups for flight tests is a lengthy but necessary task. One part of this preparation is comparing newly available measurement technology with proven setups. In our case, we wanted to compare acoustic Micro-Electro-Mechanical Systems (MEMS) to large and proven surface-mounted condenser microphones. The task started with the comparison of spectra in low-speed wind tunnel environments. After successful completion, the challenge was increased to similar comparisons in a transonic wind tunnel. The final goal of performing in-flight measurements on the outside fuselage of a twin-engine turboprop aircraft was eventually achieved using a slim array of 45 MEMS microphones with additional large microphones installed on the same carrier to drawn on for comparison. Finally, the array arrangement of MEMS microphones allowed for a complex study of fuselage surface pressure fluctuations in the wavenumber domain. The study indicates that MEMS microphones are an inexpensive alternative to conventional microphones with increased potential for spatially high-resolved measurements even at challenging experimental conditions during flight tests.

Keywords: microphone array; turbulent boundary layer; Kapton foil; flight test; wave number decomposition; propeller; spatial resolution; flexible circuit board

Citation: Ahlefeldt, T.; Haxter, S.; Spehr, C.; Ernst, D.; Kleindienst, T. Road to Acquisition: Preparing a MEMS Microphone Array for Measurement of Fuselage Surface Pressure Fluctuations. *Micromachines* **2021**, *12*, 961. https://doi.org/10.3390/mi12080961

Academic Editor: Libor Rufer

Received: 16 July 2021
Accepted: 9 August 2021
Published: 14 August 2021

1. Introduction

The characteristics of pressure fluctuations on the airplane fuselage govern the vibro-acoustic excitation of surface panels exposed to the boundary layer flow. These pressure fluctuations are caused either by the Turbulent Boundary Layer (TBL) itself or acoustic fluctuations induced by-for instance-the engine or by airframe noise. Experimental data of these pressure fluctuations are difficult to obtain in ground-based tests due to the high Reynolds numbers required. Flight tests are an appropriate method to obtain knowledge about the characteristics of these pressure fluctuations under realistic conditions.

Sensors suitable for this task require a high overload point to resolve the high-amplitude pressure peaks within the turbulent boundary layer. The size of the sensor influences the spatial resolution of the measurement, i.e. the size of detectable pressure structures and the measurable coherence between any two sensors. Surface microphones that are currently commercially available and for in-flight testing are comparably large and thus focus on providing acoustic pressure fluctuations. Small sensors suitable for capturing both acoustic and hydrodynamic pressure fluctuations are to our knowledge not very common.

This paper aims to present the experience gained from the design, application, and evaluation of a flight test with a phased microphone array to measure the fuselage excitation by the acoustic as well as hydrodynamic wavelength. The array developed in-house for this task consists of MEMS pressure sensors, concealed beneath a protective layer.

Commercially available surface microphones typically have a sensitive surface size between 1/2 inch and 1/4 inch and a dynamic range reaching up to 160–175 dB. These microphones are reasonably thin (thickness between 1.4 mm to 2.5 mm) so that they do not excessively protrude into the TBL during the measurement. They are suitable for the measurement of the acoustic excitation of an aircraft fuselage by (acoustic) wavelengths larger than the sensitive surface size (acoustic wavelengths of frequencies up to 20.00 kHz).

The measurement of small, high-energy TBL structures connected to fuselage excitation is a complex task compared to the assessment of acoustic pressure fluctuations. In the acoustically relevant frequency range, the wavelengths of typical hydrodynamic pressure fluctuations are small compared to acoustic wavelengths [1]. Thus, the acquisition of data in the flight test scenarios requires prior knowledge of the measurement environment and a careful design of the MEMS array.

The discourse in this paper will go as follows: after a brief introduction to the state of the art, the challenges of measuring in a flow using sensors of different sizes are described in Section 2. The in-house MEMS pressure sensor array is introduced in Section 3 followed by the ground tests conducted with and without flow (Section 4). The setup for the final in-flight measurements is presented in Section 5 including selected results demonstrating the feasibility and performance of the introduced MEMS microphone array.

State of the Art

Currently, the state-of-the-art sensors for measuring these fluctuating pressure distributions are small (pinhole-mounted) piezoresistive pressure transducers (e.g., Kulite) as shown for in-flight measurements [2,3]. These sensors typically are rather expensive. Moreover, using this type of sensor has several drawbacks. The inherent noise level may mask the lower level acoustic excitation, which can nevertheless contribute to the fuselage excitation due to its large spatial coherence. The space required for mounting the sensors limits possible locations for measurements on the fuselage mostly to passenger windows replaced by dummy windows. Furthermore, the mounting condition of the sensors governs the achievable results by possible flow-induced self-noise, pressure damping within a sound hole, or the minimum distance between two sensors (see Section 2).

Alternatives to these pressure sensors are MEMS microphones due to their small size and low cost. In aeroacoustic testing, one must distinguish between in-flow sensors, which are mounted on any kind of surface that is exposed to the flow-field, and out-of-flow sensors, which are mounted out of the flow-field of an open jet wind tunnel. A recent example with out-of-flow sensors by Zhou et al. [4] shows the application of a 256 MEMS microphone array in an open jet wind tunnel for aeroacoustic measurements. In the current paper, only the in-flow placement of acoustic sensors and their unique problems are discussed.

In 1999 Sheplak et al. [5] developed the first custom aeroacoustic MEMS microphone. They also defined the critical performance parameters of an aeroacoustic MEMS microphone which state, that the dynamic range should be 60 dB to 160 dB and the upper frequency limit should be higher than 50 kHz. The resulting sensor implementation offered an upper frequency limit of 6 kHz (predicted 300 kHz) with a noise floor of the packaged sensor of 92 dB/$\sqrt{\text{Hz}}$ (predicted 63 dB/$\sqrt{\text{Hz}}$). The maximum sound pressure level was 155 dB.

Several publications on custom aeroacoustic MEMS microphones were published since then (see Table 8-2 in [6]). A good summary of these developments and details on the fabrication process are the Ph.D. theses of Williams [7] and Reagan [6]. Both authors developed new custom MEMS microphones which are meeting the requirements stated above. However, these sensors are not available on the open market. Therefore, implementations of MEMS microphones in experimental acoustic setups are based on commercially available sensors.

In comparison to surface microphones, commercially available MEMS microphones have a limited Acoustic Overload Point (AOP) of 135 dB maximum. While being sufficient for the consumer market, the small-scale hydrodynamic excitation within the TBL can exceed this limit already at moderate flow speeds of 30–50 m/s (see Section 4.2.1) and overload the sensors. Microphones in general show a nonlinear attenuation of the output amplitude response once the input amplitude has an order of magnitude similar to the AOP. Depending on how this compression of the signal is implemented in the sensor for very high input amplitudes, an undetectable and irrecoverable shaping of the signal, a high nonlinear distortion or sensor malfunctioning may occur.

Shams et al. [8] developed a 128 MEMS microphone array for aeroacoustic testing in 2004 and successful aeroacoustic measurements in a wind tunnel were first conducted by Humphreys et al. [9] in 2005 using MEMS microphones from Knowles. Both authors do not mention a problem with a low acoustic overload point. Sanders et al. [10] embedded digital MEMS microphones in trailing-edge serrations in order to measure the hydrodynamic coherence length. They state that MEMS microphones with a higher acoustic overload point should be used for Reynolds numbers larger than 7×10^5 due to problems with signal clipping. Leclere et al. [11] and Salze et al. [12] used a MEMS microphone array for turbulent boundary layer measurements in a wind tunnel and turbofan duct. Both do not report any type of acoustic overload with measurements conducted at speeds from 0 m/s to 50 m/s free stream velocity.

To overcome the limitations of the low overload point, the MEMS microphones in the current study were covered with a Kapton foil of thickness 25 μm to attenuate the pressure excitation by approx. 38 dB. Covering the sensors has the positive side-effect of protecting the MEMS sensors on the fuselage from humidity while at the same time smoothing the surface by covering surface irregularities in the vicinity of the sensor (e.g., pinhole edges). The response of this Kapton foil to acoustic waves must be considered in the analysis. Calibration measurements of the foil are introduced in Section 4.

2. Sensors in Flow

When measuring pressure fluctuations underneath a turbulent boundary layer, the ratio of the size of the sensitive microphone surface relative to the "size" of the turbulent structure in the boundary layer can affect the measurement output. More specifically, the wavenumber of the turbulent structure's pressure fluctuation has to be small enough in order for the sensor to spatially resolve the fluctuations. If the sensor is not able to resolve the turbulent structure, signal attenuation can occur.

This was described by Schewe [1] who recorded signals from microphones of various sizes flush-mounted underneath a TBL. Schewe introduced a dimensionless sensor diameter $d^+ = du_\tau/\nu$ dependent on the inner wall variables, such as the wall shear velocity u_τ and the kinematic viscosity ν. This dimensionless sensor diameter was used to describe the dependence of several higher-order moments of the the normalized total fluctuating pressure, including power $p_{\mathrm{rms}}/q_\infty$ and flatness $\overline{p^4}/p_{\mathrm{rms}}^2$.

The fluctuating power was found to decrease with dimensionless sensor size and is connected to rare high-power events, indicated by a similar increase in flatness of probability density function of the pressure fluctuations. The effect is quite significant: when using a sensor of $d^+ \approx 300$, only half of the pressure power is received compared to the power estimated at an extrapolated value of $d^+ = 0$. The larger share of power measured with high-resolution sensors implies that a lot of fluctuating power is contained in the small-scale structures of the boundary layer. In addition to possibly underestimating the power content of the pressure fluctuations, the pressure applied by these small-scale structures may exceed the AOP of sensors used. Schewe found that the portion of fluctuation power not resolved by larger diameters was connected to higher frequencies of the fluctuations, thus linking the higher frequencies with higher wavenumbers.

The dampening of higher wavenumbers can be modeled by averaging a fluctuation over a sensitive surface of the desired shape. For circular and various other transducer shapes this was done by Ko [13] and will be shown below in Section 2.1. The wavenumber characteristics of TBL surface pressure fluctuations were described by Corcos [14] who formulated a spectral dampening compensation for power levels at higher frequencies. The effects of different pressure characteristics on the results are described in Section 2.2.

2.1. Sensitive Surface Averaging

Concerning the characterization of signal attenuation as a function of the wavenumber, Ko [13] modeled the effect of a cosine-shaped surface pressure fluctuation over a circular shaped transducer. The resulting attenuation was found to be a function of the size ratio $a = k \cdot r$ of sensitive surface radius r and wavenumber k of the surface pressure fluctuations. The gain $s(k, r)$, resulting from the modeled surface attenuation is given by

$$s(k, r) = \frac{2 J_1(k \cdot r)}{k \cdot r},\tag{1}$$

with $J_1(...)$ being the Bessel function of first kind of order one. The gain $s(k, r)$ resulting from Equation (1) is shown in Figure 1 in dB for a range of frequencies with exemplary propagation velocities. For TBL convection, the phase propagation wavenumber k_p is used for the relation of sensor size and wavenumber in Equation (1) It is obtained by using $k_p = f / u_p$ with u_p being the phase velocity. The phase velocity describes the speed of signal phase changes over the array and takes values of approximately $u_p \approx 0.7u$, with u being the flow velocity determined from the Mach number $M = u / c_0$. Furthermore, f describes the frequency, and $c_0 = 340$ m/s the speed of sound. With constant phase velocity, wavenumber is proportional to the frequency and thus, the stronger the attenuation effect gets. For TBL pressure fluctuations propagating over the largest sensitive surface area ($r = 12.7$ mm = 1/2 in) at a velocity of $M = 0.2$ the loss in gain does not exceed 1 dB below 10 kHz. At first glance, this appears to be sufficient for most acoustic full-scale tests with Helmholtz-number unity. However, TBL pressure fluctuations have a limited coherence length which pushes some of the wavenumber content of the convective propagation towards higher wavenumber values. This will be explained below.

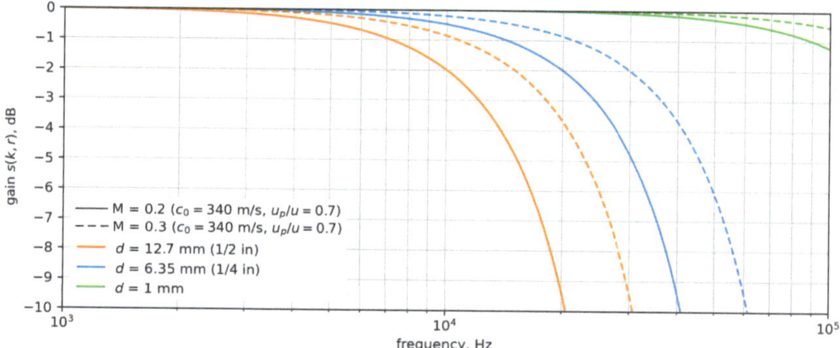

Figure 1. Modeled attenuation of hydrodynamic surface pressure fluctuations as a function of frequency at different Mach numbers and sensitive surface radii.

2.2. Influence of Limited Coherence Length

While acoustic pressure fluctuations keep their self-similarity when propagating over large distances, hydrodynamic TBL pressure fluctuations do not. The hydrodynamic propagation mechanism has a limited coherence length-a characteristic length to describe the size of a pressure patch where pressure is applied to a surface coherently. Besides the lower propagation velocity of subsonic hydrodynamic pressure fluctuations, their limitation in coherence length enlarges the wavenumber content contained in the propagation mechanism and adds to the attenuation of spectral power. The effect was described by Corcos [14]. Figure 2 shows an exemplary hydrodynamic source modeled in the wavenumber domain after Graham [15] at a frequency of 10 kHz for two velocities $M = 0.2$ in Figure 2a and $M = 0.3$ in Figure 2b. The velocity fraction for convective transport was assumed again at 0.7. fraction At a frequency of 10 kHz, this results in convection wavenumbers of $k_c \approx 210$ m^{-1} and $k_c \approx 140$ m^{-1} respectively, where the center of the convective ridge is located. Using a set of limited coherence lengths of $l_x = 5$ cm and $l_y = 5$ mm on the sources leads them to spread out in k_x- and k_y-direction rather than being a point source located at k_c. Especially due to the shorter coherence length in y-direction, the convective ridge can be observed to be spread out much more prominently in k_y-direction, resulting in a much higher wavenumber content.

In Figure 2, the -3 dB B-thresholds of sensors sizes 1/2 in, 1/4 in, and 1 mm are shown as orange, blue and green, respectively. For the case of M = 0.2 the center of the convective source is still contained within the -3 dB-line of the 1/2 in sensor. However, due to the limited coherence length, the hydrodynamic source is spread out towards higher wavenumbers and less content is contained within the passband of the spatial filter of the sensitive surface. This leads to this wavenumber content being attenuated.

Raising the flow velocity to M = 0.3 shifts the center of the convective ridge closer to the origin of the plot. However due to the large spread of the convective ridge still only a fraction of the energy is picked up by the sensor. This applies as well for the 1/2 inch sensor with its -3 dB line shown in blue, but to a lesser extent. Using a sensor of size 1 mm will yield a much larger passband of the spatial filtering as shown by the green curve. Here, almost the entire energy contained in the convective ridge is picked up.

The amount of source content contained within the -3 dB-line for each of the two sensor sizes is shown in the legend of each figure segment as a percentage of the total energy contained in the simulated source. In this simplified exemplary case, a sensor having a diameter of 1/2 inch would only capture 48% of the total energy at M = 0.2 and 55% at M = 0.3. The simulated 1/4 inch microphone would yield 75% and 77% for the two Mach numbers, and the 1 mm sensor is able to recover 99% of the energy for both simulated velocities under consideration.

Small sensors are therefore required for characterizing the TBL surface pressure fluctuations. Apparently, the difference in measured signal level due to the sole influence of velocity and convective wavenumber has a minor impact compared to the impact of a small coherence length.

The use of the -3 dB-threshold in the above example was used exemplary for illustration of the passband of the spatial filter constituted by each sensor's size. This is not to be seen as a hard cutoff, but rather as a smooth transition from passband to stopband as shown in Figure 1.

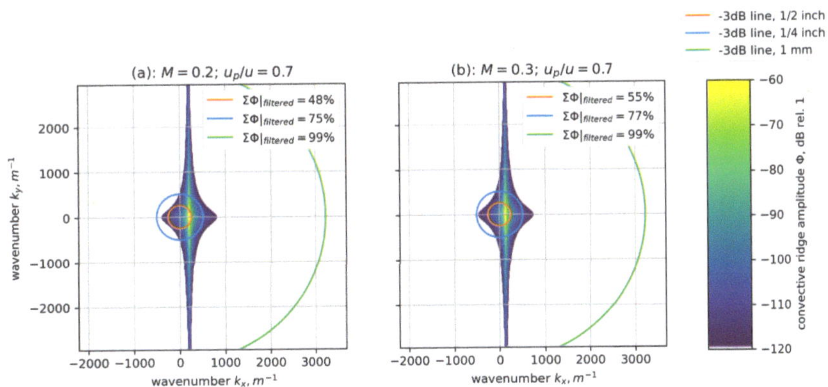

Figure 2. Exemplary attenuation caused by broadening of sources in the wavenumber domain due to limited coherence length. (**a**) $M = 0.2$ (**b**) $M = 0.3$.

2.2.1. Influence on Measurement Time

Besides having a smaller sensitive surface, smaller sensors can be placed closer together on a surface. This enables the measurement of smaller coherence lengths when measurement time is limited. A sketch of this effect is shown in Figure 3 where the signal coherence level is modeled [16] exemplary for small sensor separation distances in cross-flow direction Δy in the presence of a very small coherence length in cross-flow direction $l_y = 5$ mm. l_y takes very small values in the presence of thin boundary layers that may occur at the position of the flight deck in the front of the aircraft. The noise level is proportional to the inverse square root of the number of averages applied to the signal when evaluating and thus, very long measurement times are required to reduce the noise level. This can be circumvented to some extent by using sensors spaced closer together in a way that a higher signal coherence is still present and above the noise level even for shorter measurements. In order to better capture the proceedings underneath the boundary layer, in this region of the fuselage, the use of small sensors is advisable.

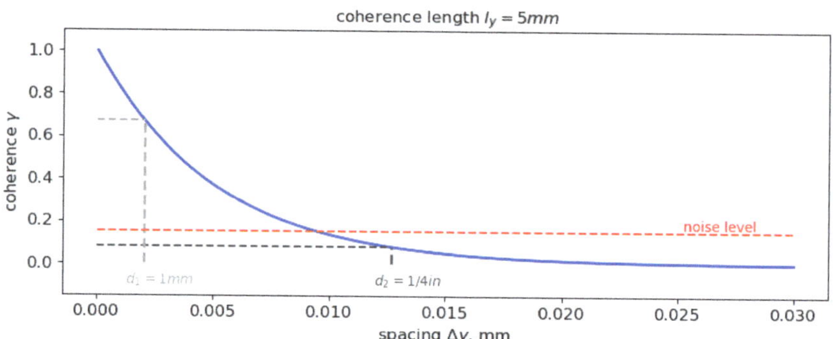

Figure 3. Different sensor sizes allow for different sensor spacings. The level of coherence between two signals from sensors spaced closer together is high, allowing for the determination of the coherence drop even at higher noise levels.

2.2.2. Requirements for Sensors in Flow

Resulting from the aforementioned characteristics of sensors in flow, several requirements can be deduced in order to optimize an array for flow measurements.

- The sensor should be able to resolve small turbulent structures of high wavenumber. This requires a small sensor. Typically, the characteristic size of the sensor should be small compared to relevant scales in the TBL (e.g., the boundary layer thickness and inner length scales)
- Sensors should be placed closely together to allow for capturing hydrodynamic signal correlation even of small turbulent structures. This requirement calls for small sensors as well.
- The sensor should have a high dynamic range (not mentioned in the section above): Acoustic and hydrodynamic pressure fluctuations typically do not exhibit the same amplitude. Often, one of the two sources has a significantly higher amplitude than the other one. In order to resolve a source with low amplitude in the presence of a source with a very high amplitude, the dynamic range of the sensors needs to be large. This applies not only to combinations of hydrodynamic/acoustic sources but also to combinations of acoustic/acoustic sources and hydrodynamic/hydrodynamic sources.

3. MEMS Array Structure

In this section, the MEMS microphone array used for the flight test is introduced, presenting its general properties and structure as well as the microphones.

With the intention in mind to perform array measurements in parallel with analogue high-precision condenser measurement microphones for reference, analogue MEMS sensors of type ICS-40617 [17] were selected for the array. For time-synchronous acquisition, all analogue sensors were recorded by the same data acquisition system.

The MEMS microphone has dimensions of 3.50 mm by 2.65 mm with a thickness of 0.98 mm. The diameter of the sensor port and the diaphragm is 1 mm. It was mainly chosen for its reasonably high dynamic range of 111 dB and an AOP with 10% THD+N at 129 dB. However, for aeroacoustic measurements the AOP at 2% is of interest to gain reliable data for the evaluation. According to the data sheet [17], at 2% the AOP is approximately around 127 dB. The sensor also shows a nonlinear amplitude output response above the AOP, i.e., signal compression, as mentioned in Section 1.

The MEMS microphone array was constructed on the basis of a flexible Printed Circuit Board (PCB) to enable an application onto curved surfaces. The PCB was manufactured in multilayer design using 6 layers with the two outer layers being used for shielding the signal lines on the inner layers. The selected dimensions for the aperture of the array were chosen to be 400 mm by 300 mm as depicted in Figure 4 (left side) with a thickness of the PCB of 0.45 mm. In combination with the thickness of the MEMS sensor a combined thickness of 1.45 mm was achieved.

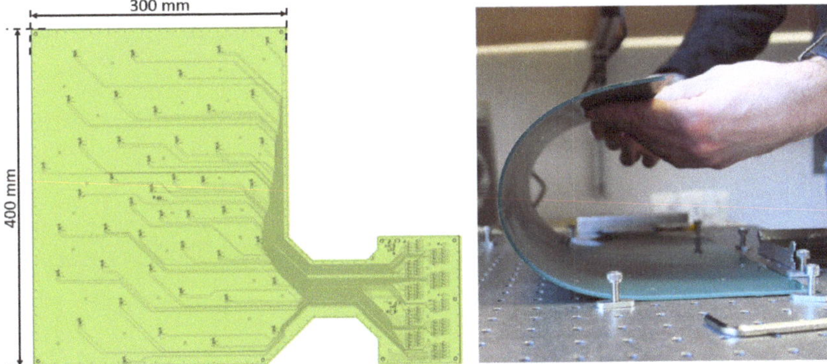

Figure 4. Left: Schematic of the printed circuit board with dimensions. Right: Bending demonstration of the array.

The sensor port of the MEMS is located at the connector side which enabled an installation on the far side of the PCB with regard to the flow. Each sensor port was aligned with a through-hole of size 1 mm in the PCB connecting it to the smooth flow side of the array. For flush mounting on smooth surfaces, for insulating the electronic parts, and for moisture and dust protection during a test, the backside of the array (with the protruding sensor) was cast with silicone to an overall and equal thickness of the array of 1.55 mm. The finished MEMS microphone array on the flexible PCB responded well to bending as illustrated in Figure 4 (right side) without damaging the solder joints on the circuit board.

In total, 45 acoustic MEMS sensors were placed on the PCB to form the array with the focus set on demonstrating their applicability in flight testing in general. The MEMS microphones were arranged in a Vogel spiral suitable for acoustic beamforming [18].

The point spread function for wavenumber space beamforming is shown in Figure 5 for selected frequencies.

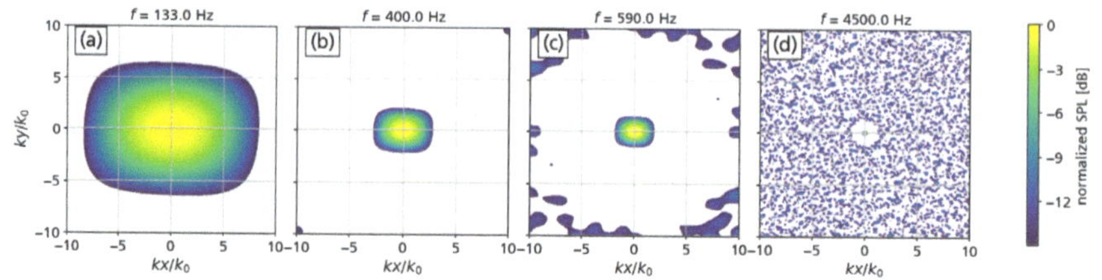

Figure 5. (**a–d**) Point spread function (PSF) of the array in wavenumber space at selected frequencies.

As expected, a unity source in the center of the map exhibits a broad main lobe. The main lobe width decreases as the evaluation frequency increases. For a dynamic range of 15 dB some grating lobes start to appear at $f = 590$ Hz in the selected focus grid range of $-10 \leq k_x/k_0, k_y/k_0 \leq 10$ approximately 9 dB below the maximum. At 4500 Hz, the width of the main lobe further decreases, yielding a higher resolution of the array. However, more grating lobes appear reaching levels of approximately 6 dB below the maximum. The analytical point spread functions are used for a DAMAS2.1 deconvolution attempt in Section 5.

The maximum difference in cable length was estimated to be the diagonal of the array area and thus 0.5 m. With an estimated delay caused by the cables of roughly 5 ns/m (corresponding to 0.67 times the speed of light), a maximum time delay of 2.5 ns is estimated between two channels of the array. This accounts for a maximum phase error of 0.09 deg at 100 kHz which is considered to be acceptable.

The MEMS required an electrical power supply voltage of 3 V which was provided by two batteries of type AA. The power was delivered to the array by utilizing two dedicated pins in the cable connectors.

3.1. Data Acquisition System

In this paper, two similar data acquisition systems were used. One system ("GBM Viper-48") is capable of recording 48 channels up to 250 kHz at a bit depth of 16 bit, allowing for a maximum dynamic range of 96 dB.

The other system ("GBM Viper-HDR") is capable of recording 64 channels up to 250 kHz at a bit depth of 24 bit, allowing for a maximum dynamic range of 144 dB.

Each channel of the two systems is equipped with its own sigma-delta A/D conversion unit and all A/D converters are synchronized and receive their clocking signal from one common clock source. Both systems possess a second-order high-pass filter with a cutoff frequency of 500 Hz which was used to reduce the influence of the low-frequency noise prior to A/D conversion and to make better use of the dynamic range of the system.

4. Preliminary Tests

The array and the sensors had to be examined extensively to ensure their functionality for the flight test. A total of four preliminary tests were performed. First, the in-situ frequency response of the individual microphones was determined without flow in an anechoic chamber. Second, a test was performed to study their behavior at low flow speeds. Observations made here lead to the introduction of Kapton foil as a treatment. Third, the influence of the foil on the frequency response was measured. Forth, a final test at comparable flight speeds in an industrial wind tunnel was performed in order to assess the influence of the foil and flow on the MEMS microphone's frequency response compared to a reference sensor.

4.1. Frequency Response

The data sheet [17] lists a frequency response that is typical for this type of microphone, showing a flat response from 200 Hz to 10 kHz. Here, the individual frequency response of the MEMS sensors for a range up to 100 kHz was measured in the anechoic chamber of the aeroacoustic laboratory of DLR Göttingen. This was performed using two loudspeakers: a calibration speaker (B&K Omnisource 4295) for frequency range up to 6 kHz and a tweeter (ELAC-4Pi-II) for the frequency range from 5 kHz to 100 kHz. The MEMS microphone array was positioned at a 3.00 m distance to the speakers. A condenser microphone (B&K Typ 4944) with known frequency response was used as reference. This microphone was positioned in the center of a metal plate that had the dimensions of the MEMS microphone array, to mimic the influence of the array itself. In order to minimize the influence of source directionality especially of the tweeter special attention was paid to exact positioning when the MEMS array was exchanged for the reference plate. Here, the 24-bit GBM Viper-HDR data acquisition system was used (see Section 3.1).

The frequency response H was then calculated by comparing the reference microphone spectrum G_Y with those of the MEMS array microphone spectra G_X:

$$H(f) = \frac{G_X(f)}{G_Y(f)} \tag{2}$$

the measurement time was 30 s at a sampling frequency of 200 kHz. The data was then processed using an overlap of 50% and a fast Fourier transform block size of 5000 samples, with a Hann window, yielding 2399 averages and a narrowband frequency resolution of 40 Hz. The calculated results were then smoothed with a Savitzky-Golay filter [19] with a window length of 51 and a poly order of three.

Figure 6 shows the frequency response which was calculated by merging both results from the two separate measurements with different speakers and frequency range from 5 kHz to 10 kHz using a linear weighted transition. For the range of 300 Hz to 10 kHz the frequency response is approximately constant. For higher frequencies, a resonance peak is visible at 22 kHz and an anti-resonance drop at 44 kHz. The standard deviation of the frequency responses of all microphones is almost negligible with somewhat larger values from 500 Hz to 1 kHz and around the anti-resonance drop.

In general, the measured mean frequency response shows a low standard deviation between all MEMS sensors. The frequency response from the data sheet shows an almost perfect flat response from 10 Hz to 17 kHz. The comparison shows somewhat lower values of the measured frequency response below 1 kHz and somewhat higher values above 1 kHz. These deviations are probably due to inaccuracies in the measurement setup (influence of the measuring room and setup).

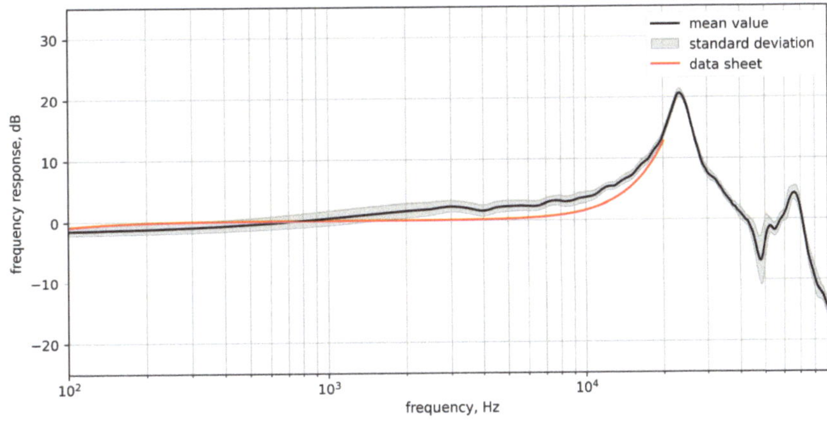

Figure 6. Mean frequency response of all array sensors with standard deviation in comparison to the data sheet.

4.2. Tests in Flow Conditions

4.2.1. First Tests (AKG)

The first flow application tests were performed in the aeroacoustic wind tunnel facility (AKG) of the aeroacoustic laboratory of DLR Göttingen. This small wind tunnel of Göttingen type has a 0.4 m by 0.4 m open test section with a length of approximately 2 m and can be operated at flow speeds up to $u = 50$ m/s.

The microphone array was flush-mounted in a plate that was used as a extension of one wind tunnel wall with its surface aligned with and extending the bottom wall of the nozzle of the wind tunnel as shown in Figure 7. The tests were performed at different flow speeds ranging from $u = 5$ m/s to $u = 40$ m/s.

Here, the 24-bit GBM Viper-HDR data acquisition system was used (see Section 3.1). To lower the required dynamic range for recording the signals, which is mainly caused by the dominant low-frequency wind tunnel noise and hydrodynamic pressure, the 500 Hz high-pass filter was applied.

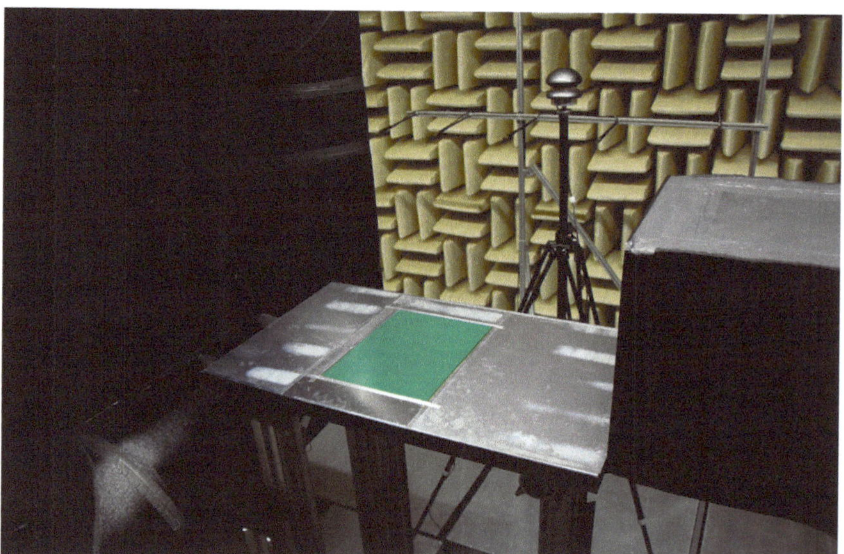

Figure 7. Photo of the measurement setup for the first tests under flow conditions. The MEMS array is mounted on the plate that is used as an extension of the bottom wind tunnel wall.

The individual MEMS measurements were investigated as time signals and spectra. While observing the live-feed from an ongoing measurement, it appeared that some online spectra would freeze and not change their shape even after turning off the wind tunnel. The observation was made at flow speeds from $u = 20$ m/s and above. Therefore, the time signals of the microphones were investigated in detail. Figure 8 shows a segment of a recorded time signal from an exemplary microphone at a flow speed of $u = 40$ m/s (blue).

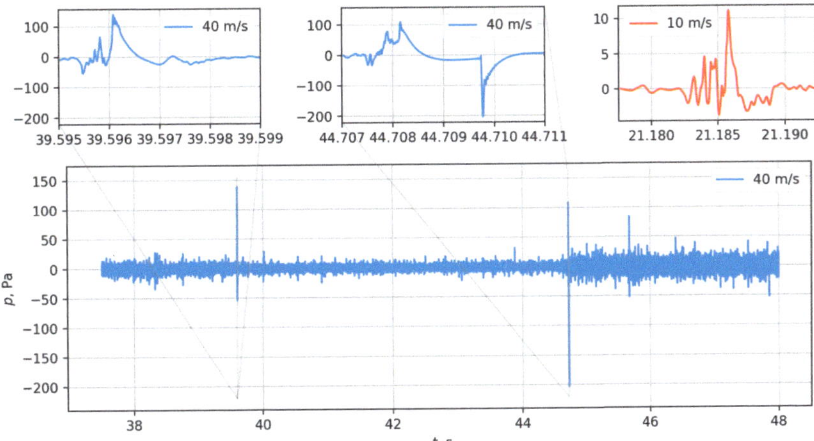

Figure 8. Time series of a chosen MEMS microphone at 40 m/s (blue) with zoomed-in sub-figures of peculiar events. For comparison, a zoomed-in result is shown at 10 m/s (red).

Several isolated peaks are conspicuous, exceeding the predominant fluctuation amplitude by far. An interesting observation can be made after the second peak in Figure 8 at $t \approx 44.7$ s. The fluctuation intensity takes different levels before and after the peak, with the higher level continuing throughout the end of the graph. At a closer inspection after zooming in, the isolated peaks exhibit peculiar details. For the first event at $t \approx 39.6$ s, a quick increase of fluctuation energy can be observed leading up to a strong peak exceeding 140 Pa (corresponding to 137 dB). After the peak, an exponential decay of the signal can be observed, leading to an apparent regular signal again after one to two milliseconds. The second event shows a similar behavior, with the first exponential decay being followed by a steep and sudden drop (-200 Pa within 1 sample) in level and a subsequent second exponential mitigation.

The exponential decay after the high-amplitude events is caused by a clipping of the signals in the sensor and the signals being recorded with a 500 Hz high-pass filter applied to them.

Contrary to the first event, the overall fluctuation amplitude appears to be increased by a factor of two. The signal of this channel persisted to fluctuate even after turning off the wind tunnel. After a waiting time of several minutes, the sensor behavior returned to normal. This behavior occurred in approximately 10% of the sensors when subjected to flow velocities exceeding 20 m/s. Sensors exhibiting this behavior were located exclusively within the shear layer on the right and left sides of the array, produced by the corresponding nozzle edges. Both, the sensor behavior returning back to normal after some time, as well as the highly localized occurrence of the phenomenon in the experiment lead to the conclusion that it was not just a bad batch of sensors causing the observed behavior.

For comparison, a comparable signal from a sensor exhibiting the strange signal behavior is shown in Figure 8 for a lower flow velocity of $u = 10$ m/s (red). Although an event is visible exhibiting a similar fluctuation behavior leading up to a main peak, no exponential drop is observed after the main peak. For the case at $u = 10$ m/s, this type of event was found significantly more often at locations where previously at higher speeds the peculiar signal behavior was observed. The signals are likely caused by footprints of instabilities from the widening shear layer.

Probably due to the small sensor size (see Section 2) the sensor is able to capture hydrodynamic structures with a small coherence length and/or high wavenumber. It is likely that the amplitude of these small structures takes very high values which were not fully captured due to the integration effect of the sensitive surface of larger microphones.

It appears that this signal cannot be fully captured because of the limitation set by the AOP of the microphones. This leads to a significant overload which can also lead to a temporary malfunction.

The question arises, why this peculiar behavior of the sensors has not been observed or reported in prior studies (see Section 1). There, implementations of signal compression could result in such events not being observed due to undetectable shaping of the signal. For the sensors used in the present study, one explanation could be that the implementation of signal compression leads to the observed peculiar behavior (nonlinear distortion or malfunction).

The findings in the previous section can be summarized into three points:

- due to the small sensor size the MEMS microphone is capable of capturing small (short coherence length and high wavenumber) hydrodynamic structures.
- These structures may sometimes exhibit a very high pressure amplitude which leads to exceeding the AOP of the MEMS microphones.
- This results in signal artifacts being imprinted onto the signal for a limited but significantly long duration and sometimes temporary malfunction.

Therefore, either sensors with a higher AOP must be used or measures are required to increases the AOP value of the employed sensors. Currently, sensors with higher AOP are not available to the authors and therefore, the application of Kapton foil for increasing the AOP value is presented in the following section.

4.2.2. Surface Treatment: Application of Kapton Foil

To overcome the limitations of the MEMS microphones observed in the first tests under flow conditions, a one-sided adhesive Kapton foil of 25 μm thickness was applied in order to attenuate the pressure excitation covering all the port holes on the PCB. As a positive side-effect, this coverage provides protection to the MEMS microphones on the fuselage from humidity and dust particles.

The influence of this treatment on frequency response was measured in the anechoic chamber of the aeroacoustic laboratory of DLR Göttingen. This was performed using a rectangular plate with the surface-mounted MEMS microphone array and a surface-mounted reference condenser microphone (B&K type 4944). As a sound source, an Adam A3X speaker was used in the frequency range of 50 Hz to 60 kHz. The Kapton foil response was achieved by comparing the measurement with foil to the one without foil with respect to the reference microphone. The reference microphone was used to take into account the influence of the setup such as imperfections of the anechoic environment and differences in the excitation from the speaker. Thus, the transfer function of the Kapton foil H_K was calculated in the frequency domain by comparing the cross-spectrum R between the case of MEMS microphones with Kapton applied X_K to the case of MEMS microphones without Kapton $X_{\overline{K}}$ by means of a reference microphone Y.

$$H_K = \frac{R_{X_K,Y}(f)}{R_{X_{\overline{K}},Y}(f)} \tag{3}$$

for each MEMS microphone. Here, the 24-bit GBM Viper-HDR data acquisition system was used (see Section 3.1). The measurement was performed for 60 s at a sampling frequency of 200 kHz and processed using an overlap of 50% and a fast Fourier transform block size of 5000 samples, with a Hann window, yielding 4799 averages and a narrowband frequency resolution of 40 Hz. The calculated result was then smoothed with a Savitzky-Golay filter [19] with a window length of 51 and a poly order of three.

In order to investigate the dependence on the application procedure, Kapton foil was applied a second time onto the MEMS microphone array and the measurement procedure was repeated. In the second application, a felt card was used as a tool to apply the contact pressure. It should be noted that measurement #1 was performed with the very exact Kapton foil applied to the MEMS microphone array as used in the final flight test (Section 5), without removing the foil in between.

Figure 9 shows the resulting frequency response with the confidence interval for both measurements. Both results show an approximately flat attenuation of -40 dB to -38 dB (measurement #1) and -34 dB (measurement #2). After a minimum at 23 kHz, the attenuation drops to about -10 dB to -5 dB at 36 kHz. Thereafter, it further increases and decreases in the attenuation range from -10 dB to -25 dB with different courses for both measurements up to 60 kHz.

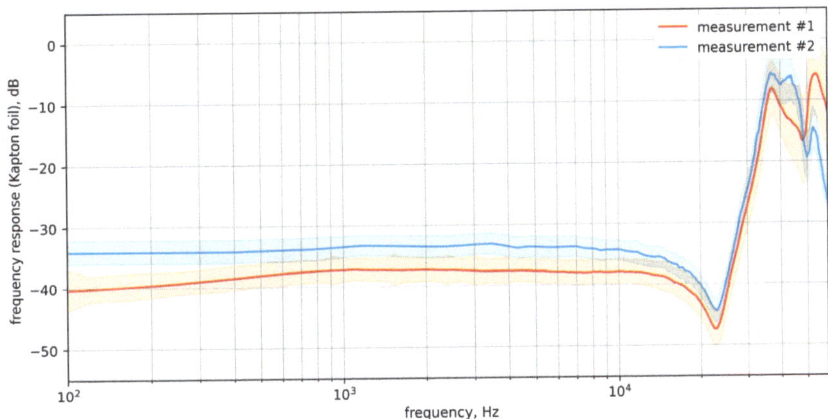

Figure 9. Mean frequency response of all array sensors treated with Kapton foil with standard deviation for two different applications of the coating with the Kapton foil (measurement #1: exact treatment as used in this paper). The respective standard deviation is depicted by the envelopes.

As an important outcome, the comparison of both results shows that the application of the Kapton foil can lead to different results, dependent on the method of application (usage of pressure tool, applied contact pressure). Thus, the transfer function should be measured individually for each Kapton foil application.

4.2.3. Validation Measurements (TWG)

The validation measurements at increased Mach numbers were performed in the Transsonic Wind Tunnel Göttingen (DNW-TWG) facility, which is a DNW research and development facility at the DLR Göttingen site. This wind tunnel of Göttingen type has a 1.0 m by 1.0 m closed test section and can be operated for Mach numbers from of M = 0.3 to M = 2.2 under vacuumed and pressurized conditions.

Here, embedded in another experiment, the MEMS microphones used in this paper treated with Kapton foil and a 1/2 inch B&K microphone (type 4134) were compared at a Mach number of 0.3. The microphones were flush-mounted into the wind tunnel wall as shown in Figure 10. For this measurement, the 16-bit GBM Viper-48 data acquisition system was used (see Section 3.1) using a 500 Hz high pass filter. The results were later corrected with regard to the filter response.

Figure 10. Photos of the microphones on the wind tunnel wall. Left: MEMS microphone array with Kapton foil. Right: flush mounted reference 1/2 inch microphone.

The spectra of a Kapton-foil-treated MEMS microphone and a reference microphone are presented in Figure 11. The black line shows the spectrum of the reference microphone. To illustrate the influence of the frequency response corrections explained in Section 4.1 (frequency response of the MEMS microphone itself) and Section 4.2.2 (Kapton foil) three different spectra are shown for the data acquired using the MEMS. The uncorrected spectrum is shown by the red dotted line, whereas the spectrum corrected for the Kapton foil is shown by the red dashed line. The fully corrected spectra are shown by the red solid line. The uncertainty range, which results from the combined standard deviations of the two correction measurements, is depicted by the orange envelope.

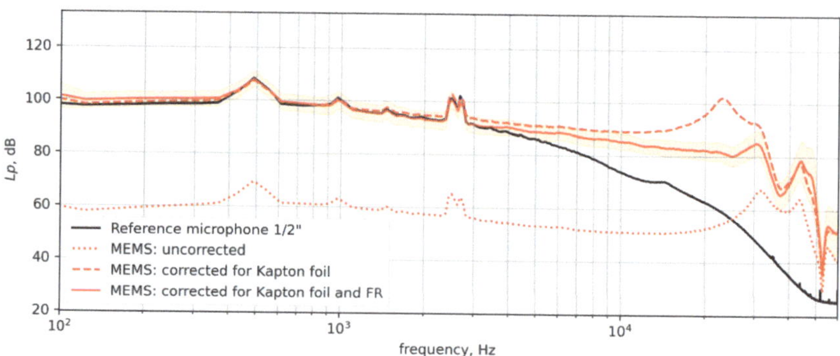

Figure 11. Comparison of the MEMS microphone spectrum and the reference microphone spectrum in the TWG at a Mach number of 0.3. The orange envelope illustrates the combined error margin obtained from the Kapton foil and frequency response measurements.

First, the fully corrected MEMS microphone spectrum is compared to the spectrum of the reference microphone. Second, the application of the frequency response corrections is assessed.

From 100 Hz up to a frequency of 3 kHz both spectra collapse very well, showing an slight downward slope of about 10 dB. Within this frequency range, some peaks appear at 500 Hz and 1 kHz and a double peak at about 2.5 kHz. After 3 kHz the reference microphone spectrum continues the downward slope until about 50 kHz where it transitions to a constant floor level of 25 dB at 60 kHz. The MEMS microphone spectrum, however, continues with a much less significant downward slope up to a frequency of 25 kHz. Then the spectrum decreases with superimposed ripples in the downward slope down to a level of 50 dB.

For the interpretation of the spectra, it is assumed that the broadband spectra are dominated mainly by the boundary layer noise. In the wind tunnel, no artificial sound source was present, thus only the peaks at 500 Hz and 1 kHz (wind tunnel drive, 1st and 2nd Blade Passing Frequency (BPF) order) and the peaks around 2.5 kHz are assumed to be of acoustic nature. The frequency where the strong downward slope starts (3 kHz vs. 25 kHz) is similar to the modeled attenuation for the correspondent sensor sizes in Figure 1, Section 2.1.

The estimated amplitude attenuation due to surface integration shown in Figure 1 exhibits significantly less attenuation than can be seen between the MEMS microphone signal and the reference microphone signal in Figure 11. The attenuation difference between a 1 mm MEMS microphone and a 1/2 inch microphone due to surface integration (Section 2.1) was estimated to be approximately -1 dB at 10 kHz in Figure 1. However, at 10 kHz in Figure 11, a difference in power level of approximately 10 dB is observed in the measurement. The remaining difference is due to the strong influence of a short coherence length of the TBL pressure fluctuations as discussed in Section 2.2.

For validation, the application of the frequency response corrections on the MEMS microphone is discussed. Applying the Kapton foil correction leads to an increase of the spectrum by approximately 40 dB. With the foil correction alone, the MEMS spectrum's overall amplitude coincides well with the reference microphone below 3 kHz. At 22 kHz an additional hump is now observed which coincides with the resonance frequency of the MEMS microphones already known from the calibration in Figure 6. Applying the MEMS calibration curve to the measured spectrum removes the hump at 22 kHz and leaves a smooth spectrum up to 26 kHz. At higher frequencies, the spectrum still shows a wavy response which already is observed in the uncorrected spectrum. It can be concluded that in the frequency range greater than 26 kHz, the corrections cannot be meaningfully applied.

5. Application for In-Flight Measurements

5.1. Setup

Experiments were conducted on a DLR test carrier, the Dornier 228-101 propeller aircraft as shown in Figure 12. One window bank on the left-hand side of the aircraft was equipped with a dummy window. The dummy window allowed the passage of the cabling from the MEMS array which was attached on the fuselage close to the window and directly opposite of the propeller.

Figure 12 also shows a sketch of the MEMS microphone array mounting. The fuselage was coated with a flight-approved aluminum tape. As described in Section 3, the backside of the PCB of the MEMS microphone array was coated with silicone. Thus, the MEMS microphone array was applied via adhesion between the tape and the silicone. In order to attenuate the pressure excitation as motivated in Section 4.2.2 the MEMS microphone array was also coated with a Kapton foil. For the final fixation, the array was completely secured with the flight-approved tape around the edges.

Additionally, an array of 1/2 inch aerospace surface microphones (Brüel&Kjær type 4948) was mounted on the fuselage. Surface microphones in the vicinity of the MEMS microphone array will be used as a reference here. All signals were routed to the data acquisition system inside the aircraft. Here, the 24-bit GBM Viper-HDR data acquisition system was used (see Section 3.1) using a 500 Hz high pass filter. The results were later corrected with regard to the filter response. The microphone signals of the MEMS microphone array and the surface microphones were recorded simultaneously with a sampling frequency of 250 kHz. Spectra will be shown up to a frequency of 5 kHz. To reduce the influence of the low-frequency boundary layer noise, high-pass filter with a cutoff frequency of 500 Hz was used. The acquired data were corrected with regard to the filter response. The data were processed using an overlap of 50% and a fast Fourier transform block size of 65,536 samples, with a Hann window, yielding a narrowband frequency resolution of 3.81 Hz.

Figure 13 depicts the positions of all MEMS microphones and the surface microphones chosen as a reference.

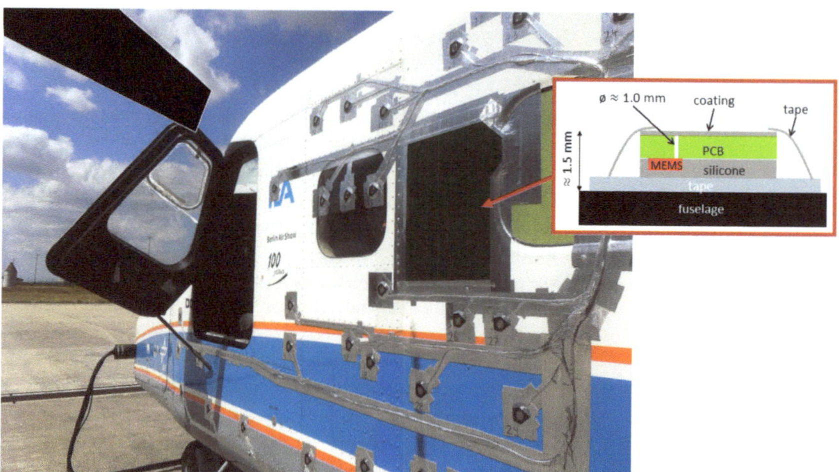

Figure 12. Photo of the propeller aircraft with the MEMS microphone array mounted on the fuselage and a sketch depicting the MEMS microphone array mounting.

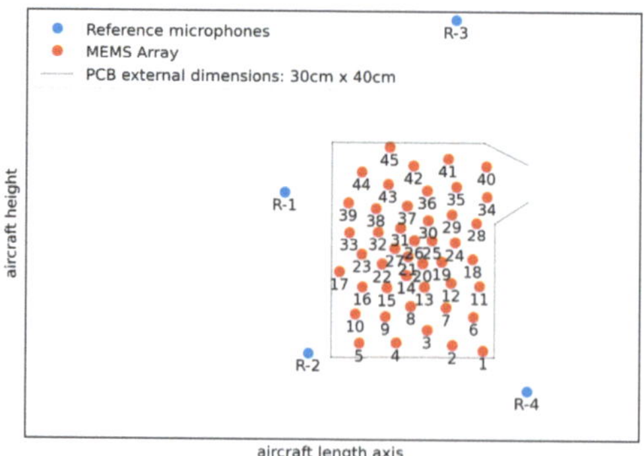

Figure 13. Position of all MEMS microphones (red) and the chosen reference surface microphones (blue).

5.2. Results

The spectral content of the signals from both the reference microphones and the MEMS microphones in the form of an array yield valuable information about the characteristics of the surface pressure fluctuations. This information can be deduced from the autospectra and by looking at the wavenumber maps computed from the signals of the MEMS microphone array.

5.2.1. Autospectra of MEMS and Reference Microphones

Figure 14 compares the mean spectra of the reference microphones to the MEMS microphones obtained from one in-flight measurement.

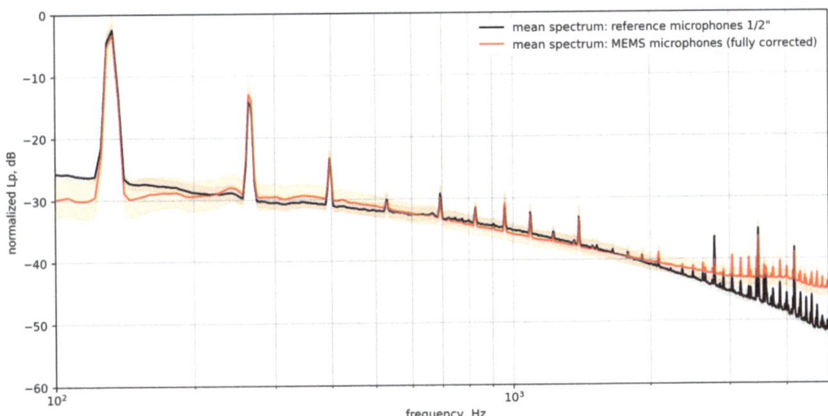

Figure 14. Comparison of the mean spectra of the MEMS microphones and the reference microphones with standard deviation.

In general, both spectra show a general broadband decrease over the frequency and a dominant peak at approx 133 Hz with its n-th order harmonics. It can be assumed, that the dominant structures are of acoustic nature. The dominant peak at 133 Hz and its n-th order harmonics are related to the propeller. The speed of the propeller was measured to be $n \approx 1590$ rpm. Taking into account the five propeller blades this leads to a BPF $f_{BPF} \approx 26.5$ Hz $\times$ 5 $\approx$ 133 Hz which equals the measured peak frequency.

While the dominant tonal components (which are assumed to be of acoustic nature as they coincide with multiples of the BPF) measured with both the reference microphone and the MEMS microphone are very similar, some differences are visible for the broadband shape of the signal.

Above 2 kHz the reference microphone exhibits a stronger drop in signal compared to the MEMS microphone. This drop in signal amplitude at higher frequencies is due to the surface integration of the larger microphone modeled in Section 2.1, shown in Figure 1. Here, the 1/2 inch microphone exhibits a strong level decrease at about 2 to 3 kHz.

5.2.2. Wavenumber Maps

Wavenumber maps were calculated by applying a beamforming algorithm with planar wave steering vectors without removing the diagonal elements of the cross-spectral matrix as done by Haxter and Spehr [3]. In the present case, the normalized focus grid was of size 255 by 255, extending over a normalized wavenumber range of $-10 \leq k_{x,y}/k_0 \leq 10$. A DAMAS 2.1 [20] deconvolution algorithm using $N = 10^5$ iterations was applied to the maps for enhancement.

Resulting maps at are shown in Figure 15 for the first BPF of 133.5 Hz in Figure 15a,e, the third BPF of 400.5 Hz in Figure 15b,f, and selected broadband ranges in Figure 15c,g as well as Figure 15d,h. The top row in Figure 15 shows the direct beamformer output (or "dirty map") without the DAMAS 2.1 enhancement algorithm applied to it. The bottom row shows the output of the DAMAS 2.1 algorithm for the respective frequency or frequency range of the map shown above. The solid circle in the center of each map represents the domain of the acoustic propagation mechanism. Sources located within this circle propagate at the speed of sound or above. The dashed circle shown is the integration map used to separate the acoustic surface pressure fluctuations from all other pressure fluctuations. To compensate for the main lobe width of sources even after a deconvolution attempt, the acoustic domain was increased to $3k_0$ which should incorporate the remaining imprecision in source appearance. The flight speed was sufficiently low for the convective ridge to be located reasonably far away from the acoustic domain. This gap allowed for the increase in acoustic domain threshold with no overlap of enlarged acoustic domain and convective ridge occurring.

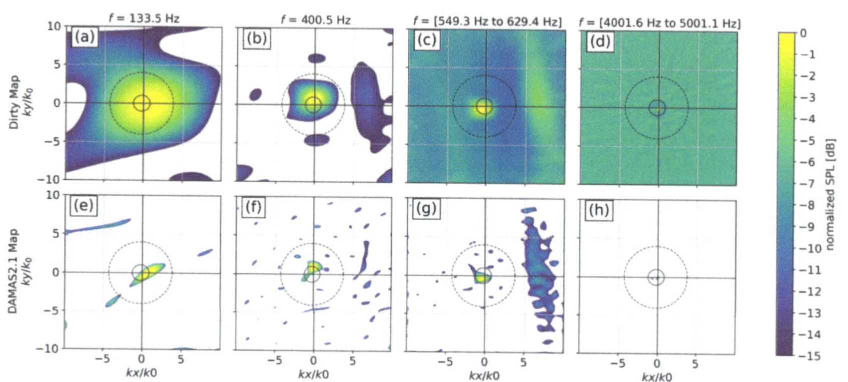

Figure 15. Wavenumber dirty maps and deconvolved maps at selected frequencies, plotted with a dynamic range of 15 dB. The acoustic domain is shown as a solid circle in the center of the map. The expanded integration border for separation of acoustic and other types of pressure fluctuations is shown as a dashed line surrounding the acoustic domain. (**a**)/(**e**): narrow band dirty map/deconvolved map at 133.5 Hz; (**b**)/(**f**): narrow band dirty map/deconvolved map at 400.5 Hz; (**c**)/(**g**): integrated dirty maps/deconvolved maps from 549.3 Hz to 629.4 Hz; (**d**)/(**h**): integrated dirty maps/deconvolved maps from 4001.6 Hz to 5001.1 Hz

At different frequencies, different sources can be seen to dominate the spectrum. At the first BPF at $f \approx 133$ Hz a single source is visible in the center of the map in Figure 15a,e. In the upper Dirty Map, the source is located in the center of the map but is spread out over a very large wavenumber region due to the poor resolution at low frequencies. Applying the DAMAS 2.1 deconvolution in the lower map reduces the source size, but also introduces some ghost sources at various positions in the map. However, source integration over the acoustic domain is enhanced since the size of the central source is reduced, enabling a sharper separation. Relative to the acoustic domain, the center of the dominant source appears to be located on the right-hand side of the acoustic domain, indicating a propagation direction from front to back over the array.

For the third BPF at $f \approx 400$ Hz in Figure 15b, a dominant source occurs within the acoustic domain as well. Its location relative to the acoustic domain appears to be on the upper edge of the domain, indicating a propagation at the speed of sound from the lower end of the array towards the top. Besides this acoustic source, there are four other appearances in the wavenumber map worth mentioning. The first appearance is a large elongated source on the right-hand side representing the TBL pressure fluctuations at subsonic speeds. As mentioned in Section 2.2, this convective ridge inherits its shape from the limited coherence lengths of the TBL pressure fluctuations. With the source being located to the right of the map origin and its main elongation being in k_y-direction it can be concluded that the flow is passing over the array from front to back (as expected) and that the coherence length in cross-flow direction is smaller than the coherence length in in-flow direction. Two other sources of interest are located above and below the center acoustic source at $k_x/k_0 \approx 0$ and $5 \geq k_y/k_0 \geq 6$. With the propeller passing in an upward motion on its side facing the array, the upper source location represents the direct near-field pressure fluctuation passing over the array. The lower source is possibly caused by a reflection of the near-field pressure fluctuations at the bottom side of the wing, thus propagating over the array in the opposite direction from top to bottom. It appears that the bottom source is located closer to the origin of the spectrum than the upper source, indicating that the wavenumber of the presumably reflected source is slightly smaller.

The fourth source is located towards the left of the acoustic domain, indicating an upstream propagation below the speed of sound. We could not find an explanation for this source so far. Whether or not this source is real or is just an artifact of signal processing remains an open question. Other sources in the corners of the map are likely side lobes from real sources in the map.

With DAMAS 2.1-enhancement applied to the source map at 400.5 Hz in Figure 15f, the size of the center acoustic source is significantly reduced. However, the shape of the convective ridge on the right-hand side has been dissolved. Many ghost sources have appeared over the entire map. While the two sources from the near-field of the propeller above and below the acoustic domain are still present and even have increased in relative amplitude with enhancement, their presence does not stand out any more relative to the ghost sources.

In the frequency range 549.3 Hz $\geq f \geq$ 629.4 Hz in Figure 15c both an acoustic source in the bottom left part of the acoustic domain, as well as the convective ridge on the right-hand side, are visible. A high background noise level is present in the map. The DAMAS 2.1 enhancement is able to remove most of the background noise with the convective ridge keeping its characteristic shape. Several ghost sources are visible within the confinement of the extended acoustic domain. Regarding the near-field sources visible at lower frequency, there is only a small dot visible in the enhanced bottom map at the lower position. This however might just be a spurious processing artifact.

The last frequency range is shown in Figures 15d,h from 4001.6 Hz to 5001.1 Hz. Only two distinct acoustic sources are present: a dominant source in the center and a secondary source on the the left hand side of the acoustic domain. The remaining noise in the map exhibits a slight radial pattern centered around this acoustic source which is a characteristic feature of distinct source maps integrated over several frequencies. The image enhancement by DAMAS 2.1 is able to singularize this source, leaving only thin traces of noise behind. Remarkably, there is no convective ridge visible which is likely due to the small coherence length of TBL pressure fluctuations at these high frequencies. The array was not designed to measure this kind of pressure fluctuations and thus the convective ridge vanishes.

5.2.3. Separation of Spectra

Using the extended acoustic domain as a separation between the propagation types, the deconvolved source maps can be integrated over each separated region to obtain the separated spectra. For a noise threshold of -15 dB the separated spectra are shown in Figure 16. The spectrum obtained from integrating the extended acoustic domain is shown in red. It exhibits very distinct peaks at the BPF and higher harmonics. With increasing frequency, the broadband noise level appears to be connected to the level of hydrodynamic pressure fluctuations shown in blue. This is likely due to ghost sources resulting from incomplete source map enhancement by the DAMAS2.1 leaking into other parts of the spectrum.

The hydrodynamic spectrum is shown in blue in Figure 16 and it exhibits a rather complex appearance. At lower frequencies, indicated by the dashed blue line, a distinct drop in power is visible. This is very likely due to the turbulent structures breaking down at low frequencies as their size is limited by the boundary layer thickness. This diminishes the coherent pressure footprint and considerably reduces the signal correlation between different microphones which is required for wavenumber analysis. The acoustic peaks mostly remain unchanged as the acoustic propagation still leaves a coherent signal on all transducers in the array. The increased hydrodynamic power at the BPF is likely caused by both, the limited dynamic range of the analysis as well as near field effects as observed in Figure 15b being located outside the extended acoustic domain used for source separation.

An increased constant noise floor would likely not be reconstructed by the DAMAS 2.1 algorithm as such but rather be approximated by distributed small sources that would explain the noise. Additionally, the likelihood of phase mismatches between measurement and Green's function is increased at higher frequencies which could explain a drop in efficiency of the enhancement algorithm. The close resemblance of the acoustic spectrum by the hydrodynamic spectrum at high frequencies is again likely due to the limited dynamic range of the analysis.

In the mid-frequency range from ≈400 Hz to ≈2.2 kHz the hydrodynamic spectrum exhibits a slow decrease of fluctuation power with intermediate peaks at the higher harmonics of the BPF which are likely caused by the near-field of the propeller. At slightly higher frequencies, the influence of the peaks at BPF increasingly dominates the hydrodynamic spectrum. Whether or not this is a real phenomenon remains uncertain at this time.

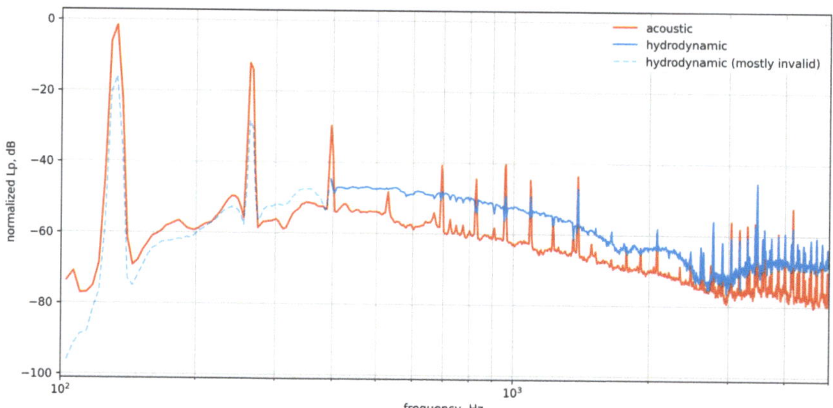

Figure 16. Separated spectra of acoustic pressure fluctuations and hydrodynamic pressure fluctuations. The performance of separation is best in the mid-frequency range as indicated by the solid line for the hydrodynamic pressure fluctuations.

Overall, the wavenumber spectra provide valuable insights into the origins and characteristics of the surface pressure fluctuations when measured by an array of surface-mounted sensors.

6. Summary

The paper describes the process of preparing MEMS microphones within a flexible circuit board array for the measurement of unsteady pressure fluctuations on an airplane fuselage during flight tests.

The MEMS sensors have a small sensitive surface area which allowed them to capture the presence of high-energy small-scale structures within the turbulent boundary layer when pre-testing them in a low-speed wind tunnel environment. Although the relationship between the sensitive surface area size of the sensor and an increasingly high measured amplitude is known, it can easily be overlooked if such high amplitudes are not expected. As measurements nowadays are performed more frequently using small-scale MEMS sensors, higher measured amplitudes are bound to occur more frequently as well.

With the high measured amplitudes resulting from the matched size ratio of the turbulent boundary layer pressure fluctuations to the sensitive surface area of the sensor, it is possible to elicit nonlinear behavior in the response of the MEMS sensors due to them temporarily exceeding the upper limit of their dynamic range.

Some makes of MEMS sensors respond to exceeding this limit by either clipping or compressing the signal, which may cause an inaccurate underestimation of the amplitude of the small-scale hydrodynamic pressure fluctuations. If compression is used by the internal signal conditioning of the sensor, such a distortion may go unnoticed.

To overcome these limitations of current MEMS microphones and be able to apply them in flight tests, the sensitivity of the MEMS array was decreased by the application of Kapton foil of 25 μm thickness. The damping resulting from the applied foil was measured and revealed a mean insertion loss of 34 to 40 dB with a variation of ±3 dB between single microphones in the frequency range of interest.

After a low-speed test, the Kapton-covered MEMS were reviewed further in a high-speed wind tunnel test. The test was aimed at comparing the signals from a reference 1/2 inch microphone to the signal from the MEMS microphones with frequency response correction and Kapton foil response correction applied. The analysis of the high-speed lead to the conclusion that the application of both frequency responses to the MEMS microphone signals leads to a very good agreement with the reference signals with expected differences due to sensitive surface size.

Finally, flight tests were conducted on the DLR Dornier 228 propeller aircraft with the MEMS microphone array mounted on the left-hand side of the fuselage, opposing the left-hand-side propeller. The comparison with 1/2 inch microphones mounted in the vicinity of the MEMS microphone array shows a good general signal agreement in the expected frequency range when applying the appropriate frequency corrections. Apart from the good general agreement, some differences were visible as well. The mean broadband spectra of the MEMS microphones revealed more fluctuating energy especially in the range above 2 kHz. Using a wavenumber analysis it was determined that these increased levels were caused by hydrodynamic pressure fluctuations.

The results indicate that the Kapton foil covering the MEMS microphones was able to damp the pressure fluctuations as expected, while still preserving the critical phase information required for wavenumber analysis.

A great advantage of MEMS microphones lies in their small size and the possibility to arrange them closely-spaced and flush-mounted in a very thin array that can be applied to follow the shape of the underlying fuselage structure. The close spacing and small sensitive surface area of the sensors allows for analyzing the wavenumber spectra of both, the acoustic pressure fluctuations as well as the overlaying turbulent boundary layer. The results of the wavenumber analysis from the flight test measurements reveal the presence of both the acoustic pressure fluctuations induced by propeller and engine and the convective ridge of the hydrodynamic pressure fluctuations at different relative amplitudes depending on frequency.

It can be concluded that MEMS microphones are an inexpensive alternative to conventional microphones even at challenging experimental conditions during flight tests. Due to their small size the MEMS exhibit an increased potential for spatially high-resolved measurements which larger sensors are not capable of.

Author Contributions: Conceptualization and methodology: T.A., S.H. and C.S.; investigation: T.A., S.H., C.S., D.E., and T.K.; software and formal analysis: T.A. and S.H.; hardware development: T.K.; data curation: D.E.; writing—original draft preparation: T.A., S.H., C.S. and D.E.; writing—review and editing: S.H. and T.A.; visualization: T.A.; supervision: C.S.; project administration: S.H. and T.A.; funding acquisition: C.S. All authors have read and agreed to the published version of the manuscript.

Funding: This research has received funding through three different projects: (1) MEMS microphone usage in aeroacoustic applications: project *ADEC (LPA)* from the Clean Sky 2 Joint Undertaking (JU) under grant agreement No. 945583. The JU receives support from the European Union's Horizon 2020 research and innovation programme and the Clean Sky 2 JU members other than the Union.; (2) MEMS microphone array implementation: project *VICTORIA* from the German Aerospace Center (DLR).; (3) Flight test: project *FusionProp* from the Federal Ministry of Economics and Technology (BMWi) as part of the aerospace research program (LuFO V-3), grant number 20T1731B.

Acknowledgments: The authors would like to sincerely acknowledge contributions by the following people and entities: Stefan Kommallein and the team from DLR Flight Experiments Facility (FX); airplus Maintenance GmbH; Armin Goudarzi, Hans-Georg Raumer, Florian Philipp and Carsten Fuchs from DLR Experimental Methods.

Conflicts of Interest: The authors declare no conflict of interest.

Abbreviations

The following abbreviations are used in this manuscript:

A/D	Analog-to-digital converter
AOP	Acoustic Overload Point
BPF	Blade Passing Frequency
DAMAS	Deconvolution Approach for the Mapping of Acoustic Sources
DLR	Deutsches Zentrum für Luft-und Raumfahrt (German Aerospace Center)
DNW	Deutsch-Niederländische Windkanäle (German-Dutch Wind Tunnels)
FR	Frequency response
MEMS	Micro Mechanical System
PCB	Printed Circuit Board
SNR	Signal-to-Noise Ratio
THD+N	Total Harmonic Distortion + Noise
TWG	Transonic wind tunnel facility

References

1. Schewe, G. On the structure and resolution of wall-pressure fluctuations associated with turbulent boundary-layer flow. *J. Fluid Mech.* **1983**, *134*, 311–328. [CrossRef]
2. Palumbo, D.L. Determining correlation and coherence lengths in turbulent boundary layer flight data. *J. Sound Vib.* **2021**, *331*, 3721–3737. [CrossRef]
3. Haxter, S.; Spehr, C. Wavenumber Characterization of Surface Pressure Fluctuations on the Fuselage During Cruise Flight. In *Flinovia—Flow Induced Noise and Vibration Issues and Aspects-III*; Ciappi, E., De Rosa, S., Franco, F., Hambric, S.A., Leung, R.C.K., Clair, V., Maxit, L., Totaro, N., Eds.; Springer International Publishing: Cham, Switzerland, 2021; pp. 157–180.
4. Zhou, Y.; Valeau, V.; Marchal, J.; Ollivier, F.; Marchiano, R. Three-dimensional identification of flow-induced noise sources with a tunnel-shaped array of MEMS microphones. *J. Sound Vib.* **2020**, *482*, 115459. [CrossRef]
5. Sheplak, M.; Seiner, J.; Breuer, K.; Schmidt, M.A. MEMS microphone for aeroacoustics measurements. In Proceedings of the 37th Aerospace Sciences Meeting and Exhibit, Reno, NV, USA, 11–14 January 1999.
6. Reagan, T.N. MEMS on a Plane: A Flush-Mount MEMS Piezoelectric Microphone for Aircraft Fuselage Arrays. Ph.D. Thesis, University of Florida, Gainesville, FL, USA, 2017.
7. Williams, M.D. Development of a MEMS Piezoelectric Microphone for Aeroacoustic Applications. Ph.D. Thesis, University of Florida, Gainesville, FL, USA, 2011.
8. Shams, Q.A.; Graves, S.S.; Bartram, S.M.; Sealey, B.S.; Comeaux, T. Development of MEMS microphone array technology for aeroacoustic testing. *J. Acoust. Soc. Am.* **2004**, *116*, 2511–2511. [CrossRef]
9. Humphreys, W.; Shams, Q.; Graves, S.; Sealey, B.; Bartram, S.; Comeaux, T. Application of MEMS microphone array technology to airframe noise measurements. In Proceedings of the 11th AIAA/CEAS Aeroacoustics Conference. American Institute of Aeronautics and Astronautics, Monterey, CA, USA, 23–25 May 2005.
10. Sanders, M.P.; de Santana, L.D.; Azarpeyvand, M.; Venner, C.H. Unsteady surface pressure measurements on trailing edge serrations based on digital MEMS microphones. In Proceedings of the 24th AIAA/CEAS Aeroacoustics Conference, Atlanta, GA, USA, 25–29 June 2018.
11. Leclere, Q.; Dinsenmeyer, A.; Salze, E.; Antoni, J. A Comparison between Different Wall Pressure Measurement Devices for the Separation and Analysis of TBL and Acoustic Contributions. In *Flinovia—Flow Induced Noise and Vibration Issues and Aspects-III. FLINOVIA 2019*; Ciappi, E., Rosa, S.D., Franco, F., Hambric, S.A., Leung, R.C.K., Clair, V., Maxit, L., Totaro, N., Eds.; Springer: Cham, Switzerland, 2021; doi:10.1007/978-3-030-64807-7_9. [CrossRef]
12. Salze, E.; Jondeau, E.; Pereira, A.; Prigent, S.L.; Bailly, C. A new MEMS microphone array for the wavenumber analysis of wall-pressure fluctuations: Application to the modal investigation of a ducted low-mach number stage. In Proceedings of the 25th AIAA/CEAS Aeroacoustics Conference, Delft, The Netherlands, 20–23 May 2019.
13. Ko, S.H. Performance of various shapes of hydrophones in the reduction of turbulent flow noise. *J. Acoust. Soc. Am.* **1993**, *93*, 1293–1299. [CrossRef]
14. Corcos, G.M. Resolution of Pressure in Turbulence, *J. Acoust. Soc. Am.* **1963**, *32*, 192–199. [CrossRef]
15. Graham, W.R. A Comparison of Models for the Wavenumber-Frequency Spectrum of Turbulent Boundary Layer Pressures. *J. Sound Vib.* **1997**, *206*, 541–565. [CrossRef]

16. Haxter, S.; Spehr, C. Comparison of model predictions for coherence length to in-flight measurements at cruise conditions. *J. Sound Vib.* **2017**, *390*, 86–117. [CrossRef]
17. ICS-40618 Datasheet. Available online: https://invensense.tdk.com/wp-content/uploads/2016/02/DS-000044-ICS-40618-v1.0.pdf (accessed on 24 February 2021).
18. Sarradj, E. A Generic Approach to Synthesize Optimal Array Microphone Arrangements. In Proceedings of the BeBeC 2016, Berlin, Germany, 29 February–1 March 2016. Available online: http://www.bebec.eu/Downloads/BeBeC2016/Papers/BeBeC-2016-S4.pdf (accessed on 12 August 2021).
19. Savitzky A.; Golay, M.J.E. Smoothing and Differentiation of Data by Simplified Least Squares Procedures. *Anal. Chem.* **1964**, *36*, 1627–1638. [CrossRef]
20. Haxter, S. Extended Version: Improving the DAMAS 2 Results for Wavenumber-Space Beamforming. In Proceedings of the BeBeC 2016, Berlin, Germany, 29 February–1 March 2016. Available online: http://www.bebec.eu/Downloads/BeBeC2016/Papers/BeBeC-2016-D8.pdf (accessed on 12 August 2021).

Article

Piezoelectric MEMS Acoustic Transducer with Electrically-Tunable Resonant Frequency

Alessandro Nastro [1,*], Marco Ferrari [1], Libor Rufer [2], Skandar Basrour [2] and Vittorio Ferrari [1]

1   Department of Information Engineering, University of Brescia, 25123 Brescia, Italy; marco.ferrari@unibs.it (M.F.); vittorio.ferrari@unibs.it (V.F.)
2   CNRS, Grenoble INP, TIMA, University Grenoble Alpes, 38000 Grenoble, France; libor.rufer@univ-grenoble-alpes.fr (L.R.); skandar.basrour@univ-grenoble-alpes.fr (S.B.)
*   Correspondence: alessandro.nastro@unibs.it

Abstract: The paper presents a technique to obtain an electrically-tunable matching between the series and parallel resonant frequencies of a piezoelectric MEMS acoustic transducer to increase the effectiveness of acoustic emission/detection in voltage-mode driving and sensing. The piezoelectric MEMS transducer has been fabricated using the PiezoMUMPs technology, and it operates in a plate flexural mode exploiting a 6 mm × 6 mm doped silicon diaphragm with an aluminum nitride (AlN) piezoelectric layer deposited on top. The piezoelectric layer can be actuated by means of electrodes placed at the edges of the diaphragm above the AlN film. By applying an adjustable bias voltage V_b between two properly-connected electrodes and the doped silicon, the d31 mode in the AlN film has been exploited to electrically induce a planar static compressive or tensile stress in the diaphragm, depending on the sign of V_b, thus shifting its resonant frequency. The working principle has been first validated through an eigenfrequency analysis with an electrically induced prestress by means of 3D finite element modelling in COMSOL Multiphysics®. The first flexural mode of the unstressed diaphragm results at around 5.1 kHz. Then, the piezoelectric MEMS transducer has been experimentally tested in both receiver and transmitter modes. Experimental results have shown that the resonance can be electrically tuned in the range $V_b = \pm 8$ V with estimated tuning sensitivities of 8.7 ± 0.5 Hz/V and 7.8 ± 0.9 Hz/V in transmitter and receiver modes, respectively. A matching of the series and parallel resonant frequencies has been experimentally demonstrated in voltage-mode driving and sensing by applying $V_b = 0$ in transmission and $V_b = -1.9$ V in receiving, respectively, thereby obtaining the optimal acoustic emission and detection effectiveness at the same operating frequency.

Keywords: MEMS; piezoelectric; PiezoMUMPs; acoustic transducer; tunable; resonant frequency; finite element modelling

Citation: Nastro, A.; Ferrari, M.; Rufer, L.; Basrour, S.; Ferrari, V. Piezoelectric MEMS Acoustic Transducer with Electrically-Tunable Resonant Frequency. *Micromachines* **2022**, *13*, 96. https://doi.org/10.3390/mi13010096

Academic Editor: Jose Luis Sanchez-Rojas

Received: 20 December 2021
Accepted: 6 January 2022
Published: 8 January 2022

Publisher's Note: MDPI stays neutral with regard to jurisdictional claims in published maps and institutional affiliations.

1. Introduction

Acoustic transducers based on micro electro-mechanical systems (MEMS) represent a lively research field and, at the same time, provide a significant number of concrete solutions and commercial devices. Specifically, thanks to the advantages provided by MEMS technology such as compact sizes, low production costs and high compatibility with IC technology [1], acoustic MEMS transducers have been extensively employed in different applications. In biomedical fields they have been exploited to monitor heart and lungs sounds [2,3] and for cochlear implants [4]. In the livestock sector MEMS acoustic transducers have been used to estimate the state of health of animals [5] while in industrial fields they have been used for noise and vibration measurements [6], as resonant photoacoustic combustion gas monitors [7] or as hydrophones for pipeline leak detection [8]. In recent years MEMS acoustic sensors and actuators have been extensively produced for consumer applications as microphones for wearable devices [9,10] and voice controllable systems [11,12] or as microspeakers for in-ear devices [13,14]. MEMS acoustic transducers rely on the conversion of energy between mechanical/acoustic and electrical domains which can be achieved by different transduction mechanisms. The most commonly used are the electrostatic [15–17], piezoresistive [18] and piezoelectric [19–23] mechanisms. Specifically, the piezoelectric transduction mechanism compared to other principles has higher energy density and does not require polarization voltages [24]. A crucial parameter for the development of an acoustic piezoelectric MEMS transducer is the resonant frequency of the MEMS structure, since it influences the frequency response of the device [25] and also represents the condition around which the transducer has the maximum transmission and receiving effectiveness in narrowband operation. Usually, a piezoelectric transducer near resonance can be described by an equivalent electrical circuit composed of two parallel branches, i.e., the motional and the electrical branches [26]. The mechanical properties such as the effective mass, mechanical damping, and stiffness of the resonator are modelled by the motional branch, while the electrical branch is associated to the capacitance arising from the dielectric nature of the piezoelectric material [27]. Therefore, the frequency response of the device is characterized by two resonances which differ in value named the series resonant frequency f_s and parallel resonant frequency f_p [28]. According to such an equivalent circuit, a piezoelectric acoustic transducer, under voltage excitation, exhibits the highest transmitting output at the series resonance, while, under voltage readout, it displays the highest receiving sensitivity at the parallel resonance, shifted to a higher frequency [29]. However, when only a single piezoelectric element is used as a transceiver for both transmitting and receiving acoustic signals at resonance, a dynamic frequency tuning would be desirable to obtain the maximum transmitting-receiving effectiveness [30]. Typically, to reach that goal, additional electrical tuning circuits or matching networks can be dynamically added to the transducer [31–34]. However, in systems with several transducers such as in arrays, multiple networks are required to connect all transducer elements thus increasing the complexity of the overall system. Other solutions rely on the application of DC voltages to change mechanical properties of the transducer allowing to increase the bandwidth merging two closely-spaced resonance modes [35].

In this context, the present work proposes a technique to obtain an electrically-tunable matching between the series and parallel resonant frequencies of a single piezoelectric MEMS acoustic transducer to increase the effectiveness of acoustic emission/detection in voltage-mode driving and sensing. A DC bias voltage is applied to the piezoelectric layer inducing a controllable stress thus leading to a matching of the series and parallel resonant frequencies in transmitter and receiver modes. At first, the working principle has been validated through a 3D finite element modelling employing a parametric two-step study to compute the first flexural mode of the structure considering the influence of a static electric field applied across the piezoelectric layer. Then, the proposed technique has been experimentally verified by configuring the fabricated MEMS acoustic transducer in both emitter and receiver modes with applied DC bias. Finally, the matching of the acoustic emission and detection characteristics with the same operating frequency in voltage-mode driving and sensing has been experimentally achieved.

The paper is organized as follows: fabrication technology and device design (Section 2), finite element analysis of the piezoelectric MEMS device (Section 3), experimental results (Section 4) and conclusions (Section 5).

2. Fabrication Technology and Device Design

The top view of the proposed piezoelectric MEMS device, taken from the graphic design system (GDS) file, is reported in Figure 1. The proposed 9×9 mm MEMS device exploits a 6×6 mm highly doped silicon diaphragm with an aluminum nitride (AlN) piezoelectric layer deposited on top that can be actuated by eight interdigital transducers (IDTs), each composed by two interlocking metal comb-shaped arrays of twenty equally spaced fingers. The IDTs are placed on the inner and outer edges of the diaphragm and disposed symmetrically with respect to its centre. The doped silicon layer can be electrically connected employing the four metal pads located at the die corners. The layout of the device, and specifically of the IDTs, has been designed to create a general-purpose piezoelectric MEMS platform exploitable in different applications. In particular, the IDTs have been exploited to generate Lamb waves in the diaphragm at frequencies in the megahertz range to drive mechanical vortexes in liquids for biological applications [36].

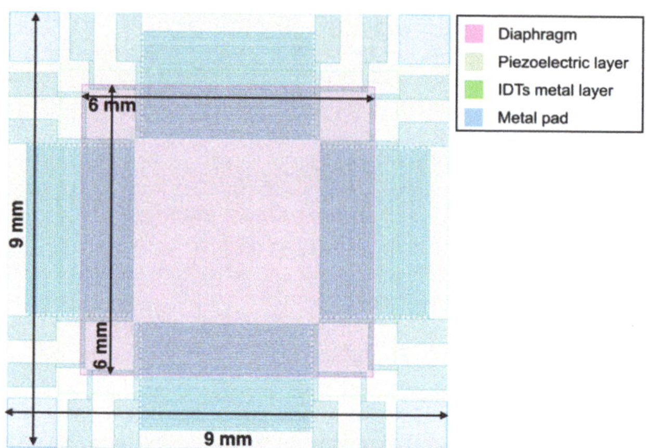

Figure 1. Top view image of the proposed piezoelectric micro electro-mechanical systems (MEMS) taken from the graphic design system (GDS) file.

In this work, on the other hand, to induce a planar static compressive or tensile stress in the diaphragm and to excite/detect acoustic signals, the IDTs have been employed as top plates over the piezoelectric layer, while the highly doped silicon layer has been used as a common bottom plate, thus configuring the electrodes to produce in all respects parallel-plate transducers. Therefore, the IDTs layout, and specifically the spacing between two consecutively electrodes, does not affect the presently proposed application.

The piezoelectric MEMS has been manufactured by employing the piezoelectric multi-user MEMS processes (PiezoMUMPs) technology developed by the MEMSCAP foundry [37]. The manufacturing process steps are illustrated in Figure 2a–f. The process employs a 150 mm <100> oriented silicon-on-insulator (SOI) wafer where the silicon, the oxide and the silicon substrate have thicknesses of 10 ± 1 μm, 1 ± 0.05 μm and 400 ± 5 μm, and are shown in Figure 2 with red, black and blue colours, respectively. A bottom side oxide layer, shown in green colour, is also present on the starting substrate. The process begins with the doping step of the wafer reported in Figure 2a. This step involves the deposition of a phosphosilicate glass (PSG) layer, shown in purple colour, and its annealing at 1050 °C for 1 h in argon. The PSG layer is subsequently removed using wet chemical etching. The piezoelectric film lift-off occurs as the second step of the manufacturing process, reported in Figure 2b. The piezoelectric film consisting of 0.5 μm of aluminum nitride (AlN), shown in cyan colour, is deposited over the wafer by reactive sputtering. The third step involves the pad metal lift-off, reported in Figure 2c. A metal stack consisting of 20 nm of chrome (Cr) and 1000 nm of aluminum (Al), shown in grey colour, is deposited by beam evaporation.

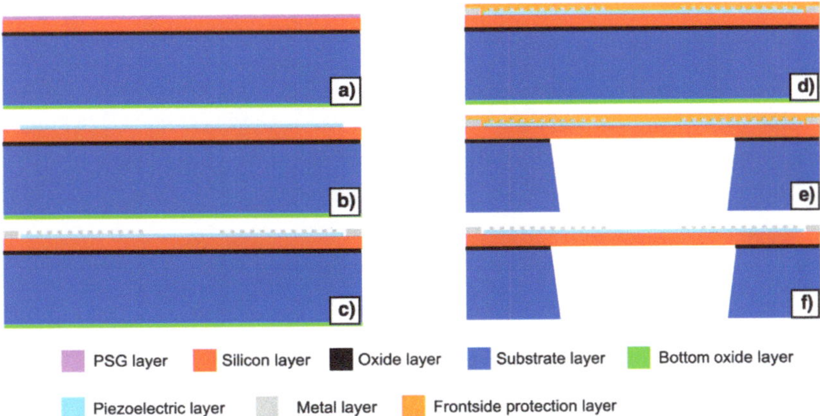

Figure 2. Manufacturing process steps (**a**–**f**) of the piezoelectric multi-user MEMS processes (PiezoMUMPs) technology involved for the fabrication of the piezoelectric MEMS device.

A front side polyamide protection coat, shown in Figure 2d in orange colour, is then applied to the top surface of the wafer. The wafer is then reversed, and the substrate layer is lithographically patterned from the bottom side, as reported in Figure 2e. Reactive ion etching (RIE) is used to remove the bottom side oxide layer while a DeepRIE (DRIE) is subsequently used to etch the substrate layer up to the silicon layer. Finally, the front side protection material is stripped during the release step, as reported in Figure 2f. Top and bottom views of the fabricated piezoelectric MEMS device are reported in Figure 3a,b, respectively. The proposed device embeds electrical terminals for each comb-shaped arrays of fingers, while four electrically shorted metal pads are placed in each corner of the device to contact the highly doped silicon layer beneath the AlN piezoelectric layer, as reported in Figure 3c,d, respectively. Each comb-shaped array of the IDTs includes fingers with width of 28 µm and pitch of 112 µm, as reported in Figure 3c.

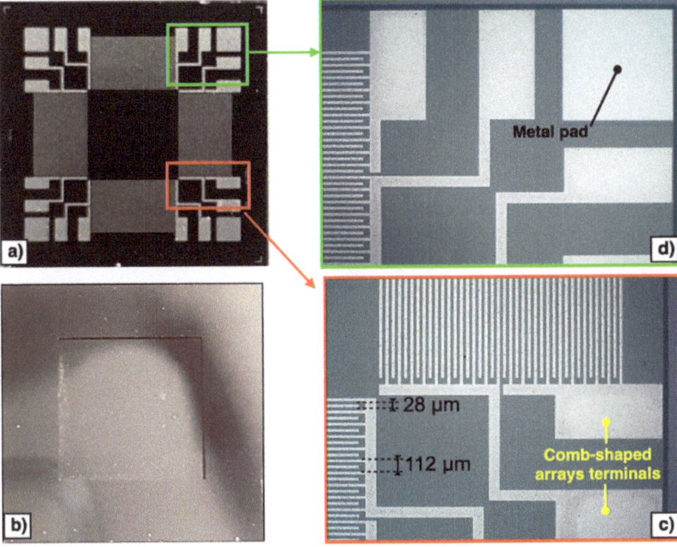

Figure 3. Top (**a**) and bottom (**b**) views of the fabricated piezoelectric MEMS device. Enlarged images of the comb-shaped arrays terminals (**c**) and metal pad (**d**).

The possibility to obtain an adjustable matching between the series and parallel resonant frequencies of the first flexural mode of the piezoelectric MEMS transducer to increase the effectiveness of acoustic emission/detection has been investigated by electrically tuning the mechanical characteristics of the diaphragm. The displacement and the section view of a fully-clamped square plate vibrating at the first flexural mode is reported in Figure 4a,b, respectively.

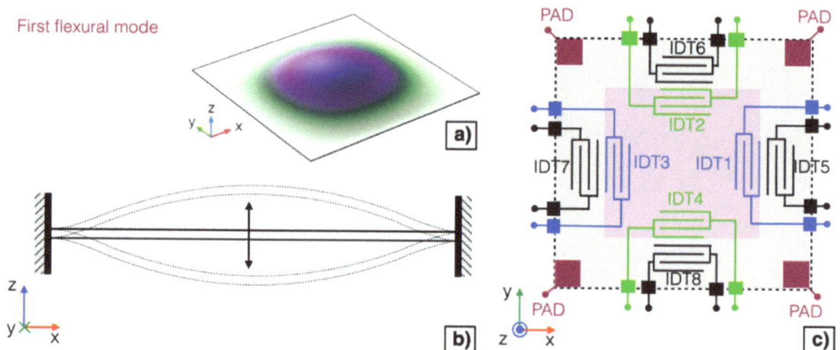

Figure 4. Displacement (**a**) and section view (**b**) of a fully-clamped square plate vibrating at the first flexural mode. Simplified schematic of the proposed piezoelectric MEMS device (**c**).

Two IDTs, namely IDT1 and IDT3, placed on the opposite inner edges of the diaphragm, shown in the simplified schematic of Figure 4c in blue colour, have been shorted and employed as a top plate over the piezoelectric layer.

The doped silicon layer, contactable by employing the metal pads shown in Figure 4c in dark red colour, has been grounded and employed as a bottom plate. The four IDTs on the outer edges of the diaphragm, i.e., IDT5–IDT8, shown in black colour, have not been employed and thus left unconnected. The remaining two IDTs, namely IDT2 and IDT4, on the inner edges of the diaphragm, shown in green colour, have been adopted for excitation or detection of acoustic signals.

By applying a DC bias voltage V_b between IDT1 shorted with IDT3 and the silicon pad, it is possible to exploit the d31 mode in the AlN film to induce a planar static compressive or tensile stress in the diaphragm, depending on the sign of the bias voltage V_b, as reported in Figure 5a,b, respectively. The application of an electrically controllable mechanical stress allows the mechanical resonant frequency f_{R0} of the diaphragm to be shifted thus leading to tunable resonant frequency. Studies have proven that a stress induced to a clamped square plate leads to variations of the frequencies of its vibrational modes, including the first flexural mode [38,39]. Considering the spacing of 28 µm between two consecutive fingers and the thickness of the AlN layer of 0.5 µm, the electric field **E** induced within the AlN layer below each finger can be assumed as in the configuration of parallel plates, where the top plate is the corresponding finger, and the bottom plate is the highly doped silicon layer. Given the piezoelectric polarization vector **P** oriented along the negative direction of the z-axis, by applying a positive bias voltage $V_b > 0$, an expansion along the z-axis and a contraction along the x-axis of the piezoelectric material is produced at each finger, as reported in the inset of Figure 5a.

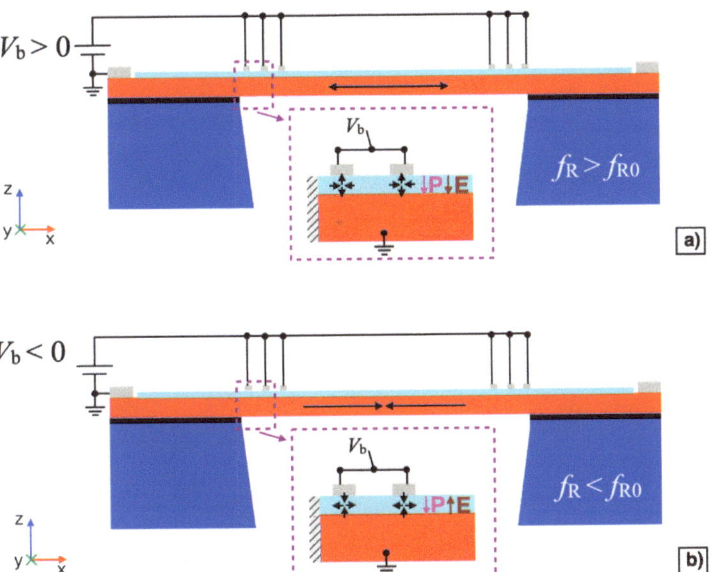

Figure 5. Cross-sectional view of the piezoelectric MEMS device with positive ($V_b > 0$ V) (**a**) and negative ($V_b < 0$ V) (**b**) bias voltage.

Consequently, a planar static tensile stress is induced into the diaphragm as indicated by the arrows, thus increasing the mechanical resonant frequency f_R, i.e., $f_R > f_{R0}$. On the contrary, by applying a negative bias voltage $V_b < 0$ a contraction along the z-axis and an expansion along the x-axis of the piezoelectric material is produced at each finger, as reported in the inset of Figure 5b. Consequently, a compressive stress will be induced into the diaphragm as indicated by the arrows, thus decreasing the mechanical resonant frequency f_R, i.e., $f_R < f_{R0}$.

3. Finite Element Analysis of the Piezoelectric MEMS Device

The electro-mechanical behaviour of the piezoelectric MEMS device described in Section 2 has been investigated by means of 3D finite element modelling in COMSOL Multiphysics®. Top and bottom views of the developed 3D model of the device are reported in Figure 6a,b, respectively.

Figure 6c reports an enlarged view of the structural layers that have been included in the 3D model. The nominal dimensions reported in Section 2 have been considered, i.e., neglecting tolerances in layer thicknesses produced by the manufacturing process. In the reported 3D model, the metal layer has been considered as made by Al, thus Cr has been neglected, since the Al thickness is 50 times higher than the Cr thickness. The four IDTs on the outer edges of the diaphragm have not been included in the model since, as described in Section 2, they have not been actuated and they do not affect the mechanical properties of the diaphragm to any significant extent. The SiO_2 has been used as the oxide layer material while Si <100> has been adopted for the substrate and the silicon layer.

The piezoelectric coefficients $d_{31} = -2.78$ pC/N and $d_{33} = 6.5$ pC/N have been specified for the AlN piezoelectric layer as reported in [22,35]. A rotation of 180 deg around the x-axis of the coordinate system has been adopted for the piezoelectric layer to correctly align the poling direction with the negative direction of the z-axis. The piezoelectric effect has been considered by including in the simulation the piezoelectric multiphysics which combines the solid mechanics with the electrostatics physics.

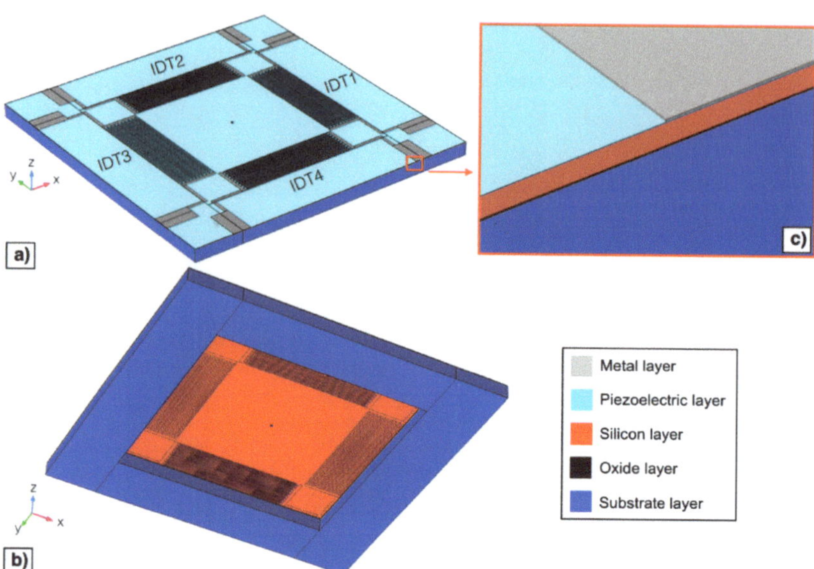

Figure 6. Top (**a**) and bottom (**b**) 3D model views of the proposed piezoelectric MEMS device. Enlarged image of the layers employed in the simulation model (**c**).

Regarding the solid mechanic physics, a fixed boundary constraint has been applied to the bottom surface of the substrate while for the piezoelectric layer a strain-charge constitutive relation has been specified including the AlN material properties. The gravity constraint has been applied to the domains of the whole structure.

Regarding the electrostatics physics, a charge conservation boundary condition has been applied to the AlN layer. Terminal constraints have been specified to the domain of each comb-shaped arrays of fingers. The metal pads placed in the device corners to contact the silicon layer beneath the piezoelectric layer have not been included in the 3D model since a ground constraint has been applied to the top surface of the silicon. The mesh domain has been carefully designed to obtain a convergent solution while reducing the computational workload. Top and bottom views of the mesh domain are shown in Figure 7a,b, respectively. Layers that compose the diaphragm have been studied with a finer mesh, while layers laid on the outer edges of the diaphragm with a coarser mesh, as reported in Figure 7c. Specifically, top surfaces of the metal and AlN layers that compose the diaphragm have been meshed with a mapped resolution distribution of 1 µm and with a free triangular minimum element size of 36 µm, respectively. Whereas, a free triangular mesh with a minimum element size of 90 µm has been applied to layers laid on the outer edge of the diaphragm and swept down to the substrate layer.

A two-step study with parametric sweep has been employed to evaluate the effect of the electrical DC bias to the resonant frequency of the diaphragm. The terminal voltage V_b of IDT1 shorted with IDT3 has been varied within the range of ± 8 V with a step size of 2 V while leaving the terminals of IDT2 and IDT4 electrically floating.

As a first step, a stationary study has been employed to analyse the mechanical effect of the electric static load, i.e., an electrically induced prestress, on the diaphragm.

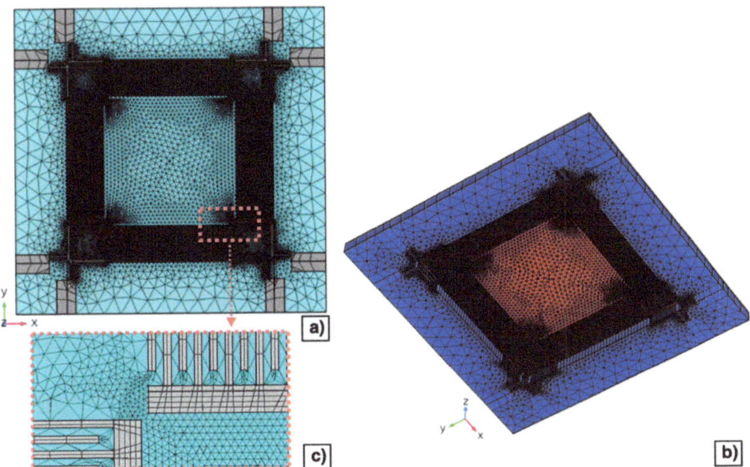

Figure 7. Top (**a**) and bottom (**b**) view of the mesh domain developed for the 3D model of the piezoelectric MEMS transducer. Enlarged view of the mesh of the comb-shaped arrays (**c**).

The stationary study results of the z-axis displacement for $V_b = 8$ V and $V_b = -8$ V have been reported with a 3D representation, not in true scale, in Figure 8a,b, respectively. It can be noticed that, as expected, the convexity of the diaphragm deflection is function of the sign of the applied bias voltage V_b due to the induced planar static compressive or tensile stress. The z-axis displacement w_p of the point laid on the top of the AlN surface in the centre of the diaphragm as a function of the bias voltage V_b is plotted in Figure 9. A displacement of 0.48 µm and -0.51 µm has been obtained at $V_b = 8$ V and $V_b = -8$ V, respectively.

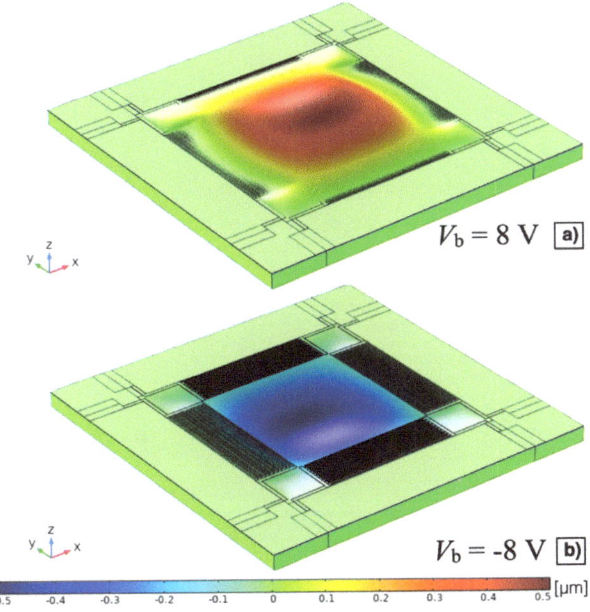

Figure 8. Z-axis displacement, not in true scale, of the MEMS device with $V_b = 8$ V (**a**) and $V_b = -8$ V (**b**).

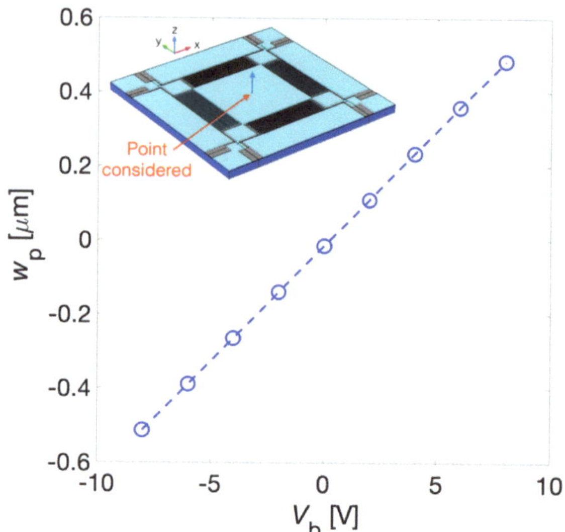

Figure 9. Z-axis displacement w_p of the point reported in the inset as a function of the bias voltage V_b.

With $V_b = 0$ V, i.e., without electrically induced prestressed, w_p is equal to -15 nm due to the gravity effect included in the simulation.

As a second step, an eigenfrequency study has been employed to compute the first flexural mode of the structure considering the influence of the electric static load previously evaluated by means of the stationary study. The simulation results of the prestressed eigenfrequency study related to the first eigenmode of the structure are reported in Figure 10.

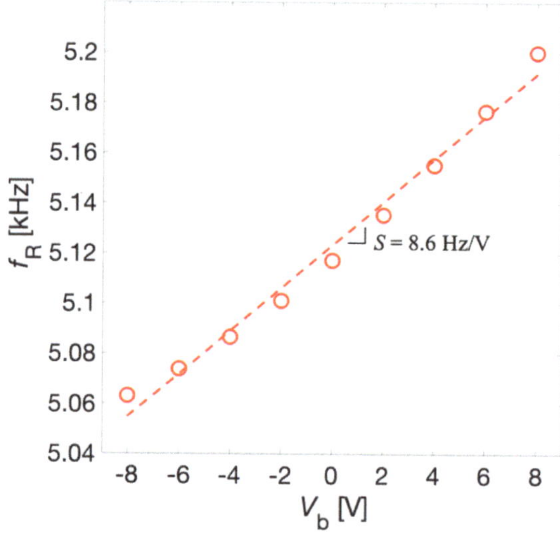

Figure 10. Simulation results of the resonant frequency f_R of the first eigenmode of the diaphragm as a function of the bias voltage V_b.

Specifically, the mechanical resonant frequency f_R of the piston-like first flexural vibrational mode of the diaphragm is plotted versus V_b. The estimated resonant frequency varies from 5.06 kHz for $V_b = -8$ V up to 5.19 kHz for $V_b = 8$ V. Therefore, by adjusting the voltage V_b it is possible to electrically tune the resonant frequency of the diaphragm. As expected, this could provide the system with the capability of reaching the coupling between the series and parallel resonant frequencies of a piezoelectric MEMS acoustic transceiver. The tuning sensitivity $S = 8.6$ Hz/V of the system defined as the linearized ratio between the resonant frequency shift and the applied bias voltage has been estimated by taking the angular coefficient of the linear fitting of simulated data shown in Figure 10.

4. Experimental Results

The possibility to improve the receiving-transmitting effectiveness through an applied DC bias voltage V_b was experimentally investigated by testing the piezoelectric MEMS device in both acoustic receiver and transmitter modes.

The block diagram of the piezoelectric MEMS device configured as acoustic receiver is reported in Figure 11a. In receiver mode the direct piezoelectric effect was exploited by measuring the voltage signal $v_{out}(t)$ at frequency near the mechanical resonant frequency. The MEMS device can be represented by the equivalent Butterworth–Van Dyke model (BVD) reported in Figure 11b, where the effective mass, mechanical damping, and elastic compliance are represented by the inductance L_m, resistance R_m, and capacitance C_m, respectively. The force induced by the impinging acoustic signal is represented by the voltage $v_a(t)$ in the mechanical branch while the parallel capacitance C_p represents the dielectric nature of the piezoelectric material. According to such an equivalent circuit, the piezoelectric acoustic device, under voltage readout, displays the highest receiving response at the parallel resonance f_p [29], defined as:

$$f_p = \frac{1}{2\pi}\sqrt{\frac{C_p + C_m}{L_m C_m C_p}} \tag{1}$$

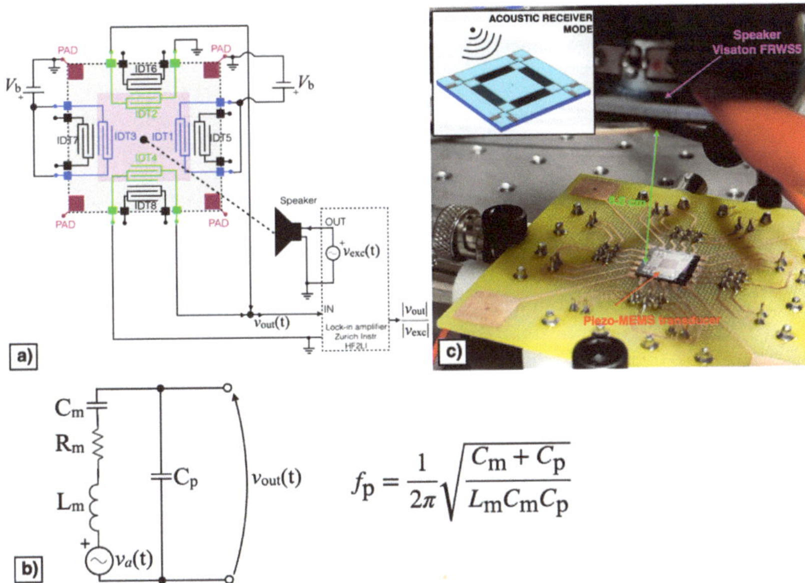

$$f_p = \frac{1}{2\pi}\sqrt{\frac{C_m + C_p}{L_m C_m C_p}}$$

Figure 11. Block diagram (**a**), equivalent Butterworth–Van Dyke (BVD) model (**b**) and experimental set-up (**c**) of the piezoelectric MEMS device in acoustic receiver mode.

A sinusoidal excitation voltage $v_{exc}(t)$ with peak amplitude $A_{exc} = 1$ V and frequency f_{exc} within the bandwidth 5.3–5.6 kHz, provided by the lock-in amplifier (HF2LI, Zurich Instruments: Zurich, Switzerland), was applied to a speaker (FRWS5, Visaton: Haan,

Germany) with a flat response in the frequency region of interest placed at 6.5 cm above the diaphragm, as shown in Figure 11c.

The output voltage signal $v_{out}(t)$ was measured across the parallel connection of IDT2 and IDT4, while the bias voltage V_b was applied between IDT1 shorted with IDT3 and the silicon pad using a power supply (Polytec: Grenoble, France). The acquired voltage $v_{out}(t)$ was synchronously demodulated with the excitation signal $v_{exc}(t)$ by the lock-in amplifier, thus providing the magnitude ratio $|v_{out}|\,/\,|v_{exc}|$ of the resulting receiving transfer function which is plotted as a function of f_{exc} for different values of V_b in Figure 12.

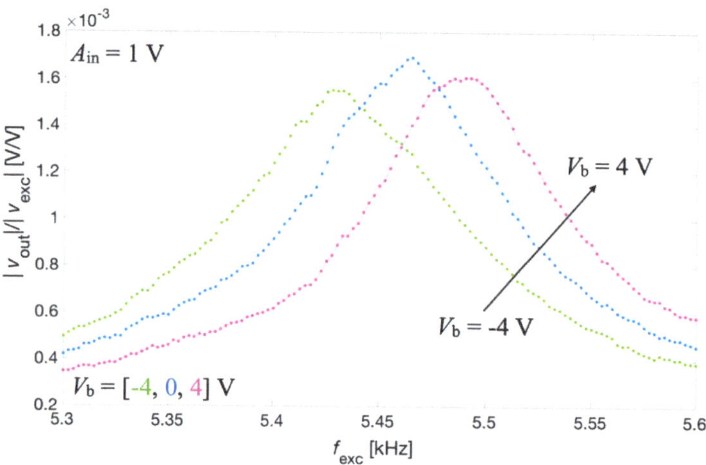

Figure 12. Measured magnitude ratio $|v_{out}|\,/\,|v_{exc}|$ as a function of the frequency f_{exc} for the acoustic receiver mode at different values of the bias voltage V_b.

The results of Figure 12 show that by acting on V_b it is also possible to electrically tune the resonant frequency of the piezoelectric MEMS device configured as an acoustic receiver.

The tuning sensitivity S was derived by the linear fitting of experimental data reported in Figure 13. The uncertainty for f_P was estimated as $\sigma = 5$ Hz, and the uncertainty of S was obtained exploiting the error propagation approach [40]. The tuning sensitivity S results 7.8 ± 0.9 Hz/V for the receiver mode. Given the electrical constraints imposed for the FEM simulation reported in Section 3 the simulated mechanical resonant frequency f_R is expected to approach the parallel resonant frequency f_P defined in Equation (1). The obtained values of sensitivity show a good agreement between simulated and experimental results, demonstrating that a tunability of the parallel resonant frequency can be obtained in the explored range for V_b. Discrepancies between the simulated and experimental results of f_P are probably related to the tolerances introduced by the fabrication process of the device which were not taken into full account in the simulations.

The block diagram of the piezoelectric MEMS configured as acoustic transmitter is reported in Figure 14a. In transmitter mode the converse piezoelectric effect was exploited by applying the alternating excitation voltage $v_{exc}(t)$ at frequency near the mechanical resonant frequency. The MEMS device can be represented by the equivalent BVD model of Figure 14b. The velocity of the diaphragm causing the emitted acoustic signal is represented in electrical formalism by the current $i_a(t)$. According to such an equivalent circuit, the piezoelectric acoustic device, under voltage excitation, exhibits the highest transmitting output at the series resonance f_s [29] defined as:

$$f_s = \frac{1}{2\pi} \frac{1}{\sqrt{L_m C_m}} \tag{2}$$

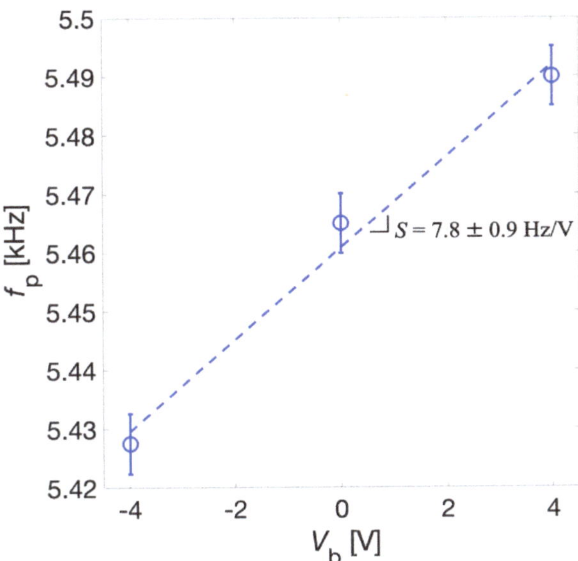

Figure 13. Measured parallel resonant frequency f_p (circles) and linear fitting (dotted line) as a function of V_b. The error bars extend one standard deviation σ on each side of the experimental data.

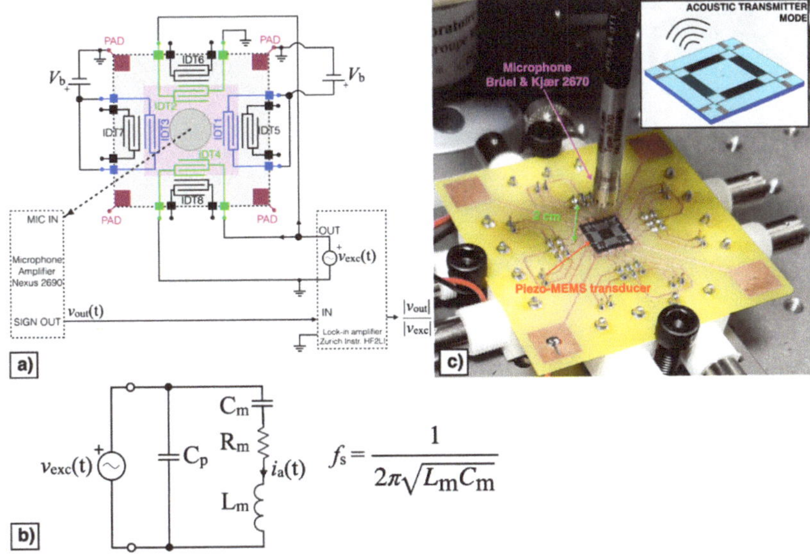

$$f_\mathrm{s} = \frac{1}{2\pi\sqrt{L_\mathrm{m}C_\mathrm{m}}}$$

b)

Figure 14. Block diagram (**a**), equivalent BVD model (**b**) and experimental set-up (**c**) of the piezoelectric MEMS device in acoustic transmitter mode.

The excitation voltage $v_\mathrm{exc}(t)$ was applied by means of the lock-in amplifier to the parallel connection of IDT2 and IDT4. The DC bias voltage V_b was applied between IDT1 shorted with IDT3 and the silicon pad and swept within the range of ± 8 V with a step size of 2 V. The generated acoustic signal was measured by a microphone (2670, Brüel & Kjaer: Nærum, Denmark) placed at 2 cm above the diaphragm, as shown in Figure 14c. The microphone output was fed to an amplifier (Nexus 2690, Brüel & Kjaer: Nærum, Denmark) set with a sensitivity of 1 V/Pa. The measured output signal $v_{out}(t)$ was fed to the lock-in amplifier input for synchronous demodulation with the excitation signal. The magnitude ratio $|v_{out}|/|v_{exc}|$ of the resulting transmitting transfer function is reported as a function of f_{exc} for different values of V_b in Figure 15.

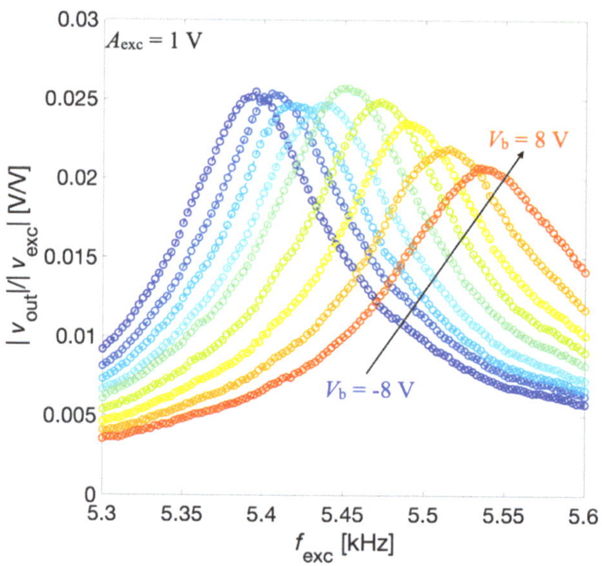

Figure 15. Measured magnitude ratio $|v_{out}|/|v_{exc}|$ as a function of f_{exc} for the acoustic transmitter mode at different values of V_b.

The results of Figure 15 show that by acting on the prestress caused by the bias voltage V_b it is possible to electrically tune the resonant frequency of the piezoelectric MEMS also in transmitter mode. Considering the maximum of the magnitude of the transmitting transfer function v_{out}/v_{exc}, the series resonant frequency f_s was estimated.

The tuning sensitivity S of the system was derived by the linear fitting of experimental data reported in Figure 16. The uncertainty for f_S, as for the receiver mode, was estimated as $\sigma = 5$ Hz. The tuning sensitivity S results 8.7 ± 0.5 Hz/V for the transmitter mode. The reported data demonstrates that a tunability of about 130 Hz can be obtained in the explored range for V_b. The obtained experimental values of S in the receiver and transmitter modes, taking into account their uncertainties, are compatible with each other in metrological sense [40] and closely approach the simulated value.

The measured tuning sensitivities and frequency shifts obtained in both receiver and transmitter modes demonstrate that matching of the series resonant frequency with the parallel resonant frequency can be obtained by acting on the bias voltage in either one of the two working modes. A comparison between the receiver and the transmitter modes in terms of the normalized measured magnitude ratio as a function of the frequency f_{exc} without and with the applied tuning by V_b is reported in Figure 17a,b, respectively.

Specifically, a bias voltage $V_b = -1.9$ V was applied to the device configured as receiver to match the resonant frequency of the device configured as transmitter. Therefore, by electrically tuning V_b it is possible to finely control the resonant frequency of the device, thus obtaining the optimal acoustic emission and detection characteristics with the same operating frequency in both voltage-mode driving and sensing.

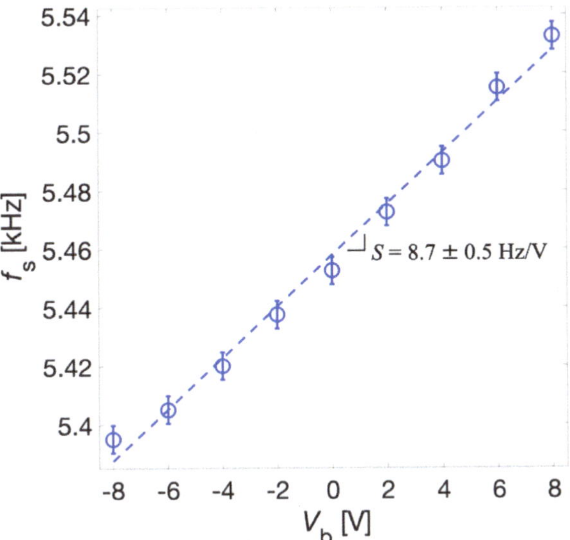

Figure 16. Measured series resonant frequency f_s (circles) and linear fitting (dotted line) as a function of V_b. The error bars extend one standard deviation σ on each side of the experimental data.

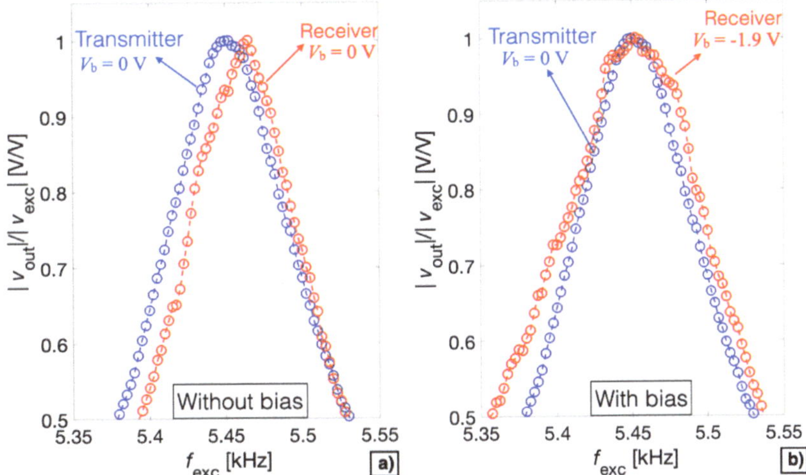

Figure 17. Comparison between the receiver and transmitter mode in terms of the normalized measured magnitude ratio as a function of the frequency f_{exc} without (**a**) and with (**b**) the tuning effect induced by V_b.

5. Conclusions

This work has presented a technique to electrically tune the resonant frequency of a piezoelectric MEMS acoustic transducer to obtain matching between the series and parallel resonant frequencies. The piezoelectric MEMS device has been fabricated with the PiezoMUMPs technology exploiting a doped silicon diaphragm with an AlN piezoelectric layer deposited on top. Electrodes disposed symmetrically with respect to the centre of the diaphragm allow for actuating and sensing. By applying a bias voltage V_b between the bottom doped silicon layer and top electrodes on the AlN layer, an electrically-controllable stress can be induced into the diaphragm, thus leading to the tuning of the resonant frequency.

The working principle of the proposed technique has been studied by 3D finite element modelling in COMSOL Multiphysics® and experimentally verified configuring the piezoelectric acoustic transducer in both receiver and transmitter modes.

Experimental results have shown a tuning sensitivity $S = 7.8 \pm 0.9$ Hz/V in receiver mode, whereas a frequency shift of 130 Hz for $V_b = \pm 8$ V and a tuning sensitivity $S = 8.7 \pm 0.5$ Hz/V have been reached in transmitter mode. A comparison between the receiver and the transmitter modes has been performed by applying a bias voltage $V_b = -1.9$ V to the device configured as receiver to match the resonant frequency of the device configured as transmitter, thus obtaining the optimal acoustic emission and detection characteristics with the same operating frequency in voltage-mode driving and sensing.

Taking advantage of the non-directional response in the low-frequency range, the proposed device can be employed in pulsed-echo mode as a proximity/presence, or gesture detector. Furthermore, the proposed technique can be transferred to a properly down-scaled structure to obtain a tunable piezoelectric micromachined ultrasound transducer (PMUT).

Author Contributions: Conceptualization, V.F.; Data curation, M.F. and L.R.; Formal analysis, A.N.; Investigation, A.N., M.F. and L.R.; Methodology, A.N. and M.F.; Supervision, S.B. and V.F.; Validation, A.N.; Visualization, A.N.; Writing—original draft, A.N.; Writing—review and editing, A.N., M.F., L.R., S.B. and V.F. All authors have read and agreed to the published version of the manuscript.

Funding: This research received no external funding.

Conflicts of Interest: The authors declare no conflict of interest.

References

1. Bryzek, J.; Roundy, S.; Bircumshaw, B.; Chung, C.; Castellino, K.; Stetter, J.R.; Vestel, M. Marvelous MEMs: Advanced IC sensors and microstructures for high volume applications. *IEEE Circuits Devices Mag.* **2006**, *22*, 8–28. [CrossRef]
2. Qu, M.; Yang, D.; Chen, X.; Li, D.; Zhu, K.; Xie, J. Heart sound monitoring based on a piezoelectric mems acoustic sensor. In Proceedings of the IEEE 34th International Conference on Micro Electro Mechanical Systems (MEMS), Virtual Conference, 25–29 January 2021; pp. 59–63.
3. Liu, H.; Liu, S.; Shkel, A.A.; Tang, Y.; Kim, E.S. Multi-band MEMS resonant microphone array for continuous lung-Sound monitoring and classification. In Proceedings of the IEEE 33rd International Conference on Micro Electro Mechanical Systems (MEMS), Vancouver, BC, Canada, 18–22 January 2020; pp. 857–860.
4. Udvardi, P.; Radó, J.; Straszner, A.; Ferencz, J.; Hajnal, Z.; Soleimani, S.; Schneider, M.; Schmid, U.; Révész, P.; Volk, J. Spiral-shaped piezoelectric MEMS cantilever array for fully implantable hearing systems. *Micromachines* **2017**, *8*, 311. [CrossRef]
5. Cheng, Y.; Narusawa, K.; Iijima, S.; Nakayama, M.; Ishimitsu, S.; Ishida, A.; Mikami, O. Fundamental research of an early detection system to find respiratory diseases for Pigs using body-conducted sound. *ICIC Express Lett. Part B Appl.* **2019**, *10*, 737–742.
6. Sahdom, A.S. Application of micro electro-mechanical sensors (MEMS) devices with Wifi connectivity and cloud data solution for industrial noise and vibration measurements. *J. Phys. Conf. Ser.* **2019**, *1262*, 012025. [CrossRef]
7. El-Safoury, M.; Dufner, M.; Weber, C.; Schmitt, K.; Pernau, H.-F.; Willing, B.; Wöllenstein, J. On-board monitoring of SO2 ship emissions using resonant photoacoustic gas detection in the UV range. *Sensors* **2021**, *21*, 4468. [CrossRef]
8. Yang, D.; Yang, L.; Chen, X.; Qu, M.; Zhu, K.; Ding, H.; Li, D.; Bai, Y.; Ling, J.; Xu, J.; et al. A piezoelectric AlN MEMS hydrophone with high sensitivity and low noise density. *Sens. Actuators A Phys.* **2021**, *318*, 112493. [CrossRef]
9. Grigoriev, D.M.; Generalov, S.S.; Polomoshnov, S.A.; Nikiforov, S.V.; Amelichev, V.V. Condenser MEMS microphone. *Russ. Microelectron.* **2020**, *49*, 37–42. [CrossRef]
10. Liu, Z.-Y.; Chen, R.-Z.; Ye, F.; Guo, G.-Y.; Li, Z.; Qian, L. Time-of-arrival estimation for smartphones based on built-in microphone sensor. *Electron. Lett.* **2020**, *56*, 1280–1283. [CrossRef]
11. Berol, D. The Advantages of Using Piezoelectric MEMS Microphones in Your Alexa-Enabled Product. 2017. Available online: https://developer.amazon.com/it/blogs/alexa/post/bcef47f5-f1ca-4614-8a57-6af7eabfd1eb/the-advantages-of-using-piezoelectric-mems-microphones-in-your-alexa-enabled-product (accessed on 7 January 2022).
12. Oh, T.; Aiken, W.; Kim, H. "Hey Siri—Are you there?" Jamming of voice commands using the resonance effect (work-in-progress). In Proceedings of the International Conference on Software Security and Assurance (ICSSA), Seoul, Correa, 26–27 July 2018; pp. 73–76. [CrossRef]
13. Cheng, H.-H.; Huang, Z.-R.; Wu, M.; Fang, W. Low frequency sound pressure level improvement of piezoelectric mems microspeaker using novel spiral spring with dual electrode. In Proceedings of the 20th International Conference on Solid-State Sensors, Actuators and Microsystems & Eurosensors, Berlin, Germany, 23–27 June 2019; pp. 2013–2016. [CrossRef]

14. Clarke, P. Infineon is Moving the Growing Microphone Market. 2020. Available online: https://www.eenewsanalog.com/news/infineon-moving-growing-microphone-market (accessed on 7 January 2022).

15. Anzinger, S.; Bretthauer, C.; Tumpold, D.; Dehé, A. A non-linear lumped model for the electro-mechanical coupling in capacitive MEMS microphones. *J. Microelectromech. Syst.* **2021**, *30*, 360–368. [CrossRef]

16. Zawawi, S.A.; Hamzah, A.A.; Majlis, B.Y.; Mohd-Yasin, F. A review of MEMS capacitive microphones. *Micromachines* **2020**, *11*, 484. [CrossRef] [PubMed]

17. Anzinger, S.; Bretthauer, C.; Manz, J.; Krumbein, U.; Dehé, A. Broadband acoustical MEMS transceivers for simultaneous range finding and microphone applications. In Proceedings of the 20th International Conference on Solid-State Sensors, Actuators and Microsystems & Eurosensors XXXIII (TRANSDUCERS & EUROSENSORS XXXIII), Berlin, Germany, 23–27 June 2019; pp. 865–868. [CrossRef]

18. Nebhen, J.; Savary, E.; Rahajandraibe, W.; Dufaza, C.; Meillère, S.; Haddad, F.; Kussener, E.; Barthélémy, H.; Czarny, J.; Walther, A. Low-noise CMOS analog-to-digital interface for MEMS resistive microphone. In Proceedings of the IEEE 20th International Conference on Electronics, Circuits, and Systems (ICECS), Abu Dhabi, United Arab Emirates, 8–11 December 2013; pp. 445–448. [CrossRef]

19. Wang, Q.; Ruan, T.; Xu, Q.; Yang, B.; Liu, J. Wearable multifunctional piezoelectric MEMS device for motion monitoring, health warning, and earphone. *Nano Energy* **2021**, *89*, 106324. [CrossRef]

20. Rahaman, A.; Kim, B. Sound source localization by *Ormia ochracea* inspired low–noise piezoelectric MEMS directional microphone. *Sci. Rep.* **2020**, *10*, 9545. [CrossRef]

21. Prasad, M.; Aditi, A.; Khanna, V.K. Development of MEMS acoustic sensor with microtunnel for high SPL measurement. *IEEE Trans. Ind. Electron.* **2021**, *69*, 3142–3150. [CrossRef]

22. Kabir, M.; Kazari, H.; Ozevin, D. Piezoelectric MEMS acoustic emission sensors. *Sens. Actuators A Phys.* **2018**, *279*, 53–64. [CrossRef]

23. Nastro, A.; Rufer, L.; Ferrari, M.; Basrour, S.; Ferrari, V. Piezoelectric Micromachined acoustic transducer with electrically-tunable resonant frequency. In Proceedings of the 20th International Conference on Solid-State Sensors, Actuators and Microsystems & Eurosensors XXXIII (TRANSDUCERS & EUROSENSORS XXXIII), Berlin, Germany, 23–27 June 2019; pp. 1905–1908. [CrossRef]

24. Yi, S.H.; Kim, E.S. Micromachined piezoelectric microspeaker. *Jpn. J. Appl. Phys.* **2005**, *44*, 3836. [CrossRef]

25. Liechti, R.; Durand, S.; Hilt, T.; Casset, F.; Dieppedale, C.; Verdot, T.; Colin, M. A piezoelectric MEMS loudspeaker lumped and FEM models. In Proceedings of the 22nd International Conference on Thermal, Mechanical and Multi-Physics Simulation and Experiments in Microelectronics and Microsystems (EuroSimE), St. Julian, Malta, 19–21 April 2021; pp. 1–8. [CrossRef]

26. Pillai, G.; Li, S.-S. Piezoelectric MEMS resonators: A review. *IEEE Sens. J.* **2021**, *21*, 12589–12605. [CrossRef]

27. Arnau, A. *Piezoelectric Transducers and Applications*; Springer: Berlin/Heidelberg, Germany, 2004.

28. Queiros, R.; Girao, P.S.; Serra, A.C. Single-mode piezoelectric ultrasonic transducer equivalent circuit parameter calculations and optimization using experimental data. *IMEKO TC4 Symp.* **2005**, *2*, 468–471.

29. Getman, I.; Lopatin, S. Matching of series and parallel resonance frequencies for ultrasonic piezoelectric transducers. In Proceedings of the 12th IEEE International Symposium on Applications of Ferroelectrics (ISAF 2000), Honolulu, HI, USA, 21 July–2 August 2000; Volume 2, pp. 713–715. [CrossRef]

30. Emeterio, J.L.S.; Ramos, A.; Sanz, P.T.; Ruiz, A. Evaluation of impedance matching schemes for pulse-echo ultrasonic piezoelectric transducers. *Ferroelectrics* **2002**, *273*, 297–302. [CrossRef]

31. Shuyu, L. Study on the parallel electric matching of high power piezoelectric transducers. *Acta Acust. United Acust.* **2017**, *103*, 385–391. [CrossRef]

32. Ens, A.; Reindl, L.M. Piezoelectric transceiver matching for multiple frequencies. *J. Sens. Sens. Syst.* **2015**, *4*, 9–16. [CrossRef]

33. Yang, Y.; Wei, X.; Zhang, L.; Yao, W. The effect of electrical impedance matching on the electromechanical characteristics of sandwiched piezoelectric ultrasonic transducers. *Sensors* **2017**, *17*, 2832. [CrossRef] [PubMed]

34. Garcia-Rodriguez, M.; Garcia-Alvarez, J.; Yañez, Y.; Garcia-Hernandez, M.-J.; Salazar, J.; Turo, A.; Chavez, J.A. Low cost matching network for ultrasonic transducers. *Phys. Procedia* **2010**, *3*, 1025–1031. [CrossRef]

35. Kusano, Y.; Wang, Q.; Luo, G.L.; Lu, Y.; Rudy, R.Q.; Polcawich, R.G.; Horsley, D.A. Effects of DC bias tuning on air-coupled PZT piezoelectric micromachined ultrasonic transducers. *J. Microelectromech. Syst.* **2018**, *27*, 296–304. [CrossRef]

36. Demori, M.; Baù, M.; Ferrari, M.; Basrour, S.; Rufer, L.; Ferrari, V. MEMS device with piezoelectric actuators for driving mechanical vortexes in aqueous solution drop. In Proceedings of the 20th International Conference on Solid-State Sensors, Actuators and Microsystems & Eurosensors XXXIII (Transducers & Eurosensors XXXIII), Berlin, Germany, 23–27 June 2019; pp. 2318–2321. [CrossRef]

37. Cowen, A.; Hames, G.; Glukh, K.; Hardy, B. *PiezoMUMPs Design Handbook*; revision 1.3 ed.; MEMSCAP Inc.: Durham, NC, USA, 2014.

38. Weinstein, A.; Chien, W. On the vibrations of a clamped plate under tension. *Q. Appl. Math.* **1943**, *1*, 61–68. [CrossRef]

39. Leissa, W.A. *Vibration of Plates*; Scientific and Technical Information Division, Office of Technology Utilization, National Aeronautics and Space Administration: Washington, DC, USA, 1969.
40. Taylor, R.J. *An Introduction to Error Analysis, the Study of Uncertainties in Physical Measurements*; University Science Books: Sausalito, CA, USA, 1997.

MDPI AG
Grosspeteranlage 5
4052 Basel
Switzerland
Tel.: +41 61 683 77 34

Micromachines Editorial Office
E-mail: micromachines@mdpi.com
www.mdpi.com/journal/micromachines